T0323400

Modern Applications of Graph Theory

Modern Applications of Graph Theory

VADIM ZVEROVICH

OXFORD
UNIVERSITY PRESS

OXFORD

UNIVERSITY PRESS

Great Clarendon Street, Oxford, OX2 6DP,
United Kingdom

Oxford University Press is a department of the University of Oxford.
It furthers the University's objective of excellence in research, scholarship,
and education by publishing worldwide. Oxford is a registered trade mark of
Oxford University Press in the UK and in certain other countries

© Vadim Zverovich 2021

The moral rights of the author have been asserted

First Edition published in 2021

Impression: 3

All rights reserved. No part of this publication may be reproduced, stored in
a retrieval system, or transmitted, in any form or by any means, without the
prior permission in writing of Oxford University Press, or as expressly permitted
by law, by licence or under terms agreed with the appropriate reprographics
rights organization. Enquiries concerning reproduction outside the scope of the
above should be sent to the Rights Department, Oxford University Press, at the
address above

You must not circulate this work in any other form
and you must impose this same condition on any acquirer

Published in the United States of America by Oxford University Press
198 Madison Avenue, New York, NY 10016, United States of America

British Library Cataloguing in Publication Data

Data available

Library of Congress Control Number: 2020952669

ISBN 978–0–19–885674–0

DOI: 10.1093/oso/9780198856740.001.0001

Printed and bound by
CPI Group (UK) Ltd, Croydon, CR0 4YY

Links to third party websites are provided by Oxford in good faith and
for information only. Oxford disclaims any responsibility for the materials
contained in any third party website referenced in this work.

To my wife, Svetlana, and our daughters, Alexandra and Katherine. Hopefully, not too many goodnight kisses had been missed during the completion of this book.

Contents

Preface

This book discusses many modern, cutting-edge applications of graph theory, such as traffic networks, navigable networks and optimal routing for emergency response, placement of electric vehicle charging stations and graph-theoretic methods in molecular epidemiology. Because of the rapid growth of research in this field, the focus of the book is on the up-to-date development of the aforementioned applications.

The proposed book will be ideal for researchers, engineers, transport planners and emergency response specialists who are interested in the recent development of graph theory applications. Moreover, this book can be used as teaching material for postgraduate students because, in addition to up-to-date descriptions of the applications, it includes exercises and their solutions. Some of the exercises mimic practical, real-life situations. Advanced students in graph theory, computer science or biology may use the problems and research methods presented in this book to develop their final-year projects, master's theses or doctoral dissertations; however, to use the information effectively, special knowledge of graph theory would be required.

The chapters of this book are independent and can be read in any order, giving the reader freedom and control over their reading experience. Another feature is the inclusion of an introductory chapter, which provides an overview of many important applications of graph theory and basic graph-theoretic algorithms. It is the author's hope that this publication of the recent development of graph theory applications will instigate further research in the field.

The Author

Dr Vadim Zverovich (MSc, 1989; PhD, 1993; Docent, 2000) is the Head of the Mathematics and Statistics Research Group at the University of the West of England (UWE) in Bristol. An accomplished researcher, he is also a Fellow of the United Kingdom Operational Research Society, and was previously a Fellow of the prestigious Alexander von Humboldt Foundation in Germany. In 2016, Dr Zverovich was awarded Higher Education Academy fellowship status, and the Faculty of Environment and Technology of the UWE named him its researcher of the year in 2017. Dr Zverovich's research interests include graph theory and its applications, networks, probabilistic methods, combinatorial optimization and emergency responses. With 30 years of research experience, he has published many research articles and two books on the above subjects and established an internationally recognized academic track record in the mathematical sciences covering both theoretical and applied aspects.

Email: vadim.zverovich@uwe.ac.uk

Disclaimer

Any statements in this book might be fictitious and they represent the author's opinion.

1

Introduction

In this chapter, we will give a brief overview of selected applications of graph theory, many of which gave rise to the development of graph theory itself. A range of such applications extends from puzzles and games to very serious scientific and real-life problems, thus illustrating the diversity of applications. The first section is devoted to the six earliest applications of graph theory. The next section introduces so-called scale-free networks, which include the web graph, social and biological networks. The last section describes some of the basic graph-theoretic algorithms which can be used to tackle a number of interesting applications and problems of graph theory.

1.1 Early Applications of Graph Theory

In this section, we will briefly discuss early applications of graph theory, which are well described in the existing literature. A *graph* G is an ordered pair (V, E), where V is a non-empty and finite set of *vertices*, and E is a set of some unordered pairs of V, which are called *edges*. The vertex set and the edge set of a graph G may also be denoted by $V(G)$ and $E(G)$ to avoid confusion. When dealing with applications of graph theory, the terms 'graphs', 'vertices' and 'edges' can be called 'networks', 'nodes' and 'links', respectively. In some cases, directed edges can be considered—this will always be obvious from the context.

1.1.1 The Königsberg Bridge Problem and Eulerian Graphs

In 1736, Euler solved the famous Königsberg Bridge Problem [17], which was the very first problem formulated and solved in terms of graph theory. The problem was to find a round tour in Königsberg where each of the seven bridges was visited exactly once (see Figure 1.1). The initial important step, which is left to the reader as an exercise at the end of this chapter, is to represent the city with seven bridges as a graph. In terms of graph theory, a round tour

Modern Applications of Graph Theory. Vadim Zverovich, Oxford University Press (2021). © Vadim Zverovich.
DOI: 10.1093/oso/9780198856740.003.0001

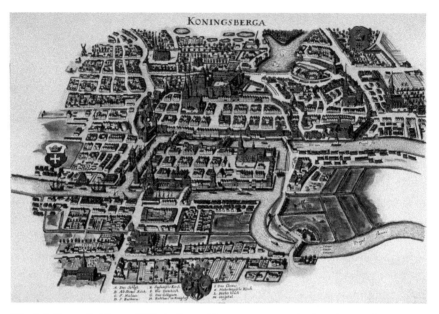

Fig. 1.1 Seven bridges of Königsberg. (Copper engraving by M. Merian, Photo: akg-images)

in a graph is a cycle; that is, a closed walk where vertices may be repeated. Nowadays, if a graph has a cycle that visits each edge exactly once, then such a cycle is called *eulerian* and the graph itself is called an *eulerian graph*. Here is the very first result in graph theory, which is attributable to Euler:

Theorem 1.1 [17] *A connected graph G is eulerian if and only if every vertex of G has even degree.*

The well-known Fleury algorithm [18] and Hierholzer algorithm [26] find an eulerian cycle in an eulerian graph, but the latter is more efficient. Note that eulerian cycles can be used in DNA fragment assembly and other practical applications such as mail carriers routing and floor designs.

1.1.2 Kirchhoff's Laws for Electrical Networks

More than 100 years after Euler's article, in 1847, Kirchhoff [28] modelled electrical networks by their underlying graphs. Interestingly, Kirchhoff was born in Königsberg, where he graduated from a local university. Kirchhoff developed the theory of trees, which are a subclass of all graphs, to describe

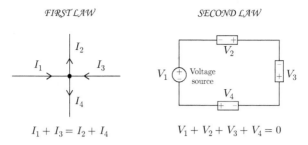

Fig. 1.2 Kirchhoff's laws.

the current around every cycle in an electrical network. More precisely, his construction was based on the independent cycles of a graph, which can be found by means of a spanning tree. Kirchhoff's first law is devoted to conservation of current:

Kirchhoff's First Law [28] *At any node v in an electrical network, the sum of the currents entering v is equal to the sum of the currents leaving the node v.*

Kirchhoff's second law deals with conservation of energy:

Kirchhoff's Second Law [28] *The algebraic sum of all voltages around any cycle in an electrical network is equal to zero.*

Kirchhoff's first and second laws are also called Kirchhoff's Current Law and Voltage Law, respectively. These laws are illustrated in Figure 1.2.

1.1.3 The Four-Colour Problem

The Four-Colour Problem has definitely been the most famous problem in graph theory: is it true that any map on a plane can be coloured with at most four colours in such a way that adjacent countries have different colours? This problem was discussed by Guthrie and De Morgan in ca. 1850, and it is possible that Möbius knew the problem even earlier. The Four-Colour Problem can easily be reformulated in terms of graph theory if countries are replaced by vertices, and two vertices are adjacent (i.e. connected by an edge) if the corresponding countries share a boundary. Thus, the problem is to colour the vertices of a planar graph with at most four colours in such a way that adjacent vertices have different colours. There is a long history of attempts to prove that

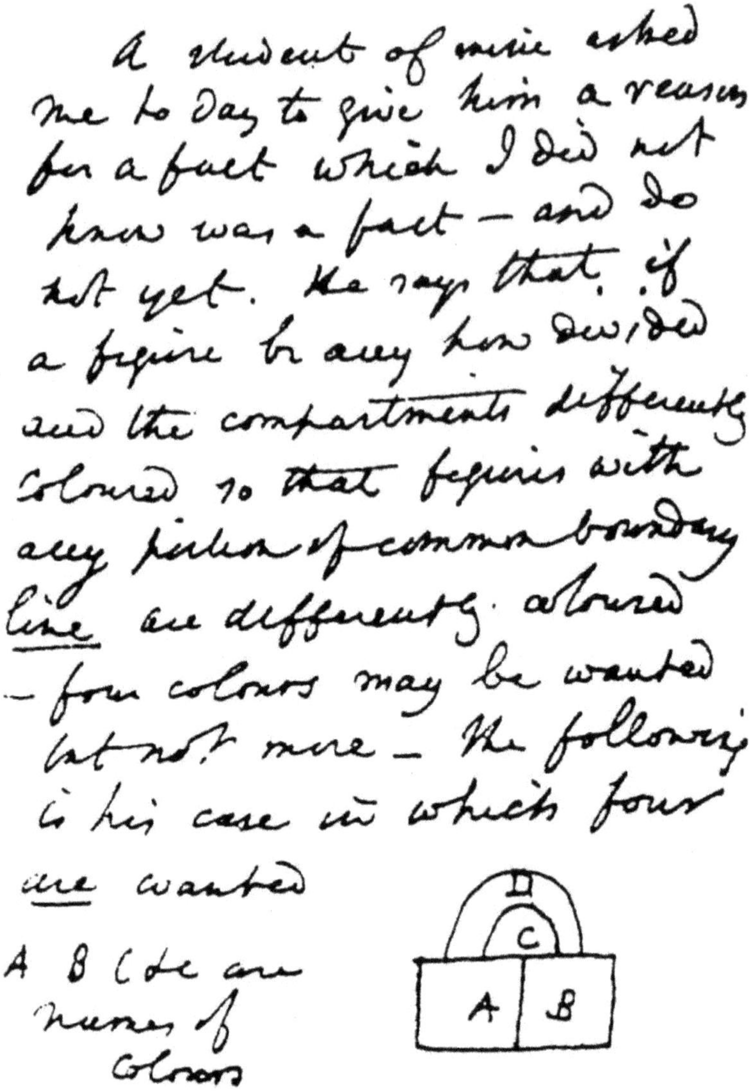

A student of mine asked
me to Day to give him a reason
for a fact which I did not
know was a fact — and do
not yet. He says that, if
a figure be any how divided
and the compartments differently
coloured so that figures with
any portion of common boundary
line are differently coloured
— four colours may be wanted
but not more — the following
is his case in which four
are wanted

A B C &c are
names of
colours

Fig. 1.3 Extract from De Morgan's letter to Hamilton about the Four-Colour Problem (1852, Dublin).

four colours are enough or find a counterexample. Finally, in 1977, Appel and Haken gave a long 'algorithmic' proof with the aid of a computer program, but a reasonably short and elegant proof of the Four-Colour Theorem is yet to be found.

Four-Colour Theorem [3, 4, 5] *Every planar graph is four-colourable.*

There are many interesting applications of graph colouring: mobile radio frequency assignment, scheduling/timetabling, register allocation in compiler optimization etc.

1.1.4 Hamilton's Games and Hamiltonian Graphs

Sir William Rowan Hamilton [22, 23] in ca. 1857 invented the Icosian Game, which included 15 example puzzles and was based on a two-dimensional (2D) representation of a dodecahedron with 20 vertices (see Figure 1.4a).

Another version of the game with simplified rules was called the Traveller's Dodecahedron. In this game, the dodecahedron has the form of a semi-sphere with a handle, which looks like a wooden mushroom, and its 20 vertices represent cities (see Figure 1.4c). The idea of the puzzle was to travel around the world (i.e. the semi-sphere) and attempt to find a round tour which would visit each city exactly once. In terms of graph theory, such a round tour in a graph is a spanning simple cycle. In other words, all vertices are included in the cycle and they are not repeated, apart from the first and the last. Such a cycle is called *hamiltonian*, and a graph is called *hamiltonian* if it has a hamiltonian cycle.

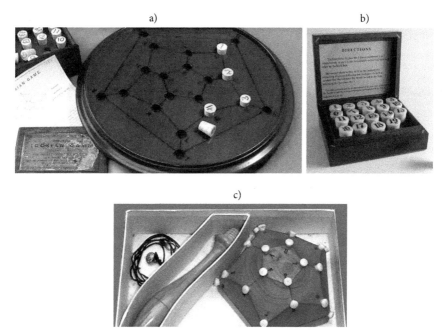

Fig. 1.4 a) The Icosian Game; b) ceramic pegs; c) the Traveller's Dodecahedron. (© 2018 The Puzzle Museum/JCD. http://puzzlemuseum.org)

There are many results devoted to hamiltonian graphs, but a complete characterization of such graphs is unknown. The following sufficient condition for a graph to be hamiltonian is due to Dirac:

Theorem 1.2 [14] *Let G be a graph with n vertices, $n \geq 3$. If every vertex of G has degree at least $n/2$, then G is hamiltonian.*

Hamiltonian cycles are closely related to the travelling salesman problem (TSP), which has numerous important applications. In fact, given a weighted graph, the TSP seeks to find a hamiltonian cycle of the smallest possible weight. The weight of an edge might be defined as the distance between its end-vertices, the cost of travel between them etc. Applications of the TSP include job scheduling, vehicle routing and computer wiring, to name a few.

1.1.5   Chemical Isomers and Enumeration of Trees

In 1857, Cayley [10] studied an application of graphs in organic chemistry, which is based on trees. The problem was to count the number of isomers of the *saturated hydrocarbons* C_nH_{2n+2}, also called *paraffins*, where n is the number of carbon atoms. For example, C_1H_4 represents methane, and C_3H_8 stands for propane. Butane and isobutane have the same formula C_4H_{10}, see Figure 1.5. Thus, the problem was to enumerate trees in which every vertex has degree 1 or 4. The starting point was to simplify the problem and count labelled trees:

Theorem 1.3 [9] *There are n^{n-2} labelled trees with n vertices.*

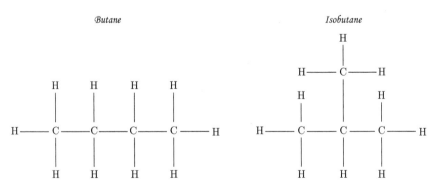

Fig. 1.5 Butane and isobutane (C_4H_{10}).

This formula also gives the number of spanning trees in a complete labelled graph. The following theorem provides the number of labelled trees with a given degree sequence:

Theorem 1.4 [33, 34, 40] *The number of labelled trees with a degree sequence* $(d_1, d_2, ..., d_n)$ *is equal to the multinomial coefficient*

$$\binom{n-2}{d_1 - 1, ..., d_n - 1}.$$

It may be pointed out that counting unlabelled trees (up to isomorphism) is a more challenging problem—no exact formula is known yet.

1.1.6 Chessboard Puzzles

We conclude this section with the famous eight queens puzzle, which was published in 1848 by the German chess player Bezzel [7]. The problem is to place eight queens on a standard 8×8 chessboard in such a way that no two queens attack each other. This means that any two queens cannot be in the same column, row or diagonal. It is not too difficult to find a particular solution to this puzzle, so the problem was later extended to finding all possible solutions. Two years later, in 1850, Nauck independently proposed the extended version of the puzzle [35] and also published all 92 solutions [36]. Notice that actually there are 12 basic solutions if symmetry (reflection and rotation) is taken into account. Interestingly, Gauss had also been involved in solving the puzzle, see [38] for further details.

Although the eight queens puzzle is not formulated in terms of graph theory, the chessboard can be represented by the *queen graph*, where vertices are squares of the chessboard and two different vertices are adjacent if the corresponding squares are in the same column, row or diagonal. Thus, it is necessary to find all independent sets of size 8 in the queen graph. In 1972, Dijkstra [11] developed a depth-first backtracking algorithm to solve this problem. Similar questions can be posed for other chess pieces: king, rook, bishop and knight; and they can be extended to an $n \times n$ chessboard.

In 1862, de Jaenisch [12] considered the problem of finding the smallest dominating set in the queen graph; that is, the smallest collection of queens which would threaten all unoccupied squares. Similar to the eight queens puzzle, the domination problem was generalized for an $n \times n$ chessboard.

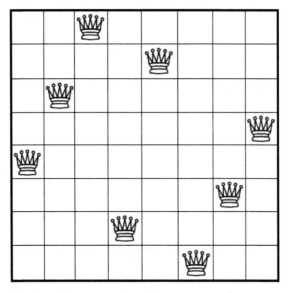

Fig. 1.6 One of twelve solutions of the eight queens puzzle.

1.2 Scale-Free Networks

In this section, we will discuss some networks whose degree distributions follow a power law. Such networks are called *scale-free networks*, and they are typically very large, sparse graphs with a small world structure. These properties are explained below.

In a given graph G with n vertices, let n_k denote the *number of vertices of degree k*, and let the sequence

$$(n_k, \ 0 \leq k \leq n - 1)$$

be the *degree distribution* of G. The degree distribution of G is said to follow a *power law* if, for every degree $k > 0$,

$$\frac{n_k}{n} \sim \frac{1}{k^\gamma}, \tag{1.1}$$

where $\gamma > 1$ is a fixed constant. The number

$$\mathbb{P}(k) = \frac{n_k}{n}$$

is the proportion of vertices of degree k and it can be viewed as a probability that a vertex is of degree k. The asymptotic relationship (1.1) is applicable for infinite graphs, and for finite graphs it means 'approximately equal to': $\mathbb{P}(k) \approx k^{-\gamma}$. In the case when G is a directed graph, the power laws for the in-degree distribution and the out-degree distribution are defined analogously.

An unusual feature of the power law distribution is its broad tail representing a polynomial decay when k tends to infinity, as opposed to typical distributions with exponential decay. This means that in massive power-law graphs there are vertices of high degree, which are called *hubs*. It may also be mentioned that in real networks the power law may have some discrepancies for vertices of very small and large degrees.

Although the power law distribution is typically applicable to massive graphs, let us consider a small graph G with 17 vertices. If G has degree sequence $(1, 1, 1, 1, 1, 1, 1, 1, 1, 1, 2, 2, 2, 2, 3, 3, 4)$, then its degree distribution is $(0, 10, 4, 2, 1, 0, 0, 0, 0, 0, 0, 0, 0, 0, 0, 0, 0)$. We can see that, apart from some discrepancy for vertices of degree 1, the graph G approximately follows a power law with exponent $\gamma = 2$:

$$\mathbb{P}(2) = \frac{4}{17} \approx \frac{1}{4}, \quad \mathbb{P}(3) = \frac{2}{17} \approx \frac{1}{9}, \quad \mathbb{P}(4) = \frac{1}{17} \approx \frac{1}{16}.$$

1.2.1 The Web Graph and Similar Technological Networks

The web graph is a relatively new phenomenon, which appeared after the birth of the internet in the early 1980s. Notice that the internet is a more general term, which includes e-mail, hardware and other components. In the *web graph* $\mathcal{W}$, vertices represent web pages and edges are the hyperlinks between web pages. Although the hyperlinks are directed edges, in some cases $\mathcal{W}$ can be viewed as an undirected graph. The web graph $\mathcal{W}$ is a real-world graph, which is continuously developing over time. Also, there are dynamic web pages (e.g. an online clock) and the number of such pages can be viewed as infinite. Hence, the web graph $\mathcal{W}$ may be considered as either finite or infinite.

Let us briefly discuss some important properties of $\mathcal{W}$. Firstly, it is an immense graph with billions of vertices. Secondly, $\mathcal{W}$ is a relatively sparse graph because the number of edges in $\mathcal{W}$ is much less than the number of all possible edges. More precisely, the average degree of a fixed vertex in $\mathcal{W}$ does not exceed a constant [6]. Thirdly, the web graph has many 'communities'. A *community* is a set of web pages having a common interest. There are several approaches for defining communities mathematically. One may define a community as a subgraph having more internal edges than external. Another definition is based on dense directed bipartite subgraphs, which are called *bipartite cores*. An example of a bipartite core is a directed complete bipartite graph $K_{3,3}$, where the directed edges go from one part to another, and some directed edges may be added to the parts.

One of the most important properties of W is the power law degree distribution. It implies that most vertices have a small number of incident edges, whereas a few vertices (hub nodes) have a large number of incident edges. There is some evidence in the existing literature [2, 8] that the in-degree distribution of the web graph W follows a power law with $\gamma = 2.1$, whereas $\gamma = 2.5$ or 2.7 for the power law out-degree distribution.

The web graph W possesses the so-called small-world property introduced in [41]. Let $d(u, v)$ denote the distance between vertices u and v in an undirected graph G, and let X consist of all pairs of different vertices for which $d(u, v)$ is finite. The *average distance* of G is defined as follows:

$$L(G) = \sum_{u,v \in X} \frac{d(u, v)}{|X|}.$$

For a connected graph G with the vertex set V, this definition can be rewritten in the following way:

$$L(G) = \binom{n}{2}^{-1} \sum_{u,v \in V} d(u, v).$$

A similar parameter, $L_d(G)$, can be defined for a directed graph G, where the distance between two vertices is the number of edges in the shortest directed path between those vertices. The small-world property says that $L(G)$ (or $L_d(G)$) should be much smaller than the number of vertices in G:

$$L(G) \leq c \ln \ln n,$$

where c is some small constant. For the web graph W, it was reported in the literature [2, 8] that $L(W) = 6.8$, whereas $L_d(W) = 16$ or 19.

Another characteristic of a small-world graph reflects its local density, which should be relatively large compared with standard random graphs having the same number of vertices and edges. If $N(v)$ denotes the set of vertices adjacent to a vertex v, then $|E\langle N(v)\rangle|$ is the number of edges in the graph induced by $N(v)$. This number is actually the number of triangles (i.e. cycles C_3) passing through v. The *clustering coefficient of a vertex* v is defined as follows:

$$C(v) = \frac{|E\langle N(v)\rangle|}{\binom{deg(v)}{2}}.$$

If v has degree k and there are q edges in the neighbourhood $N(v)$, then

$$C(v) = \frac{2q}{k(k-1)}.$$

Now we can define the *clustering coefficient of* G:

$$C(G) = \frac{1}{n}\sum_{v\in V}C(v).$$

Based on a large sample of the web graph, it was reported in [1] that $C(\mathcal{W}) = 0.1078$, whereas $C(R) = 0.0002$ for a random graph R with the same number of vertices and edges. For the directed interpretation of the web graph and the clustering coefficient, it was also reported that $C_d(\mathcal{W}) = 0.081$ and $C_d(R) = 0.001$. This confirms the claim that the clustering coefficient of the web graph is much larger compared with a random graph having the same number of vertices and edges.

The internet has different structural levels, hence the *internet graph* $\mathcal{I}$ can be defined differently. For example, vertices of $\mathcal{I}$ may represent domains and edges are connections between domains. Alternatively, vertices of $\mathcal{I}$ can be routers and edges are connections between routers. Both of these internet graphs follow a power law with exponent $\gamma = 2.5$. The *blog graph* is an induced subgraph of the web graph, where vertices represent web blogs and directed edges are the links between blogs. This graph follows a power law with exponent $\gamma = 2.1$ for both in-degree and out-degree distributions, and it has a rich community structure. In the *call graph*, vertices represent telephone numbers and two vertices are connected if there was a telephone call from one number to the other over, say, one day. The directed call graph also follows a power law with exponent $\gamma = 2.1$.

Further interesting properties of the aforementioned graphs, such as betweenness and orders of connected components, can be found in the existing literature (e.g. see [15]).

1.2.2 Social Networks

One of the first social networks was the so-called *acquaintance graph*, where people are represented by vertices and two vertices are connected by an edge if the corresponding individuals are acquainted. Let us recall that the *diameter* of a graph G is defined as follows:

$$\text{diam}(G) = \max_{u,v \in V} d(u, v).$$

Milgram [32] investigated the acquaintance graph and concluded that its diameter is about six, which gave rise to the commonplace truisms 'six degrees of separation' and 'it's a small world'.

In the well-known acquaintance puzzle, it is necessary to prove that at a party with six people, there are three individuals who are either mutually acquainted or mutually not acquainted. For example, if a six-vertex acquaintance graph H is isomorphic to a cycle C_6, then no three people are mutually acquainted, however we can easily find three individuals who are not mutually acquainted—this set of people is a three-vertex independent set.

Theorem 1.5 *Let H be a graph with six vertices. Then, H contains either a triangle K_3 or a three-vertex independent set.*

The acquaintance puzzle is closely related to Ramsey numbers, which play an important role in extremal graph theory. The *Ramsey number $R_{l,m}$* is equal to the smallest number such that each graph with $R_{l,m}$ vertices contains a complete graph K_l or an m-vertex independent set. The well-known Ramsey's theorem says that the Ramsey numbers exist. Also, the following upper bound can be deduced.

Theorem 1.6 *The Ramsey numbers satisfy:*

$$R_{l,m} \leq \binom{l + m - 2}{l - 1}.$$

In the *collaboration graph*, vertices represent researchers in some subject area and edges connect researchers who collaborate with each other. Newman [37] studied collaboration graphs for different groups of scientists and concluded that a typical collaboration graph F has the following small-world properties:

$$\text{diam}(F) \approx 20, \quad L(F) \approx 6 \quad \text{and} \quad C(F) \geq 0.3.$$

Some social networks are modelled by directed graphs if the relationship between given objects is ordered; for instance, if one person is shorter than another. A further example is a directed *graph of citations* for all publications in some subject area. Let us model kinship in a given group of people as follows: the vertices of G are individuals from the group and two individuals x and y are connected by a directed edge xy if x is a parent of y. If A is an adjacency matrix of G, then

$$A^2 = A \times A$$

is the adjacency matrix of a directed graph H. It can be proved mathematically that H represents the grandparent–grandchild relationship in the original group of people. Notice that H might be a multi-digraph if inbreeding is not forbidden. This example can be further generalized for the adjacency matrix A^k of a graph representing the ancestral relationship through k generations.

The degree distributions of social networks typically follow a power law or, in some cases, a power law with exponential cut-off.

1.2.3 Biological Networks

In general, a biomolecular network consists of vertices, which represent biomolecules, and edges are interactions between them. The biomolecules can be large molecules (e.g. proteins, genes) or small molecules (e.g. nucleic acids, sugars). For instance, in a protein interaction network, the edges represent the physical contacts between proteins in a cell.

Graph-theoretic methods have been applied for analysing structural properties of biomolecular networks. Similar to the web graph, many genetic, metabolic and protein interaction networks possess the small world property. It was reported that for some samples, the average distance is between 3 and 5, the diameter is small and the clustering coefficient is relatively high.

Some nodes in a network are more important than others; for instance, particular proteins in interaction networks can be structurally more significant. The so-called centrality measures play an important role in identifying the nodes having a 'high rank'. One important aspect is degree distribution, which was shown to follow a power law for many biological networks, for example metabolic networks and protein interaction networks. Transcriptional regulatory networks, which are a subset of gene regulatory networks, have power law out-degree distribution, however in-degrees follow an exponential distribution. There is some evidence to suggest that the degree of a vertex in a protein interaction network points out the likelihood of a protein to be essential: non-essential proteins typically have lower degrees compared with the degrees of essential proteins.

For a connected graph G, the *eccentricity* of a vertex u is defined as follows:

$$e(u) = \max_{v \in V} d(u, v).$$

Then, the *centre* of G is the set

$$\{u \in V : e(u) = \min_{v \in V} e(v)\}.$$

In other words, the centre of a graph consists of vertices with smallest eccentricity. In metabolic networks, the centre can represent some key components of the the organism's metabolism and it might be viewed as a 'bottleneck'.

Betweenness centrality is another measure for quantifying the importance of a vertex—the important vertex should belong to many shortest paths between other vertices, thus controlling their 'communication'. Let σ_{vw} denote the number of shortest paths between vertices v and w in a graph G, and $\sigma_{vw}(u)$ be the number of shortest (v, w)-paths that pass through a vertex u, where u, v and w are all distinct. The *betweenness centrality* of u is given by

$$B(u) = \sum_{\substack{v, w \in V \\ v \neq u \neq w}} \frac{\sigma_{vw}(u)}{\sigma_{vw}}.$$

For example, in yeast protein interaction networks, the average betweenness centrality for essential proteins is much higher than for non-essential proteins.

Finally, the idea that an important vertex should be adjacent to other important vertices underlies the concept of eigenvector centrality. Let A be the adjacency matrix of a connected graph G, and λ be the spectral radius of A; that is, the largest absolute value of its eigenvalues. Because G is connected, the Perron–Frobenius Theorem implies that the multiplicity of the spectral radius λ is one. Hence, λ uniquely specifies the corresponding normalised eigenvector w:

$$Aw = \lambda w.$$

If $V = \{v_1, v_2, ..., v_n\}$, then the *eigenvector centrality* of v_i is the i-th coordinate of the eigenvector w. In some cases, eigenvector centrality performs better than other centrality measures. This concept will be further explored in Chapter 6.

1.3 Graph-Theoretic Algorithms

In this section, we will discuss the most basic algorithms of graph theory: the Breadth-First Search (BFS) method, the Depth-First Search (DFS) method, Kruskal's algorithm and Dijkstra's algorithm. These methods can be used to tackle miscellaneous real-life applications, and the corresponding algorithms are described below. There are also a few other important graph algorithms, which should be mentioned here for completeness.

In some applications modelled by graphs, for example in chemistry, it is important to find a *matching*, which is a set of non-adjacent edges. A maximum matching in a graph can be found by Edmonds' blossom algorithm [16]. The Hungarian algorithm [31] solves the very well-known assignment problem. In terms of graph theory, the problem is to find a minimum weight perfect matching in a weighted bipartite graph. This can be extended to a network flow model, the so-called transshipment problem, where the objective function has to be minimized. The maximum flow problem in a given network can be solved by the Ford–Fulkerson algorithm [20].

Planar graphs are important in many real-life applications, for example for the facilities layout problem and printed circuit board design. The problem of recognizing whether a given graph is planar can be solved by the Hopcroft–Tarjan planarity testing algorithm [27]. However, this algorithm does not find some planar graph embeddings, which was fixed by Kocay [29].

Furthermore, a sequence $\underline{d}$ of non-negative integers is called *graphic* [42] if there exists a simple graph whose degree sequence is $\underline{d}$. The Havel–Hakimi algorithm [21, 25] determines whether a given sequence $\underline{d}$ is graphic. In addition, for a graphic sequence $\underline{d}$, the algorithm recursively constructs a simple graph whose degree sequence is $\underline{d}$.

1.3.1 Breadth-First Search Algorithm

When analysing a graph, in many cases we want to explore its vertices or edges step by step by traversing the graph layerwise. We start traversing at a given vertex u and examine all vertices adjacent to u. Then, a similar procedure is applied to the vertices found in the previous step, thus exploring the next layer, and so on. This is the main idea of the BFS method.

The BFS method can be used to find the shortest path between two vertices u and v in a graph G (in terms of the number of edges), and hence the distance $d(u, v)$ between the vertices. Also, it can help in determining a connected component, or all connected components, of G. Another application is to decide whether G is bipartite; that is, if the vertex set of G can be split into two sets A and B, so that there are no edges in part A and no edges in part B. In a directed graph, the BFS algorithm can find all vertices reachable from a given vertex. A more specific application of BFS will be explored in Chapter 3. This method will be used for finding the minimum number of obstructions between a given location in a building and a given epicentre of an extreme event (e.g. fire).

For a graph $G = (V, E)$, the time complexity of BFS is $O(|V| + |E|)$.

Algorithm 1.1: Breadth-first search.

Input: A graph G and a starting vertex u.
Output: The graph G with some (or all) vertices labelled by $0, 1, 2, \ldots$

(1) Start with the vertex u and label it by 0.
(2) Label by 1 all vertices in the neighbourhood of the vertex u. Denote by L_1 the set of vertices with labels 1.
(3) Label by 2 all unlabelled vertices in the neighbourhoods of the vertices labelled by 1. Denote by L_2 the set of vertices with labels 2.
(4) Repeat the following labelling procedure:
Let the set L_{k+1} consist of all unlabelled vertices in the neighbourhoods of the vertices labelled by k. If L_{k+1} is not empty, label by $k + 1$ all vertices in L_{k+1}. If L_{k+1} is empty, the algorithm stops.

1.3.2 Depth-First Search Algorithm

The DFS method is another technique for traversing a graph. In contrast to the BFS approach where we explore vertices layerwise, the DFS algorithm examines each 'branch' to the maximal depth before backtracking and starting to explore another part of the graph. Thus, the BFS process stays at a given vertex/layer as long as possible to explore all unlabelled neighbours in 'breadth', whereas the DFS method quickly moves in 'depth' from a given vertex to the first unlabelled neighbour.

The DFS approach can be used for finding spanning trees, bridges and fundamental cycles. Also, it can help in determining connected components, 2-connected and 3-connected components, and strongly connected components. DFS is used in many algorithms for providing a useful labelling of vertices and a spanning arborescence (i.e. a directed rooted tree where every edge is directed away from the root). For simplicity of presentation, the output of the DFS algorithm below includes a spanning rooted tree, not an arborescence, however the former can easily be transformed into the latter. Some further applications of DFS include planarity testing, web-crawling and artificial intelligence.

Algorithm 1.2: Depth-first search.

Input: An undirected connected graph G and a starting vertex u.

Output: A spanning rooted tree T of the graph G with their vertices labelled by $0, 1, 2, \ldots$

(1) Start with the vertex u and label it by 0. The vertex u will be the root of the tree T. Find an edge uv, let $uv \in E(T)$ and mark it as 'examined'. Move to the vertex v and label it by 1.

(2) In general, suppose that the current vertex is v and the largest assigned label is k:

If there are no unexamined edges incident to v, then go to Step 3. Otherwise, choose an unexamined edge vw. There are two possibilities:

- If the vertex w has a label, then we remain at v.
- If the vertex w has no label, then we move to w and label it by $k + 1$. Also, we set $vw \in E(T)$.

Mark the edge vw as 'examined' and repeat Step 2.

(3) Backtrack from v to the previous vertex in T by moving towards the root u. If the previous vertex is the starting vertex u and all edges incident to u are examined, then the algorithm stops. Otherwise, go to Step 2.

Note that, in general, the input for the DFS algorithm may be a disconnected graph, in which case DFS can be applied to each connected component producing a spanning forest and the corresponding labellings. Also, it may be mentioned that in large or massive graphs, incomplete DFS can be applied; that is, the search is terminated at a certain depth. However, such an incomplete search may be biased because the preference would be given to vertices of high degree.

For traversing an entire graph $G = (V, E)$, which may be disconnected, the time complexity of DFS is $O(|V| + |E|)$.

1.3.3 The Minimum Weight Spanning Tree Problem and Kruskal's Algorithm

Assume that we want to design a new gas pipeline network connecting a number of cities so that any two cities are connected—not necessarily directly, they may be connected via other cities. In addition, we want to minimize the total construction cost assuming that the construction cost of a link between any two cities is known. It is easy to see that such a network N is a connected graph, and there are no cycles because of cost restrictions. Formally speaking, the network N is a tree connecting all the cities; that is, N is a spanning tree. Typically, there may be many such trees, so the challenge is to find a tree with the smallest construction cost.

Many applications that are based on real-life networks can be modelled by graphs. The edges of such graphs often have some characteristics like distances, travel costs, construction costs, times etc. Such graphs are called *weighted graphs*. A typical problem in weighted graphs is to find a connected spanning subgraph of the smallest possible weight. As discussed above, such a subgraph is a tree because it cannot have a cycle. The problem of finding a spanning tree of the smallest possible weight can arise in design of electric power grids, railway networks, communication networks, water supply networks, natural gas networks etc. This problem can be efficiently solved by Kruskal's algorithm [30] or Prim's algorithm [39]. Here is the former:

Kruskal's Algorithm: The minimum weight spanning tree.

Input: A weighted connected graph G.
Output: A spanning tree T of minimal weight in G.

(1) Let T be an empty graph with the vertex set $V(G)$.

(2) Add to T an edge of minimal weight in G.

(3) Repeat the following until T becomes a tree:
   - Among unexamined edges of G, find an edge of minimal weight (if it is not unique, settle this randomly).
   - Add this edge to T if it does not create a cycle in T.

(4) Algorithm stops.

In Prim's algorithm [39], which is similar to Kruskal's algorithm, we start with a two-vertex tree T' consisting of the edge of minimal weight. Then, step by step, we add to T' an edge of minimal weight that connects the tree T' with one of the remaining vertices. Hence, there is no need to check that such an edge creates a cycle, but we must make sure that exactly one end-vertex of the edge belongs to T'.

For a weighted connected graph $G = (V, E)$, the time complexity of Kruskal's algorithm is $O(|E| \log |V|)$. The time complexity of Prim's algorithm is $O(|V|^2)$ if the graph is stored as an adjacency matrix/list. However, for sparse graphs this complexity can be improved if binary or Fibonacci heaps are used.

1.3.4 The Shortest Path Problem and Dijkstra's Algorithm

In the shortest path problem, we want to find a path of minimum weight between a starting vertex s and a target vertex t in a weighted connected graph G. The *weight of a path* is the sum of the weights of all edges belonging to the path. This should not be confused with the length of a path, which is the number of edges in the path—it is used for defining the distance between two vertices in a graph.

There are countless applications of the shortest path problem. The simplest example is to find the fastest route between two locations in a road network, where travel time is known for each road segment. Alternatively, if we know the length of every road segment, then we might want to find the shortest route between two locations, where the actual length (in kilometres) of a route is minimized. Another application is to determine the safest path in a building, where an extreme event is occurring—this will be explored in Chapter 3.

The shortest path algorithm presented below is due to Dijkstra [13]. In this algorithm, the set Q denotes the set of *unvisited* vertices, hence $V - Q$ consists of *visited* vertices. The label $D(x)$ is the weight of the shortest (s, x)-path if x was visited; this label is called *permanent*. If x is unvisited, then $D(x)$ is the weight of the shortest (s, x)-path found so far; this label is *temporary*. Also, each visited vertex $x \neq s$ has a label $p(x)$, which is the vertex preceding x in the shortest (s, x)-path.

A general iteration of the algorithm is to find the vertex $q^* \in Q$ with the smallest label $D(q^*)$. The vertex q^* becomes 'visited' and $D(q^*)$ becomes 'permanent'; that is, $D(q^*)$ is the minimum weight of a path from s to q^*. Next, we recalculate $D(y)$ for all neighbours of q^* in the set Q and redefine the label $p(y)$. The algorithm stops when we reach the vertex t (i.e. when $q^* = t$). The (s, t)-path of minimum weight can be found by backtracking from the vertex t using the labels $p(t)$, $p(p(t))$ etc. The weight of this path is $D(t)$.

Dijkstra's Algorithm: The shortest path.

Input: A weighted connected graph G, where $w(e) \geq 0$ for every edge e of G. A starting vertex s and a target vertex t.

Output: A shortest (s, t)-path in G (i.e. the minimum weight path).

(1) Set $D(v) = \infty$ for all $v \in V, v \neq s$. Set $D(s) = 0$ and $Q = V$. For each $v \in V$, $p(v)$ is undefined.

(2) Let q^* be a vertex in Q such that

$$D(q^*) = \min_{q \in Q} D(q).$$

Remove q^* from Q. If $q^* = t$, then go to Step 4.

(3) For each neighbour y of q^* such that $y \in Q$:

Calculate $z = D(q^*) + w(q^*y)$.
If $z < D(y)$, then $D(y) = z$ and $p(y) = q^*$.

Go to Step 2.

(4) Construct the shortest (s, t)-path: $p(t)$ is the vertex in the path before t; $p(p(t))$ is the vertex before $p(t)$ and so on. The algorithm stops.

Dijkstra's algorithm finds the shortest path between two vertices. However, with some adaptation, it can determine shortest paths from the starting vertex to all other vertices. This can be achieved if the algorithm does not stop at the vertex t and examines all vertices in the set Q. Thus, the algorithm would run until Q becomes an empty set and it would produce the shortest path tree rooted at s. Note that the input of the algorithm may be a directed graph.

The time complexity of Dijkstra's algorithm is $O(|V|^2)$ for a standard implementation. This complexity can be improved to $O(|V|\log|V| + |E|)$ if Fibonacci heaps are used.

Another path search approach is the A* algorithm [24], which extends and improves Dijkstra's algorithm. The A* algorithm uses heuristics for guiding its search, thus achieving better performance. Dijkstra's algorithm only works for weighted graphs with non-negative weights of edges. If there are negative weights but there exists no cycle of negative weight, then Floyd's algorithm [19] can find shortest paths between all pairs of vertices.

1.4 Exercises

1.4.1

Represent the city of Königsberg with seven bridges (Figure 1.1) as a graph, which reflects the relationship among land masses and bridges. The resulting graph G must not have multiple edges; that is, two different vertices in G are either connected by a single edge or not connected at all.

Is the graph G eulerian? Justify your answer.

1.4.2

This exercise is devoted to Kirchhoff's laws.

(a) Find the current I flowing out of the node v in Figure 1.7a.
(b) What is the current I' in Figure 1.7b?
(c) Find the voltage V_3 in Figure 1.8.

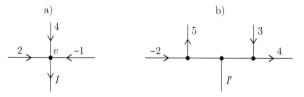

Fig. 1.7 Application of Kirchhoff's first law. The various currents are in milliamperes (mA).

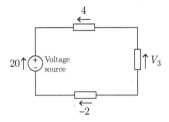

Fig. 1.8 Application of Kirchhoff's second law. The various voltages are in volts (V).

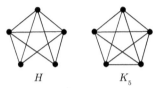

Fig. 1.9 The graphs H and K_5.

1.4.3

(a) What is the smallest number of colours needed to colour the vertices of the graphs in Figure 1.9?

(b) Is the complete graph K_5 planar? Justify your answer.

1.4.4

(a) Let D denote the dodecahedron graph of the Icosian Game in Figure 1.4a. Find a hamiltonian cycle in the graph D.

(b) The complement $\bar{D}$ of the graph D has the same vertex set, and two vertices in $\bar{D}$ are adjacent (i.e. connected by an edge) if they are not adjacent in D. Prove that the graph $\bar{D}$ is hamiltonian.

(c) Suppose that in the graph D the weights of the five outer edges are 2, and the weights of all other edges are 1. Solve the travelling salesman problem for the resulting weighted graph; that is, find a hamiltonian cycle of the smallest possible weight.

1.4.5

(a) Draw all unlabelled trees with four vertices. How many labelled trees with four vertices are there? Use your result to verify Theorem 1.3 for $n = 4$.

(b) Find all isomers of pentane C_5H_{12}. The isomers should be represented as unlabelled trees.

1.4.6

(a) There is a single solution (up to symmetry) of the six queens problem. Find this solution.
(b) Show that five queens can be placed on an 8×8 chessboard so that they attack all unoccupied squares.
(c) How can one place 14 bishops on an 8×8 chessboard so that they do not attack one another?
(d) How many knights can be placed on an 8×8 chessboard so that they do not threaten one another? (Propose a reasonable conjecture, but you do not need to prove it.)

1.4.7

(a) Describe graphs for which $C(G) = 1$. Explain when $C(G) = 0$.
(b) Find the clustering coefficient of the graph F shown in Figure 1.10.
(c) Calculate $L(Q_3)$, where Q_3 is the cube shown in Figure 1.11.

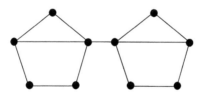

Fig. 1.10 The graph F.

Fig. 1.11 The cube Q_3.

Fig. 1.12 The acquaintance network S with 22 nodes. (Image by Gordon Johnson from Pixabay.)

1.4.8

(a) For the acquaintance network S in Figure 1.12, show that there are four individuals who are mutually acquainted and also six individuals who are not mutually acquainted.
(b) Find the diameter of the network S in Figure 1.12.
(c) Prove Theorem 1.5 without using Theorem 1.6.
(d) Show that $R_{3,3} = 6$.

1.4.9

(a) Let the path P_6 be (a, b, c, d, f, g), where the sequential vertices are adjacent. Find the centre of P_6.
(b) Calculate the betweenness centrality of each vertex in the path P_6 from part (a). Is there any relationship between the centre of P_6 and betweenness centrality of its vertices?
(c) Express the diameter of a graph G in terms of the eccentricities of its vertices.
(d) A given network N with 100 nodes has the following proportions of vertices of different degrees: $\mathbb{P}(1) = 0.70$, $\mathbb{P}(2) = 0.18$, $\mathbb{P}(3) = 0.06$, $\mathbb{P}(4) = 0.03$, $\mathbb{P}(5) = 0.02$, $\mathbb{P}(6) = 0.01$; $\mathbb{P}(k) = 0$ if $k = 0$ or $k \geq 7$. Can we conclude that N approximately follows a power law with exponent $2 \leq \gamma \leq 3$? If so, find the exponent γ (to 1 decimal place).

1.4.10

For a given graph G, explain how the BFS algorithm can be used to:

(a) Find the distance $d(u, v)$ between vertices u and v, and the shortest (u, v)-path.
(b) Determine all connected components of G.
(c) Decide whether G is bipartite. (You may assume that the graph G is connected.)

1.4.11

(a) Let G be a graph, not necessarily connected. Using the DFS approach, devise a method for deciding whether G is a tree.
(b) A vertex v is a *cut-vertex* of a graph G if $G - v$ has more connected components than G. A graph G with at least two vertices is called 2-*connected* if G is a connected graph without a cut-vertex. A maximal 2-connected subgraph of a graph is a 2-*connected component*. For example, the graph in Figure 1.10 has three 2-connected components.

    Devise an algorithm to find all 2-connected components of a graph G, which may be disconnected.
(c) Show that the time complexity of DFS is $O(|V| + |E|)$ for traversing an entire graph, which may be disconnected, and show that the time complexity is $O(|E|)$ for connected graphs.

1.4.12

(a) Apply Kruskal's algorithm to the weighted cube Q_3 of Figure 1.13, and hence find the weight of a minimum spanning tree.
(b) Let $w(T)$ be the weight of a minimum spanning tree in a weighted complete graph G, and $w(H)$ be the weight of the optimal hamiltonian cycle in G. Find the relationship between $w(T)$ and $w(H)$ and prove it.

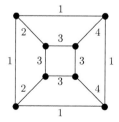

Fig. 1.13 The weighted cube Q_3.

1.4.13

(a) Using Dijkstra's algorithm, find the (s, t)-path of minimum weight in the weighted cube Q'_3 of Figure 1.14.

(b) In some cases, we do not want to use the optimal path (i.e. the minimum weight path) produced by Dijkstra's algorithm. For example, if the shortest route found by a car navigator goes through a city centre, we might want to use an alternative route to avoid congestion. Devise a method for finding an alternative near-optimal route, assuming that such a route exists.

(c) Let G be a connected graph with some negative edge weights.

   (i) Explain why Dijkstra's algorithm may fail to find a shortest path in the graph G. Give an example of such a graph G with a shortest path that cannot be found by Dijkstra's algorithm.

   (ii) Let w^* denote the smallest negative weight in G; that is,

$$w^* = \min_{e \in E} w(e).$$

Introduce new edge weights for G as follows: $w'(e) = w(e) - w^*$ for every $e \in E$. We have $w'(e) \geq 0$ for all edges in G. For given vertices s and t, run Dijkstra's algorithm for G with the updated weights $w'(e)$ to find an (s, t)-path of minimum weight in G. Explain why the resulting path is not necessarily an (s, t)-path of minimum weight in G with the original weights $w(e)$. You may want to use an example to illustrate this.

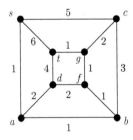

Fig. 1.14 The weighted cube Q'_3.

1.5 Solutions

1.4.1

To construct the graph G, we introduce a vertex for each land mass (large vertex) and for each bridge (small vertex). Two vertices representing a bridge and a land mass are connected by an edge if the bridge connects this land mass with another land mass. The resulting graph G is shown in Figure 1.15. This graph has vertices of odd degree. By Theorem 1.1, G is not eulerian. ■

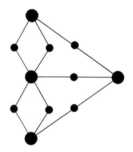

Fig. 1.15 The graph G without multiple edges that represents the city of Königsberg with seven bridges.

1.4.2

(a) Using Kirchhoff's first law, we have $2 + 4 - 1 = I$. Thus, $I = 5$ mA. ■
(b) The solution is shown in Figure 1.16. ■
(c) Using Kirchhoff's second law, we obtain $20 - 4 - V_3 + (-2) = 0$. Hence, $V_3 = 14$ V. ■

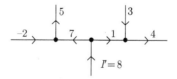

Fig. 1.16 Finding of unknown currents: 7 mA, 1 mA and $I' = 8$ mA.

1.4.3

(a) The graph H contains a complete graph with four vertices as a subgraph. Hence, we need four colours to properly colour the vertices of that subgraph. The remaining vertex can be coloured by one of those four colours because it is only adjacent to three vertices of the subgraph. We need five colours to colour the vertices of K_5 properly because it has all possible edges. ∎

(b) We know from part (a) that five colours are needed to colour the vertices of K_5. The graph K_5 is not planar, otherwise we could colour its vertices with four colours by the Four-Colour Theorem. ∎

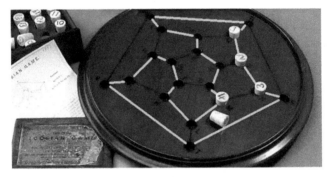

Fig. 1.17 A hamiltonian cycle of the graph D. (© 2018 The Puzzle Museum/JCD. http://puzzlemuseum.org)

1.4.4

(a) A hamiltonian cycle of the graph D is shown in Figure 1.17. ∎

(b) Every vertex of $\bar{D}$ has degree 16, and $\bar{D}$ has 20 vertices. By Theorem 1.2, $\bar{D}$ is hamiltonian. ∎

(c) Any hamiltonian cycle has 20 edges, and it must include at least three edges of weight 2. A hamiltonian cycle that contains exactly three edges of weight 2 can be seen in Figure 1.17, and its weight is $2 \times 3 + 1 \times 17 = 23$. ∎

1.4.5

(a) There exist two such trees: the path P_4 and the star $K_{1,3}$. They are shown in Figure 1.18.

Fig. 1.18 The path P_4 (left) and the star $K_{1,3}$ (right).

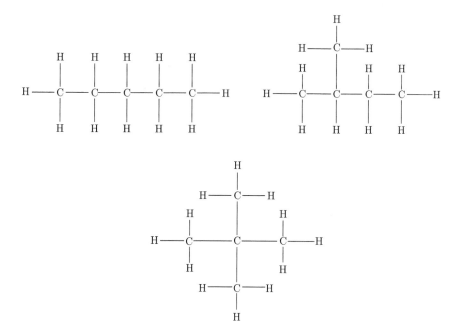

Fig. 1.19 Isomers of pentane C_5H_{12}.

There are four labellings of the star $K_{1,3}$ with numbers 1, 2, 3, 4. Also, there are twelve labellings of the path P_4. Thus, the number of labelled trees with four vertices is 16. By Theorem 1.3, this number must be $4^{4-2} = 16$. ∎

(b) The first step is to find three trees with five vertices, which would represent the 'carbon parts' of the isomers. Then, those trees can be extended to the actual isomers. They are shown in Figure 1.19. ∎

1.4.6

(a) a2, b4, c6, d1, e3, f5. ∎

(b) One possible solution: b3, c5, d2, e4, f6. ∎

(c) a1–a8, h2–h7. ∎

(d) 32 knights. They can be placed on 32 white squares. ∎

1.4.7

(a) The clustering coefficient is between 0 and 1. The clustering coefficient of a graph G is equal to 1 if and only if G is a complete graph or a disjoint union of complete graphs. $C(G) = 0$ if and only if G is a triangle-free graph. ∎

(b) $C(F) = 0.1(1 + \frac{1}{3} + \frac{1}{6} + 1 + \frac{1}{3} + \frac{1}{6}) = 0.3.$ ∎

(c) $L(Q_3) = \frac{1}{28}(1 \times 12 + 2 \times 12 + 3 \times 4) = 1\frac{5}{7}.$ ∎

1.4.8

(a) Four nodes at the right bottom corner form a complete subgraph. The following nodes form an independent set of size 6: two nodes (blue and dark brown) just under the top node, the brown node in the top right part, two green nodes in the lower part and the pink node to the left of them. ∎

(b) It can be checked directly that $\operatorname{diam}(S) = 3$. ∎

(c) Let us consider two cases. Suppose that a vertex $v \in V(H)$ is adjacent to three vertices a, b and c. If there is an edge between any two of these three vertices, then we have a triangle; otherwise $\{a, b, c\}$ is an independent set. Now suppose that v is adjacent to at most two vertices, which means that it is not adjacent to three vertices, say, x, y and z. If these three vertices are pairwise adjacent, then we have a triangle. Otherwise, some two vertices are not adjacent; for example, x is not connected to y, which implies that $\{v, x, y\}$ is an independent set. The proof is complete. ∎

(d) Let us consider a cycle C_5, which obviously does not contain a triangle and a three-vertex independent set. Therefore, $R_{3,3} \geq 6$. Taking into account the result from part (c), we conclude that $R_{3,3} \leq 6$. Thus, $R_{3,3} = 6$. ∎

1.4.9

(a) Let us calculate eccentricities for all vertices: $e(a) = e(g) = 5, e(b) = e(f) = 4$ and $e(c) = e(d) = 3$. Therefore, the centre of P_6 is $\{c, d\}$. ∎

(b) We have $B(a) = B(g) = 0, B(b) = B(f) = 4$ and $B(c) = B(d) = 6$. The set of vertices with highest betweenness centrality coincides with the centre of P_6. ∎

(c) Using the definitions of diameter and eccentricity, we obtain

$$\text{diam}(G) = \max_{v \in V} e(v).$$

∎

(d) Using Excel or by trial and error, we can conclude that N approximately follows a power law with exponent $\gamma = 2.5$:

$$\mathbb{P}(2) = 0.18 \approx 0.177 = 2^{-2.5},$$
$$\mathbb{P}(3) = 0.06 \approx 0.064 = 3^{-2.5},$$
$$\mathbb{P}(4) = 0.03 \approx 0.031 = 4^{-2.5},$$
$$\mathbb{P}(5) = 0.02 \approx 0.018 = 5^{-2.5},$$
$$\mathbb{P}(6) = 0.01 \approx 0.011 = 6^{-2.5}.$$

∎

1.4.10

(a) Run BFS with the starting vertex u. If the vertex v is labelled by k, then $d(u, v) = k$. To find the shortest (u, v)-path, we should first determine a vertex $v_{k-1} \in L_{k-1}$, which is adjacent to v. Then, we find a vertex $v_{k-2} \in L_{k-2}$, which is adjacent to v_{k-1}, and so on. The required path is $(u, v_1, v_2, ..., v_{k-1}, v)$. Note that if v is unlabelled by BFS, then u and v belong to different connected components of G, and hence $d(u, v)$ is undefined. ∎

(b) Run BFS with an arbitrary starting vertex u, and denote by C^1 all labelled vertices of G. It is easy to see that C^1 is a connected component of G. Repeat this procedure for the graph $G - C^1$ to find the second connected component C^2 and so on. ∎

(c) Run BFS with an arbitrary starting vertex u. Denote by A the set of vertices with even labels, and by B the set of vertices with odd labels. If there are no edges in A and no edges in B, then G is a bipartite graph by definition. Actually, it is enough to check that the sets L_k contain no edges. The correctness of the above procedure follows from the following observation. If a connected bipartite graph G with parts A and B is given, and $u \in A$, then, according to the BFS method, $L_1 \in B$, $L_2 \in A$ and so on. In other words, for any k, the set L_k belongs to one of the parts and hence it does not contain edges. ∎

1.4.11

(a) Let us recall that a tree is a connected acyclic graph. Apply DFS with an arbitrary starting vertex u. If all vertices of G were labelled, then G is a connected graph; otherwise G is disconnected and it cannot be a tree. Also, during the run of DFS we can count the number c of examined edges that were not included to T (the first part of Step 2). If $c > 0$, then G has a cycle and it cannot be a tree. Thus, G is a tree if all vertices of G have labels and $c = 0$. ∎

(b) We can find all connected components of G by running DFS (or BFS) many times—when a single run of DFS terminates with the highest label k, we run DFS again from any unlabelled vertex and start labelling with $k + 1$. The number of such runs will be equal to the number of connected components in G. In a similar way, let us choose a vertex x in G and determine the number of connected components in $G - x$. If the number of connected components increases, then x is a cut-vertex. Thus, we can find all cut-vertices of G and all 2-connected components. For example, is x is a cut-vertex that separates a connected component C_1 of G into three parts X, Y and Z without other cut-vertices, then $X \cup x$, $Y \cup x$ and $Z \cup x$ are 2-connected components of G. Note that for finding 2-connected components there is a more efficient algorithm based on DFS. ∎

(c) In Step 2 of the DFS algorithm we have to examine every edge incident to the vertex v exactly once. Hence, for a given vertex v (including the vertex u from Step 1), Step 2 is performed $\deg(v)$ times, and each run requires time $O(1)$. Thus, for all vertices, Step 2 needs time $O(\sum_{v \in V(G)} \deg(v)) = O(|E|)$. Also, backtracking in Step 3 is performed $|V| - 1$ times, which is $O(|V|)$. Hence, the time complexity is $O(|V| + |E|)$. Note that for a connected graph, $O(|V|)$ is dominated by $O(|E|)$ because $|V| \leq |E| + 1$. ∎

1.4.12

(a) First, we add to T three edges of weight 1 (any three edges can be chosen); the fourth edge of weight 1 cannot be added because it would create a cycle. Then, two edges of weight 2 are added to T. Finally, we add to T two edges of weight 3, which do not create a cycle (different choices are possible). An example of a minimum weight spanning tree

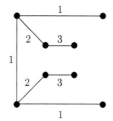

Fig. 1.20 The minimum weight spanning tree of the weighted cube Q_3.

T of the weighted cube Q_3 is shown in Figure 1.20. The weight of T is $w(T) = 13$.   ∎

(b) We will prove that $w(T)$ is a lower bound for $w(H)$. Indeed, if e is an edge of H, then $H - e$ is a path, which is a particular case of a tree. Hence,

$$w(T) \le w(H - e) \le w(H).$$

Thus, $w(T) \le w(H)$.   ∎

1.4.13

(a) (1) Set $D(v) = \infty$ for all $v \in V, v \ne s$. Set $D(s) = 0$ and $Q = V$. For each $v \in V$, $p(v)$ is undefined.

(2) $q^* = s. s \notin Q$.

(3) $D(a) = 1, p(a) = s. D(t) = 6, p(t) = s. D(c) = 5, p(c) = s$.

(2) $q^* = a. a \notin Q$, that is, $Q = \{t, d, f, g, b, c\}$.

(3) $D(d) = 3, p(d) = a. D(b) = 2, p(b) = a$.

(2) $q^* = b. b \notin Q$, that is, $Q = \{t, d, f, g, c\}$.

(3) $D(f) = 3, p(f) = b$. The labels for c are unchanged.

(2) There is a choice between d and f; any vertex may be chosen (e.g. d): $q^* = d. d \notin Q$, that is, $Q = \{t, f, g, c\}$.

(3) The labels for t and f are unchanged.

(2) $q^* = f. f \notin Q$, that is, $Q = \{t, g, c\}$.

(3) $D(g) = 4, p(g) = f$.

(2) $q^* = g. g \notin Q$, that is, $Q = \{t, c\}$.

(3) $D(t) = 5, p(t) = g$. The labels for c are unchanged.

(2) $q^* = t. t \notin Q$, that is, $Q = \{c\}$.

(4) The shortest (s, t)-path is (s, a, b, f, g, t) with weight 5.   ∎

(b) First, apply Dijkstra's algorithm to find the optimal path $P = (v_1, v_2, ..., v_k)$ between vertices v_1 and v_k. Next, remove the edge $v_1 v_2$

from the graph and find the optimal path in this updated graph using Dijkstra's algorithm. Repeat this procedure for other edges $v_2 v_3, \ldots,$ $v_{k-1} v_k$. By comparing the weights of the resulting paths, we can find the best alternative path. Note that a set of edges may be removed from the optimal path P or the original graph to find a suitable alternative path with additional properties. ∎

(c) (i) Let $V = \{s, a, b, t\}$, $E = \{sa, at, sb, bt\}$, $w(sa) = 4$, $w(at) = -4$, $w(sb) = w(bt) = 1$. It is easy to see that Dijkstra's algorithm produces the path (s, b, t) of weight 2, but the path of minimum weight is (s, a, t) of weight 0.

When there are edges of negative weights, Step 2 of Dijkstra's algorithm does not work in general. Indeed, in this step we find a vertex $q^* \in Q$ that has the smallest temporary label $D(q^*)$ and declare this label as permanent. Let us recall that $D(q^*)$ is the minimum weight of a (s, q^*)-path found so far. When all weights of edges are non-negative, there does not exist a better path to q^* through another vertex from Q because $D(q^*)$ is the smallest label in the set Q. Hence, $D(q^*)$ is the minimum weight of a (s, q^*)-path. However, if there are negative edge weights, then we cannot declare $D(q^*)$ as permanent because there might be a better (s, q^*)-path P going through a vertex $q \in Q$ to the vertex q^*. Although $D(q^*) \leq D(q)$, it might happen that $w(q q^*)$ is negative and so the weight of the path P is less than $D(q^*)$. ∎

(ii) This approach does not work in general because the number of edges in one (s, t)-path is not necessarily equal to the number of edges in another (s, t)-path. Hence, the total additional weights added to those paths might be different. (Note that in the formula $w'(e) = w(e) - w^*$, subtracting a negative number w^* means adding a positive number $|w^*|$.)

For example, let the graph G consist of three paths: $P_1 = (s, a, b, c, t)$, where every edge has weight 1; $P_2 = (s, d, t)$, where every edge has weight 3; and $P_3 = (s, f, t)$, where $w(sf) = -2$ and $w(ft) = 12$. Obviously, the path of minimum weight is P_1. Let us calculate the weights $w'(e)$. We have $w^* = -2$, hence all weights $w(e)$ should be increased by 2 to obtain $w'(e)$. Now, the new weight of P_1 is 12. Also, $w'(P_2) = 10$ and $w'(P_3) = 14$. We obtain that P_2 is the path of minimum weight in G with the new weights $w'(e)$. ∎

References

[1] L. A. Adamic, The small world web, in S. Abiteboul and A.-M. Vercoustre (eds), *Research and Advanced Technology for Digital Libraries*, Lecture Notes in Computer Science, Vol. 1696. New York, NY: Springer-Verlag, 1999, 443–452.

[2] R. Albert, H. Jeong and A. Barabási, Diameter of the world-wide web, *Nature*, **401** (1999), 130.

[3] K. Appel and W. Haken, Every Planar Map is Four Colorable. I. Discharging, *Illinois Journal of Mathematics*, **21** (3)(1977), 429–490.

[4] K. Appel, W. Haken and J. Koch, Every Planar Map is Four Colorable. II. Reducibility, *Illinois Journal of Mathematics*, **21** (3)(1977), 491–567.

[5] K. Appel and W. Haken, Solution of the Four Color Map Problem, *Scientific American*, **237** (4)(1977), 108–121.

[6] P. Baldi, P. Frasconi and P. Smyth, *Modeling the Internet and the Web, Probabilistic Methods and Algorithms*, Chichester: John Wiley & Sons, 2003.

[7] M. Bezzel (Schachfreund), *Berliner Schachzeitung*, (3)(1848), 363.

[8] A. Broder, R. Kumar, F. Maghoul, P. Raghavan, S. Rajagopalan, R. Stata, A. Tomkins and J. Wiener, Graph structure in the web, *Computer Networks*, **33** (2000), 309–320.

[9] A. Cayley, A theorem on trees, *Quarterly Journal of Pure and Applied Mathematics*, **23** (1889), 376–378.

[10] A. Cayley, On the theory of the analytical forms called trees, *Philosophical Magazine*, **13** (1857), 19–30.

[11] O.-J. Dahl, E. W. Dijkstra and C. A. R. Hoare, *Structured Programming*, London: Academic Press, 1972, 72–82.

[12] C. F. De Jaenisch, *Applications de l'Analyse Mathèmatique au Jeu des Echecs*, Vol. 1, l'Académie Impériale des Sciences, St. Petersburg, 1862.

[13] E. W. Dijkstra, A note on two problems in connexion with graphs, *Numerische Mathematik*, **1** (1959), 269–271.

[14] G. A. Dirac, Some theorems on abstract graphs, *Proceedings of the London Mathematical Society*, 3rd Ser., **2** (1952), 69–81.

[15] S. N. Dorogovtsev and J. F. F. Mendes, *Evolution of Networks: From Biological Nets to the Internet and WWW*, Oxford: Oxford University Press, 2003.

[16] J. Edmonds, Paths, trees, and flowers, *Canadian Journal of Mathematics*, **17** (1965), 449–467.

[17] L. Euler, Solutio problematis ad geometriam situs pertinentis, *Commentarii Academiae Scientiarum Petropolitanae*, (St. Petersburg Academy), **8** (1736), 128–140. (In Latin.) (English translation: *Scientific American*, **189** (1953), 66–70.)

[18] M. Fleury, Deux problemes de geometrie de situation, *Journal de Mathematiques Elementaires*, (1883), 257–261.

[19] R. W. Floyd, Algorithm 97: shortest path, *Communications of the ACM*, **5** (6)(1962), 345.

[20] L. R. Ford and D. R. Fulkerson, Maximal flow through a network, *Canadian Journal of Mathematics*, **8** (1956), 399–404.

[21] S. L. Hakimi, On realizability of a set of integers as degrees of the vertices of a linear graph. I, *Journal of the Society for Industrial and Applied Mathematics*, **10** (1962), 496–506.

[22] W. R. Hamilton, Memorandum respecting a new system of roots of unity, *Philosophical Magazine*, **12** (1856), 446.

[23] W. R. Hamilton, Account of the icosian calculus, *Proceedings of the Royal Irish Academy*, **6** (1858), 415–416.

[24] P. E. Hart, N. J. Nilsson and B. Raphael, A formal basis for the heuristic determination of minimum cost paths, *IEEE Transactions on Systems Science and Cybernetics*, **4** (2)(1968), 100–107.

[25] V. Havel, A remark on the existence of finite graphs, *Časopis pro Pěstování Matematiky*, **80** (1955), 477–480. (in Czech).

[26] C. Hierholzer and C. Wiener, Über die Möglichkeit, einen Linienzug ohne Wiederholung und ohne Unterbrechung zu umfahren, *Mathematische Annalen*, **6** (1)(1873), 30–32.

[27] J. Hopcroft and R. Tarjan, Efficient planarity testing, *Journal of the ACM*, **21** (1974), 549–568.

[28] G. Kirchhoff, Über die Auflösung der Gleichungen, auf welche man bei der Untersuchung der linearen Vertheilung galvanischer Ströme geführt wird, *Annalen der Physik und Chemie*, **72** (12)(1847), 497–508.

[29] W. Kocay, The Hopcroft–Tarjan planarity algorithm, unpublished, (1993), preprint is available at http://bkocay.cs.umanitoba.ca/G&G/G&G.html.

[30] J. B. Kruskal, On the shortest spanning subtree of a graph and the travelling salesman problem, *Proceedings of the American Mathematical Society*, **7** (1956), 48–50.

[31] H. W. Kuhn, The Hungarian method for the assignment problem, *Naval Research Logistics Quarterly*, **2** (1955), 83–97.

[32] S. Milgram, The small world problem, *Psychology Today*, **2** (1967), 60–67.

[33] J. W. Moon, The second moment of the complexity of a graph, *Mathematika*, **11** (1964), 95–98.

[34] J. W. Moon, Various proofs of Cayley's formula for counting trees, in F. Harary (ed), *A Seminar on Graph Theory*, New York, NY: Holt, Rinehart and Winston, 1967, 70–78.

[35] F. Nauck, Schach: Eine in das Gebiet der Mathematik fallende Aufgabe von Herrn Dr. Nauck, *Schleusingen Illustrirte Zeitung*, 14, No. 361, 1 June 1850, 352.

[36] F. Nauck, Briefwechseln mit Allen für Alle, *Illustrirte Zeitung*, 15, No. 377, 21 September 1850, 182.

[37] M. E. J. Newman, The structure of scientific collaboration networks, *Proceedings of the National Academy of Sciences of the USA*, **98** (2001), 404–409.

[38] C. F. Peters (ed), *Briefwechsel zwischen C. F. Gauss und H. C. Schumacher*, Vol. 6, Altona: Band, 1865.

[39] R. C. Prim, Shortest connection networks and some generalizations, *Bell System Technical Journal*, **36** (1957), 1389–1401.

[40] J. Riordan, The enumeration of labelled trees by degrees, *Bulletin of the American Mathematical Society*, **72** (1966), 110–112.

[41] D. J. Watts and S. H. Strogatz, Collective dynamics of 'small-world' networks, *Nature*, **393** (1998), 440–442.

[42] I. Zverovich and V. Zverovich, Contributions to the theory of graphic sequences, *Discrete Mathematics*, **105** (1992), 293–303.

2

Traffic Networks: Wardrop Equilibrium and Braess' Paradox

The well-known Braess' paradox illustrates situations when adding a new link to a transport network might lead to an equilibrium state in which travel times of users will increase. In this chapter, we analyse Braess' paradox and the equilibrium state in the classical network configuration introduced by Braess in 1968 [3]. This network configuration is of fundamental significance because Valiant and Roughgarden showed in 2006 that 'the "global" behaviour of an equilibrium flow in a large random network is similar to that in Braess' original four-node example' [16]. Moreover, the probability of Braess' paradox to occur in the classical network configuration will be studied, with particular emphasis on the Erlang distribution of parameters of the travel time function. This distribution is important in the context of traffic networks. However, other distributions will be analysed as well because Braess' paradox can be observed in various applied contexts such as telecommunication networks and power transmission networks, and it has been studied for an evolutionary variational inequality model of the internet [12].

2.1 Equilibrium and Braess' Paradox

2.1.1 Braess' Network Configuration and its Generalization

The classical network configuration introduced by Braess [3] consists of three paths:

$$P_1 = a - b - d, \quad P_2 = a - c - d, \quad P_3 = a - b - c - d.$$

This network is denoted by N^+ and it has four nodes and five links, where a is the origin of all travel demand, and d is the destination of all demand (see Figure 2.1). The network N is N^+ with the link (b, c) removed.

In 2006, Valiant and Roughgarden [16] showed that 'the "global" behaviour of an equilibrium flow in a large random network is similar to that in Braess'

Modern Applications of Graph Theory. Vadim Zverovich, Oxford University Press (2021). © Vadim Zverovich.
DOI: 10.1093/oso/9780198856740.003.0002

original four-node example'. Thus, Braess' network configuration is of funda-mental significance.

Let us consider a natural generalization of Braess' network, where every link of the network in Figure 2.1 is replaced by a path of arbitrary length (i.e. a route with any number of links). Thus, the generalized network 'comprises' five paths of arbitrary length: (a, b)-path, (b, d)-path, (a, c)-path, (c, d)-path and (b, c)-path. This is illustrated by the first network in Figure 2.2, where the (b, c)-path is of length 3 and other four paths have length 2. Every link (i, j) in the resulting network has a linear travel time function

$$\alpha_{ij} + \beta_{ij} f_{ij},$$

where $\alpha_{ij} \geq 0$ is the free-flow travel time for the link (i, j), $\beta_{ij} > 0$ is the delay parameter for (i, j) and $f_{ij} \geq 0$ is the flow on the link (i, j). A fixed traffic coming from outside the network is allowed. For example, in Figure 2.2 the dashed arrows represent a fixed traffic $\tilde{f}$ coming to the network and then going outside.

The assumption of a linear relationship between traffic volume on a link and the travel time on it (so-called volume-delay function) is common in the context of Braess' paradox. Although there is evidence to support such a linear approximation [17], different types of volume-delay functions (e.g. BPR functions [5]) have been used as well (see [4]).

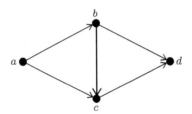

Fig. 2.1 Braess' network configuration N/N^+, where $N = N^+ - (b, c)$.

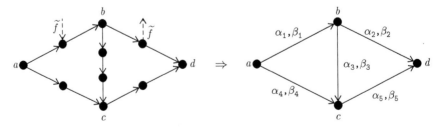

Fig. 2.2 Generalized network reduced to a four-node network N^+.

Suppose we want to decide whether Braess' paradox occurs when removing a link on the path going from b to c. Assume also that a particular link (i, j) has a fixed flow $\tilde{f}$ coming from outside the network, and the internal flow f is going from a to d through this link. Then, the travel time function of this link can be written as follows:

$$\alpha_{ij} + \beta_{ij}f_{ij} = \alpha_{ij} + \beta_{ij}(f + \tilde{f}) = (\alpha_{ij} + \beta_{ij}\tilde{f}) + \beta_{ij}f = \tilde{\alpha}_{ij} + \beta_{ij}f.$$

This function only depends on the internal flow and it is linear because the external flow $\tilde{f}$ is fixed, and hence $\tilde{\alpha}_{ij}$ is a fixed number. Thus, the first step is to update all travel time functions taking into account external flows. Further, it is easy to see that the total travel time functions for the above paths are linear functions, since the travel time functions for links are linear and all the links belonging to one of the paths share the same internal flow. For instance, if such a path P with an internal flow f consists of two links (i, j) and (j, k), then the total travel time function is as follows:

$$\alpha_{ij} + \beta_{ij}f + \alpha_{jk} + \beta_{jk}f = \alpha_{ij} + \alpha_{jk} + (\beta_{ij} + \beta_{jk})f = \alpha_P + \beta_P f.$$

Thus, if all the above paths are replaced by single links, then we obtain Braess' network with arbitrary linear travel time functions (see Figure 2.2):

$$\alpha_1 + \beta_1 f_{ab} \text{ for link } (a, b),$$

$$\alpha_2 + \beta_2 f_{bd} \text{ for link } (b, d),$$

$$\alpha_3 + \beta_3 f_{bc} \text{ for link } (b, c),$$

$$\alpha_4 + \beta_4 f_{ac} \text{ for link } (a, c),$$

$$\alpha_5 + \beta_5 f_{cd} \text{ for link } (c, d).$$

Note that in Braess' original example [3] and in many of the studies that followed it (e.g. [6]), the network is symmetric; that is, the travel time functions for the links (a, b) and (c, d) are the same as well as the travel time functions for the links (b, d) and (a, c). Also, the free-flow travel times for the links (a, b) and (c, d) are equal to zero. The occurrence of Braess' paradox in this symmetrical network configuration was described by Pas and Principio [13] (see Corollary 2.1).

We will consider a more general situation with arbitrary linear travel time functions. The existence of Braess' paradox in such a network configuration can be decided by using Theorems 2.1–2.4. Notice that the conditions for Braess' paradox to occur in the aforementioned generalized network and the corresponding reduced network are the same.

2.1.2 Equilibria and Braess' Paradox in the Classical Network Configuration

The traffic network can generally be described as a game, where a finite number of interdependent network users compete with one another by making simultaneous route choices. It is commonly assumed that network users non-cooperatively interact with one another in the traffic network to minimize their travel costs. This problem is usually modelled as an N-person non-zero-sum game (see [6]), and its solution assumes the existence of equilibrium. The concept of equilibrium in the context of transport systems had appeared in the 1950s ([2, 18]) and is based on the general assumption that network users are making adjustments to their travel choices until a state of equilibrium is reached; that is, when no individual can make a further improvement to their travel time as a result of any individual choice. Thus, at equilibrium, making an alternative path choice will not lead to a reduction in individual's travel time. It is well known that a user equilibrium always exists and in a network without capacities, it is essentially unique (e.g. see [15]).

A path P from the origin to the destination is said to have a *vanishing flow* if P carries no traffic from the origin to the destination. Note that some links in the path P may have a non-zero flow that contributes to the traffic of other paths. A path has a *non-vanishing flow* if it carries some traffic from the origin to the destination.

Definition 2.1 *A network with one origin and one destination is said to be at equilibrium if*

(a) *The travel time on paths with non-vanishing flow is the same (it is denoted by T_{Eq}) and*
(b) *The travel time on paths with no flow is at least T_{Eq}.*

This fundamental definition is, of course, a reformulation of Wardrop's first principle [18] and it can be used to determine the equilibrium time and the equilibrium flow.

The equilibrium described above is associated with aggregated strategic behaviour of all road users, described as an N-person Nash equilibrium. The concept of Nash equilibrium is related to a Wardrop equilibrium. In fact, the Nash equilibrium in a network game with a finite number of players converges to a Wardrop equilibrium when the number of players increases [8]. At equilibrium, no user can decrease their route travel time by unilaterally switching routes [18]. In other words, if a network is not at equilibrium, then some users of the network (e.g. drivers) can switch their routes in order to improve their travel times. However, if a driver decides to switch to a better route, then the travel time for this route increases, and, after a certain period of time, it will become impossible to improve drivers' travel times by switching the routes. Thus, the equilibrium describes 'stable state' behaviour in a network, and no driver has any incentive to switch routes at equilibrium because it will not improve their current travel times.

Let $Q > 0$ denote the total flow in N/N^+; that is,

$$Q = f_{ab} + f_{ac} = f_{bd} + f_{cd}.$$

Note that f_{ij} and Q are not necessarily integer numbers. Let us denote

$$\alpha_{ij} = \alpha_i + \alpha_j, \quad \beta_{ij} = \beta_i + \beta_j;$$

for example, α_{12} means $\alpha_1 + \alpha_2$. Also,

$$\alpha = \alpha_{45} - \alpha_{12}, \quad \bar{\alpha} = \alpha_4 - \alpha_{13}, \quad \hat{\alpha} = \alpha_2 - \alpha_{35},$$

and

$$\beta = \beta_{1245} = \beta_1 + \beta_2 + \beta_4 + \beta_5, \quad \beta_{ijk} = \beta_i + \beta_j + \beta_k.$$

The following identity will be used throughout the chapter:

$$\alpha = \bar{\alpha} - \hat{\alpha}.$$

Lemma 2.1 describes the equilibrium in the network N, which is N^+ with the link (b, c) removed. Note that the case (a) of this lemma corresponds to the situation when the path P_1 has a vanishing flow and P_2 has a non-vanishing flow in N. In case (b) the path P_1 has a non-vanishing flow and P_2 has a vanishing flow, and in case (c) no path has a vanishing flow. Also, the cases (a) and (b) in this lemma are mutually exclusive because one of the numbers $-\alpha/\beta_{45}$ and α/β_{12} is negative, or they both are equal to zero.

Lemma 2.1 [20] *In the network N, the travel time at equilibrium is as follows:*

(a) $T_{Eq} = \alpha_{45} + Q\beta_{45}$ *if* $0 < Q \le -\alpha/\beta_{45}$;
(b) $T_{Eq} = \alpha_{12} + Q\beta_{12}$ *if* $0 < Q \le \alpha/\beta_{12}$;
(c) $T_{Eq} = \alpha_{12} + (\alpha + Q\beta_{45})\beta_{12}/\beta$ *if* $Q > \max\{\alpha/\beta_{12}; -\alpha/\beta_{45}\}$.

Proof: Let us denote $h = f_{ab}$. Then, $f_{bd} = h$ and $f_{ac} = f_{cd} = Q - h$. We have

$$T_1 = \alpha_{12} + h\beta_{12} \quad \text{and} \quad T_2 = \alpha_{45} + (Q - h)\beta_{45},$$

where T_i is the travel time on the path P_i.

Case (a): Suppose that P_1 has a vanishing flow and P_2 has a non-vanishing flow. Then, $Q > h = 0$ and

$$T_1 = \alpha_{12} \quad \text{and} \quad T_2 = \alpha_{45} + Q\beta_{45}.$$

At equilibrium, $T_1 \ge T_2$, that is, $\alpha_{12} \ge \alpha_{45} + Q\beta_{45}$ or $Q \le -\alpha/\beta_{45}$. The travel time at equilibrium is

$$T_{Eq} = T_2 = \alpha_{45} + Q\beta_{45}.$$

Case (b): Assume that P_1 has a non-vanishing flow and P_2 has a vanishing flow. We have $Q = h > 0$ and

$$T_1 = \alpha_{12} + Q\beta_{12} \quad \text{and} \quad T_2 = \alpha_{45}.$$

At equilibrium, $T_1 \le T_2$; that is, $\alpha_{12} + Q\beta_{12} \le \alpha_{45}$ or $Q \le \alpha/\beta_{12}$. The travel time at equilibrium is as follows:

$$T_{Eq} = T_1 = \alpha_{12} + Q\beta_{12}.$$

Case (c): Suppose that no path has a vanishing flow. We have $Q > h > 0$. At equilibrium, $T_1 = T_2$; that is,

$$\alpha_{12} + h\beta_{12} = \alpha_{45} + (Q - h)\beta_{45}$$

or

$$h = (\alpha + Q\beta_{45})/\beta.$$

Therefore, the condition $Q > h > 0$ is equivalent to

$$Q > (\alpha + Q\beta_{45})/\beta > 0$$

or

$$Q > \max\{\alpha/\beta_{12}; -\alpha/\beta_{45}\}.$$

Finally,

$$T_{Eq} = T_1 = \alpha_{12} + (\alpha + Q\beta_{45})\beta_{12}/\beta.$$

■

The equilibrium in N^+ is described by seven cases in Lemma 2.2. It may be pointed out that these cases correspond to the following situations in N^+:

(a) The only path with non-vanishing flow is P_3;
(b) The only path with non-vanishing flow is P_2;
(c) The only path with non-vanishing flow is P_1;
(d) The only path with vanishing flow is P_1;
(e) The only path with vanishing flow is P_2;
(f) The only path with vanishing flow is P_3;
(g) No path has a vanishing flow.

It is not difficult to see that some of the cases in Lemma 2.2 are mutually exclusive, hence the equilibrium in a particular network N^+ is described by some of the presented seven cases. For example, if $\alpha_i = \beta_i = 1$ for $1 \leq i \leq 5$, then the equilibrium is given by just one case (f).

Let us define the Braess numbers $\mathcal{B}_i$ for $i = 1, 2, 3, 4$:

$$\mathcal{B}_1 = \beta_1\beta_5 - \beta_2\beta_4, \quad \mathcal{B}_2 = \beta_{135}\beta - \beta_{12}\beta_{45},$$

$$\mathcal{B}_3 = \beta_{45}^2\beta_{134} - \beta_4^2\beta, \quad \mathcal{B}_4 = \beta_{12}^2\beta_{235} - \beta_2^2\beta.$$

We will also need two parameters μ_1 and μ_2:

$$\mu_1 = \frac{\hat{\alpha}\beta_{14} - \alpha\beta_3}{\beta_3\beta_{45} + \beta_5\beta_{14}}, \quad \mu_2 = \frac{\bar{\alpha}\beta_{25} + \alpha\beta_3}{\beta_1\beta_{25} + \beta_3\beta_{12}}.$$

Lemma 2.2 [20] *In the network N^+, the travel time at equilibrium is as follows:*

(a) $T_{Eq}^+ = \alpha_{135} + Q\beta_{135}$ *if* $0 < Q \le \min\{\hat{\alpha}/\beta_{35}; \bar{\alpha}/\beta_{13}\}$;

(b) $T_{Eq}^+ = \alpha_{45} + Q\beta_{45}$ *if* $0 < Q \le \min\{-\alpha/\beta_{45}; -\bar{\alpha}/\beta_4\}$;

(c) $T_{Eq}^+ = \alpha_{12} + Q\beta_{12}$ *if* $0 < Q \le \min\{\alpha/\beta_{12}; -\hat{\alpha}/\beta_2\}$;

(d) $T_{Eq}^+ = \alpha_{45} + Q\beta_{45} - (\bar{\alpha} + Q\beta_4)\beta_4/\beta_{134}$ *if*

$$\max\{\bar{\alpha}/\beta_{13}; -\bar{\alpha}/\beta_4\} < Q \le \mu_1;$$

(e) $T_{Eq}^+ = \alpha_{12} + Q\beta_{12} - (\hat{\alpha} + Q\beta_2)\beta_2/\beta_{235}$ *if*

$$\max\{\hat{\alpha}/\beta_{35}; -\hat{\alpha}/\beta_2\} < Q \le \mu_2;$$

(f) $T_{Eq}^+ = \alpha_{12} + (\alpha + Q\beta_{45})\beta_{12}/\beta$ *if* $Q > \max\{\alpha/\beta_{12}; -\alpha/\beta_{45}\}$
and

$$\mathcal{B}_1 \ge \frac{\hat{\alpha}\beta_{14} + \bar{\alpha}\beta_{25}}{Q};$$

(g) $T_{Eq}^+ = \alpha_{12} + (\alpha + Q\beta_{45})\beta_{12}/\beta + g\mathcal{B}_1/\beta$, *where*

$$g = \frac{\bar{\alpha}\beta - \alpha\beta_{14} - Q\mathcal{B}_1}{\beta_3\beta + \beta_{14}\beta_{25}},$$

if $Q > \max\{\mu_1; \mu_2\}$ *and*

$$\mathcal{B}_1 < \frac{\hat{\alpha}\beta_{14} + \bar{\alpha}\beta_{25}}{Q}.$$

The proof of Lemma 2.2 can be found in [20].

The next definition is devoted to Braess' paradox [3] in the classical network configuration N/N^+; however, the same definition is valid if N/N^+ represents any network configuration. Basically, the paradox describes a situation when adding a new link to a network makes a general performance worse.

Definition 2.2 Braess' paradox *is said to occur in the network configuration* N/N^+ *for a given total flow Q if*

$$T_{Eq}^+ > T_{Eq},$$

where T_{Eq} and T_{Eq}^+ are travel times at equilibria in N and N^+, respectively.

Thus, Braess' paradox illustrates situations when adding a new link to a transport network might not reduce congestion in the network but instead increase it. This is due to individual entities acting selfishly/separately when making their travel plan choices and hence forcing the system as a whole not to operate optimally. Deeper insight into this paradox from the viewpoint of the structure and characteristics of networks may help transport planners to avoid the occurrence of Braess-like situations in real-life networks. Nagurney [11] proved that Braess' paradox disappears under higher demands, and in [9] it was shown how to avoid Braess' paradox by adding resources efficiently to a network.

Proposition 2.1 describes all possible situations when Braess' paradox may occur in N/N^+ in terms of their paths. In fact, it says that Braess' paradox may occur in N/N^+ only if, at equilibria, both P_1 and P_2 have a non-vanishing flow in N, and P_3 has a non-vanishing flow in N^+.

Proposition 2.1 [20] *Braess' paradox may occur in N/N^+ in the following cases only:*

    (a) *At equilibria, both N and N^+ have no paths with vanishing flow.*
    (b) *At equilibria, N has no path with vanishing flow, and P_3 is the only path with non-vanishing flow in N^+.*
    (c) *At equilibria, N has no path with vanishing flow, and P_1 is the only path with vanishing flow in N^+.*
    (d) *At equilibria, N has no path with vanishing flow, and P_2 is the only path with vanishing flow in N^+.*

In the following theorems, the necessary and sufficient conditions for the existence of the paradox are formulated. These theorems correspond to the four cases of Proposition 2.1.

Theorem 2.1 [20] *Suppose that at equilibria both N and N^+ have no paths with vanishing flow. Then, Braess' paradox occurs in N/N^+ if and only if the Braess number $\mathcal{B}_1$ is positive and*

$$\max\left\{\frac{\alpha}{\beta_{12}}; \frac{-\alpha}{\beta_{45}}; \mu_1; \mu_2\right\} < Q < \frac{\hat{\alpha}\beta_{14} + \bar{\alpha}\beta_{25}}{\mathcal{B}_1}.$$

Theorem 2.2 [20] *Suppose that at equilibria N has no path with vanishing flow and P_3 is the only path with non-vanishing flow in N^+. Then, Braess' paradox* *occurs in N/N^+ if and only if the Braess number $\mathcal{B}_2$ is positive and*

$$\max\left\{\frac{\alpha}{\beta_{12}}; \frac{-\alpha}{\beta_{45}}; \frac{\hat{\alpha}\beta_{45} + \bar{\alpha}\beta_{12}}{\mathcal{B}_2}\right\} < Q \le \min\left\{\frac{\hat{\alpha}}{\beta_{35}}; \frac{\bar{\alpha}}{\beta_{13}}\right\}.$$

Theorem 2.3 [20] *Suppose that at equilibria N has no path with vanishing flow and P_1 is the only path with vanishing flow in N^+. Then, Braess' paradox occurs in N/N^+ if and only if the Braess number $\mathcal{B}_3$ is positive and*

$$\max\left\{\frac{\alpha}{\beta_{12}}; \frac{-\alpha}{\beta_{45}}; \frac{\bar{\alpha}}{\beta_{13}}; \frac{-\bar{\alpha}}{\beta_4}; \frac{\bar{\alpha}\beta_4\beta - \alpha\beta_{134}\beta_{45}}{\mathcal{B}_3}\right\} < Q \le \mu_1.$$

Theorem 2.4 [20] *Suppose that at equilibria N has no path with vanishing flow and P_2 is the only path with vanishing flow in N^+. Then, Braess' paradox occurs in N/N^+ if and only if the Braess number $\mathcal{B}_4$ is positive and*

$$\max\left\{\frac{\alpha}{\beta_{12}}; \frac{-\alpha}{\beta_{45}}; \frac{\hat{\alpha}}{\beta_{35}}; \frac{-\hat{\alpha}}{\beta_2}; \frac{\hat{\alpha}\beta_2\beta + \alpha\beta_{235}\beta_{12}}{\mathcal{B}_4}\right\} < Q \le \mu_2.$$

It might be pointed out that if $\mathcal{B}_1 \ge 0$, then $\mathcal{B}_2$, $\mathcal{B}_3$ and $\mathcal{B}_4$ are positive numbers because

$$\mathcal{B}_2 = \beta_{12}\beta_{13} + \beta_{35}\beta_{45} + \mathcal{B}_1, \tag{2.1}$$

$$\mathcal{B}_3 = \beta_5^2\beta_{134} + \beta_4(\beta_3\beta_{455} + \beta_5\beta_{14} + \mathcal{B}_1), \tag{2.2}$$

$$\mathcal{B}_4 = \beta_1^2\beta_{235} + \beta_2(\beta_1\beta_{335} + \beta_2\beta_{13} + \mathcal{B}_1). \tag{2.3}$$

Moreover, Theorems 2.3 and 2.4 are mutually exclusive in the sense that they cannot provide intervals for Q simultaneously. This is true because the inequalities $\bar{\alpha}/\beta_{13} < \mu_1$ and $\hat{\alpha}/\beta_{35} < \mu_2$ are inconsistent. Note also that if, for example, Theorems 2.1–2.3 provide non-empty intervals for Q, then the interval with highest values of Q is given by Theorem 2.1, the interval with smallest values of Q is provided by Theorem 2.2, and Theorem 2.3 yields the interval with mid-range values of Q.

The original assumption $\beta_i > 0$ for all i can be relaxed by allowing $\beta_i = 0$ for some i. This can be done by introducing $+\infty$ and $-\infty$ when a non-zero number is divided by zero. For example, let us consider Arnott–Small's example [1]:

$$\alpha_1 = \alpha_5 = 0, \ \alpha_2 = \alpha_4 = 15, \ \alpha_3 = 7.5, \ \beta_1 = \beta_5 = 0.01, \ \beta_2 = \beta_3 = \beta_4 = 0.$$

Using the formulae defined before Lemmas 2.1 and 2.2, we obtain

$$\alpha = 0, \quad \bar{\alpha} = \hat{\alpha} = 7.5, \quad \beta = 0.02, \quad \mu_1 = \mu_2 = 750.$$

Now let us apply Theorems 2.1–2.4 to Arnott–Small's example, as shown in Table 2.1.

Table 2.1 Practical application of Theorems 2.1–2.4 to Arnott–Small's example.

Theorem	Braess Number	Lower Bound for Q	Upper Bound for Q	Interval for Q
2.1	$B_1 = 10^{-4}$	$\max\{0; 0; 750; 750\} = 750$	1500	$]750; 1500[$
2.2	$B_2 = 3 \times 10^{-4}$	$\max\{0; 0; 500\} = 500$	750	$]500; 750]$
2.3	$B_3 = 10^{-6}$	$\max\{0; 0; 750; -7.5/0; 0\} = 750$	750	$\emptyset$
2.4	$B_4 = 10^{-6}$	$\max\{0; 0; 750; -7.5/0; 0\} = 750$	750	$\emptyset$

Notice that the notation $]500; 750]$ in the last column of the table means an interval of real numbers from 500 (excluded) to 750 (included). Thus, by Theorems 2.1 and 2.2, Braess' paradox occurs if $500 < Q < 1500$, while Theorems 2.3 and 2.4 provide no intervals for Q. In calculating the lower bounds for Q in these theorems we have division by zero, but this problem is overcome by setting $-7.5/0 = -\infty$.

2.1.3 Symmetrical/Asymmetrical Networks

Let us consider the classical case of a symmetrical network presented by Braess [3] and discussed in Pas and Principio [13] and other articles. Using the notation introduced in the previous sections, it can be seen as a particular case of the network configuration N/N^+, where time functions are symmetrical for links that do not share nodes with each other $((a, b)$ and (c, d); (a, c) and $(b, d))$, the free-flow travel times for the links (a, b) and (c, d) are equal to zero and the delay parameter for (b, c) is equal to the delay parameter of the links (b, d) and (a, c); that is,

$$\alpha_1 = \alpha_5 = 0, \quad \alpha_2 = \alpha_4, \quad \beta_1 = \beta_5, \quad \beta_2 = \beta_3 = \beta_4.$$

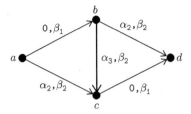

Fig. 2.3 Symmetric network M^+ ($\alpha_2 > \alpha_3$ and $\beta_1 > \beta_2$).

Also, it is assumed that $\alpha_2 > \alpha_3$ and $\beta_1 > \beta_2$. This network configuration is denoted by M/M^+ (see Figure 2.3).

Pas and Principio [13] determined the occurrence of Braess' paradox in the symmetrical network configuration M/M^+. This result follows directly from Theorems 2.1–2.4:

Corollary 2.1 [13] Braess' paradox occurs in M/M^+ if and only if

$$\frac{2(\alpha_2 - \alpha_3)}{3\beta_1 + \beta_2} < Q < \frac{2(\alpha_2 - \alpha_3)}{\beta_1 - \beta_2}.$$

The proof is left to the reader as an exercise.

Now let $\tilde{M}/\tilde{M}^+$ denote the above network configuration M/M^+ without the assumption that $\alpha_2 > \alpha_3$ and $\beta_1 > \beta_2$. A proof similar to that of Corollary 2.1 shows that Braess' paradox occurs in $\tilde{M}/\tilde{M}^+$ in the following cases only:

(a) If $\beta_1 > \beta_2$ and

$$\frac{\alpha_2 - \alpha_3}{\beta_{12}} < Q < \frac{2(\alpha_2 - \alpha_3)}{\beta_1 - \beta_2};$$

(b) If

$$\frac{2(\alpha_2 - \alpha_3)}{3\beta_1 + \beta_2} < Q \leq \frac{\alpha_2 - \alpha_3}{\beta_{12}}.$$

The both cases imply that $\alpha_2 > \alpha_3$. Another implicit relationship is obtained from (b) if we require that

$$\frac{2(\alpha_2 - \alpha_3)}{3\beta_1 + \beta_2} < \frac{\alpha_2 - \alpha_3}{\beta_{12}},$$

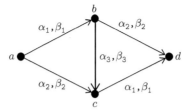

Fig. 2.4 Symmetric network S^+.

which is equivalent to $\beta_1 > \beta_2$. Thus, even though the network configuration $\tilde{M}/\tilde{M}^+$ extends M/M^+, the conditions for the occurrence of Braess' paradox are the same.

Let us further extend the network configuration $\tilde{M}/\tilde{M}^+$ by allowing any non-negative free-flow travel time α_1 for the links (a, b) and (c, d) and any positive delay parameter β_3 for the link (b, c). In other words, the symmetrical network configuration S/S^+ of Figure 2.4 is obtained from N/N^+ when travel time functions are symmetrical for links that do not share nodes with each other:

$$\alpha_1 = \alpha_5, \quad \alpha_2 = \alpha_4, \quad \beta_1 = \beta_5, \quad \beta_2 = \beta_4 \text{ (see Figure 2.4)}.$$

Corollary 2.2 [20] *Braess' paradox occurs in the symmetrical network configuration S/S^+ if and only if $\beta_1 > \beta_2$ and*

$$\frac{2(\alpha_2 - \alpha_{13})}{3\beta_1 + 2\beta_3 - \beta_2} < Q < \frac{2(\alpha_2 - \alpha_{13})}{\beta_1 - \beta_2}. \tag{2.4}$$

Proof: For the network configuration S/S^+, we have

$$\alpha = 0, \quad \bar{\alpha} = \hat{\alpha} = \alpha_2 - \alpha_{13}, \quad \mu_1 = \mu_2 = (\alpha_2 - \alpha_{13})/\beta_{13},$$
$$\mathcal{B}_1 = \beta_1^2 - \beta_2^2 = \beta_{12}(\beta_1 - \beta_2).$$

By Theorem 2.1, Braess' paradox occurs if $\mathcal{B}_1$ is positive (i.e. $\beta_1 > \beta_2$) and

$$\frac{\alpha_2 - \alpha_{13}}{\beta_{13}} < Q < \frac{2(\alpha_2 - \alpha_{13})}{\beta_1 - \beta_2}.$$

Now,

$$\mathcal{B}_2 = \beta_{131} 2\beta_{12} - \beta_{12}^2 = \beta_{12}(3\beta_1 + 2\beta_3 - \beta_2).$$

Therefore, by Theorem 2.2, Braess' paradox occurs if $3\beta_1 + 2\beta_3 > \beta_2$ and

$$\frac{2(\alpha_2 - \alpha_{13})}{3\beta_1 + 2\beta_3 - \beta_2} < Q \le \frac{\alpha_2 - \alpha_{13}}{\beta_{13}}.$$

This implies that $\alpha_2 > \alpha_{13}$. Also, it is easy to see that the lower bound is less than the upper bound only if $\beta_1 > \beta_2$, which is stronger than $3\beta_1 + 2\beta_3 > \beta_2$. Thus, the above inequalities can be written together as

$$\frac{2(\alpha_2 - \alpha_{13})}{3\beta_1 + 2\beta_3 - \beta_2} < Q < \frac{2(\alpha_2 - \alpha_{13})}{\beta_1 - \beta_2}.$$

The upper and lower bounds of Theorems 2.3 and 2.4 provide no intervals for Q. ∎

In Corollary 2.2 there is an implicit assumption that $\alpha_2 > \alpha_{13}$ because Q is a positive number (if $\alpha_2 \le \alpha_{13}$, then (2.4) provides no interval for Q). We will see in Corollary 2.6 what is happening with the times at equilibria in S/S^+ if Q exceeds the upper bound in (2.4), where $\alpha_2 > \alpha_{13}$ and $\beta_1 > \beta_2$.

The asymmetrical network configuration A/A^+ of Figure 2.5 is obtained from N/N^+ when the travel time functions for the links (a, b) and (a, c) are the same as well as the travel time functions for the links (b, d) and (c, d); that is,

$$\alpha_1 = \alpha_4, \quad \alpha_2 = \alpha_5, \quad \beta_1 = \beta_4, \quad \beta_2 = \beta_5 \text{ (see Figure 2.5)}.$$

Corollary 2.3 [20] *Braess' paradox cannot occur in the asymmetrical network configuration A/A^+.*

Proof: For the network configuration A/A^+, we have

$$\alpha = 0, \quad \bar{\alpha} = \hat{\alpha} = -\alpha_3, \quad \mathcal{B}_1 = \beta_1\beta_2 - \beta_2\beta_1 = 0, \quad \mu_1 \le 0, \quad \mu_2 \le 0.$$

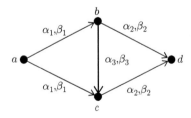

Fig. 2.5 Asymmetric network A^+.

It is easy to see that Theorems 2.1–2.4 provide no intervals, hence Braess' paradox is impossible.   ∎

Using Lemmas 2.1 and 2.2, we see that the equilibria in A/A^+ are described by the case (c) of Lemma 2.1 and the case (f) of Lemma 2.2; that is, in A no path has a vanishing flow, and in A^+ the only path with vanishing flow is P_3. Also, the flow Q is distributed evenly between P_1 and P_2, and

$$T_{Eq} = T_{Eq}^+ = \alpha_{12} + 0.5\beta_{12}Q.$$

Thus, the travel times at equilibria in A and A^+ are equal for any Q. This observation is important because adding a new link to A does not improve the general performance, even though Braess' paradox is not occurring.

Definition 2.3 *The* pseudo-paradox *occurs in the network configuration* N/N^+ *if*

$$T_{Eq}^+ = T_{Eq}$$

for an interval of values for Q.

In this definition, single values of Q are excluded when going from the situation *Braess' paradox does not occur* to *Braess' paradox occurs*. Thus, the pseudo-paradox describes a situation when adding a new link to a network does not change the general performance for a range of values of the total flow. As seen above, the pseudo-paradox occurs in the asymmetrical network configuration A/A^+ for any $Q > 0$. It seems that the pseudo-paradox is a more common phenomenon than Braess' paradox.

Corollary 2.4 [20] *For the network configuration N/N^+, the pseudo-paradox occurs if*

(a) $0 < Q \le \min\{-\alpha/\beta_{45}; -\bar{\alpha}/\beta_4\}$;
(b) $0 < Q \le \min\{\alpha/\beta_{12}; -\hat{\alpha}/\beta_2\}$;
(c) $Q > \max\{\alpha/\beta_{12}; -\alpha/\beta_{45}\}$ *and* $Q\mathcal{B}_1 \ge \hat{\alpha}\beta_{14} + \bar{\alpha}\beta_{25}$;
(d) $\mathcal{B}_1 = 0$, $\hat{\alpha}\beta_{14} + \bar{\alpha}\beta_{25} > 0$ *and* $Q > \max\{\alpha/\beta_{12}; -\alpha/\beta_{45}; \mu_1; \mu_2\}$.

The application of Corollary 2.4 (c) to the asymmetrical network configuration A/A^+ confirms the above observation:

Corollary 2.5 [20] *The pseudo-paradox occurs in the asymmetrical network configuration A/A^+ for any $Q > 0$.*

By Corollary 2.2, Braess' paradox occurs in the symmetrical network configuration if $\beta_1 > \beta_2$, $\alpha_2 > \alpha_{13}$ and the total flow Q is between the lower and upper bounds in (2.4). Corollary 2.4 (c) allows us to see what is happening with the times at equilibria if Q exceeds the upper bound:

Corollary 2.6 [20] *Suppose that Braess' paradox occurs in the symmetrical network configuration S/S^+; that is, $\beta_1 > \beta_2$ and $\alpha_2 > \alpha_{13}$. Then, S/S^+ is experiencing the pseudo-paradox for any*

$$Q \geq \frac{2(\alpha_2 - \alpha_{13})}{\beta_1 - \beta_2}.$$

Proof: We know that $\alpha = 0$, $\bar{\alpha} = \hat{\alpha} = \alpha_2 - \alpha_{13}$ and $\mathcal{B}_1 = \beta_{12}(\beta_1 - \beta_2) > 0$. Therefore, the second inequality in Corollary 2.4 (c) is equivalent to

$$Q\beta_{12}(\beta_1 - \beta_2) \geq 2(\alpha_2 - \alpha_{13})\beta_{12},$$

as required. ∎

Thus, under the conditions of Corollary 2.6, some improvement in S/S^+ is only possible if Q is less than the lower bound in (2.4), followed by Braess' paradox until Q reaches the upper bound in (2.4), followed by the pseudo-paradox for larger values of Q.

Numerical Example

Let us now consider a numerical example. Let G^+ be the generalized network shown in the left part of Figure 2.6, where free-flow travel times and delay parameters are indicated for all eleven links. As explained in Section 2.1.1, this network can be reduced to the network N^+ shown in the right part of Figure 2.6. Now suppose we want to decide whether Braess' paradox occurs in G^+ when removing any link on the path going from b to c; let us denote the resulting network by G. This is equivalent to finding the conditions for Braess' paradox to occur in the network configuration N/N^+.

Thus, the network configuration N/N^+ has the following parameters:

$$\alpha_1 = 2, \quad \alpha_2 = 36, \quad \alpha_3 = 6, \quad \alpha_4 = 40, \quad \alpha_5 = 2,$$
$$\beta_1 = 30, \quad \beta_2 = 32, \quad \beta_3 = 3, \quad \beta_4 = 8, \quad \beta_5 = 19.$$

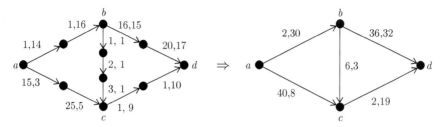

Fig. 2.6 Generalized network G^+ reduced to a four-node network N^+.

Using the formulae from the previous section, we obtain

$$\alpha = \alpha_{45} - \alpha_{12} = 4, \quad \bar{\alpha} = \alpha_4 - \alpha_{13} = 32, \quad \hat{\alpha} = \alpha_2 - \alpha_{35} = 28,$$

$$\beta = \beta_1 + \beta_2 + \beta_4 + \beta_5 = 89,$$

and

$$\mu_1 = \frac{\hat{\alpha}\beta_{14} - \alpha\beta_3}{\beta_3\beta_{45} + \beta_5\beta_{14}} = 1.31, \quad \mu_2 = \frac{\bar{\alpha}\beta_{25} + \alpha\beta_3}{\beta_1\beta_{25} + \beta_3\beta_{12}} = 0.96.$$

Note that rounded numbers to 2 dp (decimal places) are used instead of exact values.

Let us apply Theorems 2.1–2.4 as shown in the following table:

Table 2.2 Practical application of Theorems 2.1–2.4.

Th.	Braess Number	Lower Bound for Q	Upper Bound for Q	Interval for Q
2.1	$\mathcal{B}_1 = 314$	$\max\{0.06; -0.15; 1.31; 0.96\} = 1.31$	8.59	$]1.31; 8.59[$
2.2	$\mathcal{B}_2 = 2954$	$\max\{0.06; -0.15; 0.93\} = 0.93$	0.97	$]0.93; 0.97]$
2.3	$\mathcal{B}_3 = 24193$	$\max\{0.06; -0.15; 0.97; -4; 0.76\} = 0.97$	1.31	$]0.97; 1.31]$
2.4	$\mathcal{B}_4 = 116440$	$\max\{0.06; -0.15; 1.27; -0.88; 0.80\}$ $= 1.27$	0.96	$\emptyset$

Thus, Braess' paradox occurs in N/N^+ in the following cases:

$$1.31 < Q < 8.59 \text{ by Theorem 2.1,}$$

$$0.93 < Q \le 0.97 \text{ by Theorem 2.2,}$$

$$0.97 < Q \le 1.31 \text{ by Theorem 2.3.}$$

Theorem 2.4 produces no interval. Therefore, Braess' paradox occurs if and only if

$$0.93 < Q < 8.59.$$

By Corollary 2.4 (c), the pseudo-paradox happens if

$$Q \geq 8.59.$$

In other words, under this condition the travel times at equilibria in N and N^+ are the same and there is no improvement in the network.

Thus, some improvement of travel times in the network when adding the link (b, c) only occurs for small values of Q ($Q < 0.93$). The extent of this improvement and Braess' paradox can be seen from the equilibrium functions found (to 2 dp) by Lemmas 2.1 and 2.2:

$$T_{Eq} = \begin{cases} 38 + 62Q & \text{if } 0 < Q \leq 0.06, \\ 40.79 + 18.81Q & \text{if } Q > 0.06, \end{cases}$$

and

$$T_{Eq}^+ = \begin{cases} 10 + 52Q & \text{if } 0 < Q \leq 0.97, \\ 35.76 + 25.44Q & \text{if } 0.97 < Q \leq 1.31, \\ 45.10 + 18.31Q & \text{if } 1.31 < Q < 8.59, \\ 40.79 + 18.81Q & \text{if } Q \geq 8.59. \end{cases}$$

The above findings for the reduced network N/N^+ also hold for the generalized network G/G^+. Let us summarize them in the following table:

Table 2.3 Summary of findings for the networks G/G^+ and N/N^+.

Network	Improvement	Braess' Paradox	Pseudo-Paradox	Travel Times at Equilibria
N/N^+	$Q < 0.93$	$0.93 < Q < 8.59$	$Q \geq 8.59$	T_{Eq}/T_{Eq}^+
G/G^+	$Q < 0.93$	$0.93 < Q < 8.59$	$Q \geq 8.59$	T_{Eq}/T_{Eq}^+

The aforementioned equilibrium functions and Table 2.3 can be easily used to analyse the generalized network G^+ for particular values of the total demand (flow) Q. For example, suppose that the following values of Q are of interest: 0.5, 5, 10. The results are summarized in Table 2.4 (to 1 dp).

Table 2.4 Analysis of particular values of Q in the network configuration G/G^+.

Total Demand Q	Result	Travel Time T_{Eq} in G	Travel Time T_{Eq}^+ in G^+
0.5	Improvement	50.2	36.0
5	Braess' paradox	134.8	136.6
10	Pseudo-paradox	228.9	228.9

2.2 Likelihood of Braess' Paradox in Networks

We develop a new technique to show that the likelihood of Braess' paradox to occur in the classical network configuration is rather small. This is demonstrated for different distributions of parameters of travel time functions for links in a network, with particular emphasis on the Erlang distribution because of its importance for traffic networks. For example, we prove mathematically that the probability of Braess' paradox to happen does not exceed 0.129 when the parameters follow the Erlang-3 distribution. Similar estimates are true for the exponential distribution, the χ^2-distribution, the uniform and other distributions.

Our simulation results for different distributions revealed that typical probabilities for Braess' paradox to occur in the classical network configuration do not exceed 10%, and they are very low for some distributions of the parameters of travel time functions. If the classical network configuration consists of motorway sections and class A roads and the parameters of the travel time functions are modelled by the Erlang-2 distribution, then the probability of Braess' paradox to occur is 6%.

The focus of this section is on the probability of Braess' paradox to occur in the classical network configuration when a single link is added/removed, which is consistent with the original definition of the paradox. Under other assumptions, Valiant and Roughgarden [16] proved that Braess' paradox is likely to occur in a natural random network model. More precisely, they showed that in almost all networks there is a set of links whose removal improves the travel time at equilibrium for a given appropriate total flow.

Let us reformulate and simplify Theorems 2.1–2.4 by replacing the condition $\mathcal{B}_i > 0$ for $i = 2, 3, 4$ by the condition $\mathcal{B}_1 > 0$. Actually, the latter is a stronger condition as discussed after Theorem 2.4. This modification is given in Theorem 2.5. Notice that some of the intervals in this theorem may be empty. If all four intervals are empty (or $\mathcal{B}_1 \leq 0$), then there is no Braess' paradox.

Theorem 2.5 *Braess' paradox occurs in the network configuration N/N^+ if and only if the Braess number B_1 is positive and the total flow Q belongs to the following intervals:*

(A) $\max \left\{ \frac{\alpha}{\beta_{12}} ; \frac{-\alpha}{\beta_{45}} ; \mu_1 ; \mu_2 \right\} < Q < \frac{\hat{\alpha}\beta_{14} + \bar{\alpha}\beta_{25}}{B_1}$;

(B) $\max \left\{ \frac{\alpha}{\beta_{12}} ; \frac{-\alpha}{\beta_{45}} ; \frac{\hat{\alpha}\beta_{45} + \bar{\alpha}\beta_{12}}{B_2} \right\} < Q \leq \min \left\{ \frac{\hat{\alpha}}{\beta_{35}} ; \frac{\bar{\alpha}}{\beta_{13}} \right\}$;

(C) $\max \left\{ \frac{\alpha}{\beta_{12}} ; \frac{-\alpha}{\beta_{45}} ; \frac{\bar{\alpha}}{\beta_{13}} ; \frac{-\bar{\alpha}}{\beta_4} ; \frac{\bar{\alpha}\beta_4\beta - \alpha\beta_{134}\beta_{45}}{B_3} \right\} < Q \leq \mu_1$;

(D) $\max \left\{ \frac{\alpha}{\beta_{12}} ; \frac{-\alpha}{\beta_{45}} ; \frac{\hat{\alpha}}{\beta_{35}} ; \frac{-\hat{\alpha}}{\beta_2} ; \frac{\hat{\alpha}\beta_2\beta + \alpha\beta_{235}\beta_{12}}{B_4} \right\} < Q \leq \mu_2$.

Proof: Part (A) of this theorem is equivalent to the statement of Theorem 2.1. Suppose that Theorem 2.2 gives a non-empty interval for Q and $B_2 > 0$. It follows that

$$\frac{\hat{\alpha}\beta_{45} + \bar{\alpha}\beta_{12}}{B_2} < \frac{\hat{\alpha}}{\beta_{35}} \quad \text{and} \quad \frac{\hat{\alpha}\beta_{45} + \bar{\alpha}\beta_{12}}{B_2} < \frac{\bar{\alpha}}{\beta_{13}},$$

which are equivalent to

$$\hat{\alpha}\left(\frac{B_1}{\beta_{12}} + \beta_{13}\right) > \bar{\alpha}\beta_{35} \quad \text{and} \quad \bar{\alpha}\left(\frac{B_1}{\beta_{45}} + \beta_{35}\right) > \hat{\alpha}\beta_{13}.$$

Adding the last two inequalities, we obtain

$$B_1\left(\frac{\hat{\alpha}}{\beta_{12}} + \frac{\bar{\alpha}}{\beta_{45}}\right) > 0.$$

We have $\hat{\alpha} > 0$ and $\bar{\alpha} > 0$, otherwise Theorem 2.2 does not provide an interval for Q. Thus, $B_1 > 0$. According to (2.1), $B_1 > 0$ implies $B_2 > 0$, hence the latter can be dropped from the formulation.

Suppose now that Theorem 2.3 provides a non-empty interval for Q and $B_3 > 0$. It follows that

$$-\frac{\alpha}{\beta_{45}} < \mu_1 = \frac{\hat{\alpha}\beta_{14} - \alpha\beta_3}{\beta_3\beta_{45} + \beta_5\beta_{14}},$$

which is equivalent to

$$-\alpha\beta_5 < \hat{\alpha}\beta_{45}. \tag{2.5}$$

It also follows from Theorem 2.3 that

$$\frac{\bar{\alpha}\beta_4\beta - \alpha\beta_{134}\beta_{45}}{\mathcal{B}_3} < \frac{\hat{\alpha}\beta_{14} - \alpha\beta_3}{\beta_3\beta_{45} + \beta_5\beta_{14}}.$$

If we denote $L = \beta_3\beta_{45} + \beta_5\beta_{14}$, then the last inequality is equivalent to

$$(\alpha + \hat{\alpha})\beta_4\beta L - \alpha\beta_{134}\beta_{45}L < \hat{\alpha}\beta_{14}\mathcal{B}_3 - \alpha\beta_3\mathcal{B}_3$$

or

$$\alpha(\beta_4\beta L - \beta_{134}\beta_{45}L + \beta_3\mathcal{B}_3) < \hat{\alpha}(\beta_{14}\mathcal{B}_3 - \beta_4\beta L).$$

After some rearrangement and simplification of both sides, we obtain

$$-\alpha\beta_5\beta_{134}\mathcal{B}_1 < \hat{\alpha}\beta_{45}\beta_{134}\mathcal{B}_1$$

or

$$-\alpha\beta_5\mathcal{B}_1 < \hat{\alpha}\beta_{45}\mathcal{B}_1.$$

Obviously, $\mathcal{B}_1 \neq 0$. Assume that $\mathcal{B}_1 < 0$. We have

$$-\alpha\beta_5 > \hat{\alpha}\beta_{45},$$

contrary to (2.5). Therefore, $\mathcal{B}_1 > 0$. According to (2.2), $\mathcal{B}_1 > 0$ implies $\mathcal{B}_3 > 0$, which means that the latter can be dropped from the formulation.

Finally, assume that Theorem 2.4 provides a non-empty interval for Q and $\mathcal{B}_4 > 0$. It follows that

$$\frac{\alpha}{\beta_{12}} < \mu_2 = \frac{\bar{\alpha}\beta_{25} + \alpha\beta_3}{\beta_1\beta_{25} + \beta_3\beta_{12}},$$

which is equivalent to

$$\alpha\beta_1 < \bar{\alpha}\beta_{12}. \tag{2.6}$$

Moreover, the existence of a non-empty interval implies that

$$\frac{\hat{\alpha}\beta_2\beta - \alpha\beta_{235}\beta_{12}}{\mathcal{B}_4} < \frac{\bar{\alpha}\beta_{25} + \alpha\beta_3}{\beta_1\beta_{25} + \beta_3\beta_{12}}.$$

If we denote $M = \beta_1\beta_{25} + \beta_3\beta_{12}$, then the last inequality is equivalent to

$$(\bar{\alpha} - \alpha)\beta_2\beta M + \alpha\beta_{235}\beta_{12}M < \bar{\alpha}\beta_{25}\mathcal{B}_4 + \alpha\beta_3\mathcal{B}_4$$

or

$$\alpha(\beta_{235}\beta_{12}M - \beta_2\beta M - \beta_3\mathcal{B}_4) < \bar{\alpha}(\beta_{25}\mathcal{B}_4 - \beta_2\beta M).$$

We obtain after some rearrangement and simplification of both sides:

$$\alpha\beta_1\beta_{235}\mathcal{B}_1 < \bar{\alpha}\beta_{12}\beta_{235}\mathcal{B}_1$$

or

$$\alpha\beta_1\mathcal{B}_1 < \bar{\alpha}\beta_{12}\mathcal{B}_1.$$

It is obvious that $\mathcal{B}_1 \neq 0$. Let us assume that $\mathcal{B}_1 < 0$. We have

$$\alpha\beta_1 > \bar{\alpha}\beta_{12},$$

contrary to (2.6). Therefore, $\mathcal{B}_1 > 0$. According to (2.3), $\mathcal{B}_1 > 0$ implies $\mathcal{B}_4 > 0$, hence the latter can be dropped from the formulation. ∎

We will need the following technical result to show that Theorem 2.5 implies Theorem 2.6:

Lemma 2.3

(a) If $\mathcal{B}_1 > 0$, then $\frac{\alpha}{\beta_{12}} < \frac{\hat{\alpha}\beta_{14}+\bar{\alpha}\beta_{25}}{\mathcal{B}_1}$ is equivalent to $\hat{\alpha}\beta_1 + \bar{\alpha}\beta_2 > 0$;

(b) If $\mathcal{B}_1 > 0$, then $-\frac{\alpha}{\beta_{45}} < \frac{\hat{\alpha}\beta_{14}+\bar{\alpha}\beta_{25}}{\mathcal{B}_1}$ is equivalent to $\hat{\alpha}\beta_4 + \bar{\alpha}\beta_5 > 0$;

(c) If $\mathcal{B}_1 > 0$, then $\mu_1 < \frac{\hat{\alpha}\beta_{14}+\bar{\alpha}\beta_{25}}{\mathcal{B}_1}$ is equivalent to $\hat{\alpha}\beta_4 + \bar{\alpha}\beta_5 > 0$;

(d) If $\mathcal{B}_1 > 0$, then $\mu_2 < \frac{\hat{\alpha}\beta_{14}+\bar{\alpha}\beta_{25}}{\mathcal{B}_1}$ is equivalent to $\hat{\alpha}\beta_1 + \bar{\alpha}\beta_2 > 0$;

(e) $\bar{\alpha}/\beta_{13} < \mu_1$ is equivalent to $\hat{\alpha}\beta_{13} > \bar{\alpha}\beta_{35}$;

(f) $\hat{\alpha}/\beta_{35} < \mu_2$ is equivalent to $\hat{\alpha}\beta_{13} < \bar{\alpha}\beta_{35}$;

(g) $\hat{\alpha}/\beta_{35} < \mu_1$ is equivalent to $\hat{\alpha}\beta_{13} > \bar{\alpha}\beta_{35}$;

(h) $\bar{\alpha}/\beta_{13} < \mu_2$ is equivalent to $\hat{\alpha}\beta_{13} < \bar{\alpha}\beta_{35}$;

(i) $-\bar{\alpha}/\beta_4 < \mu_1$ is equivalent to $\hat{\alpha}\beta_4 + \bar{\alpha}\beta_5 > 0$;

(j) $-\hat{\alpha}/\beta_2 < \mu_2$ is equivalent to $\hat{\alpha}\beta_1 + \bar{\alpha}\beta_2 > 0$.

Let the delay parameters be arranged in a 2×2 matrix B, and let $\hat{\alpha}$ and $\bar{\alpha}$ be presented as a 2-dimensional vector $\underline{\alpha}$:

$$B = \begin{pmatrix} \beta_1 & \beta_2 \\ \beta_4 & \beta_5 \end{pmatrix}, \quad \underline{\alpha} = \begin{pmatrix} \hat{\alpha} \\ \bar{\alpha} \end{pmatrix}.$$

The next important result, which follows from Theorem 2.5, gives necessary and sufficient conditions for Braess' paradox to occur in the classical network configuration N/N^+. Although this result does not provide the interval of values for the total flow where the paradox is happening, it is very helpful for finding the probability of Braess' paradox to occur. It is interesting to note that the delay parameter β_3 for link (b, c), which is added/removed in the network configuration, plays no role in the occurrence of Braess' paradox.

Theorem 2.6 *The statements (a), (b) and (c) are equivalent:*

(a) *Braess' paradox occurs in the classical network configuration N/N^+.*

(b) *The determinant of B is positive, and the linear transformation B applied to $\underline{\alpha}$ yields a vector with positive components; that is:*

$$|B| > 0 \quad and \quad B\underline{\alpha} > \underline{0}.$$

(c) *The following inequalities are satisfied:*

$$\beta_1\beta_5 > \beta_2\beta_4, \tag{2.7}$$
$$\beta_1\hat{\alpha} + \beta_2\bar{\alpha} > 0, \tag{2.8}$$
$$\beta_4\hat{\alpha} + \beta_5\bar{\alpha} > 0. \tag{2.9}$$

Proof: It is straightforward to see that parts (b) and (c) are equivalent. Let us prove that parts (a) and (c) are equivalent. We first show that Theorem 2.5 (A) indicates that Braess' paradox occurs if and only if the inequalities (2.7)–(2.9) are satisfied. Indeed, according to Theorem 2.5 (A), Braess' paradox happens if $\mathcal{B}_1 > 0$ and there is an interval of positive values for the total flow Q. The former condition $\mathcal{B}_1 = \beta_1\beta_5 - \beta_2\beta_4 > 0$ is equivalent to (2.7), and the latter means that four inequalities must be satisfied:

$$\frac{\alpha}{\beta_{12}} < \frac{\hat{\alpha}\beta_{14} + \bar{\alpha}\beta_{25}}{\mathcal{B}_1},$$
$$-\frac{\alpha}{\beta_{45}} < \frac{\hat{\alpha}\beta_{14} + \bar{\alpha}\beta_{25}}{\mathcal{B}_1},$$
$$\mu_1 < \frac{\hat{\alpha}\beta_{14} + \bar{\alpha}\beta_{25}}{\mathcal{B}_1},$$
$$\mu_2 < \frac{\hat{\alpha}\beta_{14} + \bar{\alpha}\beta_{25}}{\mathcal{B}_1}.$$

By Lemma 2.3 (a)–(d), these inequalities are equivalent to (2.8)–(2.9).

In the next part of the proof we will show that if any of parts (B)–(D) of Theorems 2.5 provides an interval for Q and $\mathcal{B}_1 > 0$, then part (A) of this theorem gives an interval too. Thus, if $\mathcal{B}_1 > 0$, then it is enough to look at part (A) only to see whether Braess' paradox is happening or not (but without having the entire range of values for Q when the paradox occurs).

Suppose that Theorem 2.5 (C) provides an interval for Q and $\mathcal{B}_1 > 0$. It follows that

$$\max \left\{ \frac{\alpha}{\beta_{12}} ; \frac{-\alpha}{\beta_{45}} \right\} < \mu_1.$$

Hence, part (A) of Theorem 2.5 can be rewritten as follows:

$$\max \{ \mu_1 ; \mu_2 \} < Q < \frac{\hat{\alpha}\beta_{14} + \bar{\alpha}\beta_{25}}{\mathcal{B}_1}. \tag{2.10}$$

Let us prove that $\mu_1 > \mu_2$. We know that $\bar{\alpha}/\beta_{13} < \mu_1$, which is equivalent to $\hat{\alpha}\beta_{13} > \bar{\alpha}\beta_{35}$ by Lemma 2.3 (e). Then, by Lemma 2.3 (f), $\hat{\alpha}/\beta_{35} > \mu_2$, and by Lemma 2.3 (g), $\hat{\alpha}/\beta_{35} < \mu_1$. Thus,

$$\mu_1 > \frac{\hat{\alpha}}{\beta_{35}} > \mu_2.$$

It remains to prove that

$$\mu_1 < \frac{\hat{\alpha}\beta_{14} + \bar{\alpha}\beta_{25}}{\mathcal{B}_1}. \tag{2.11}$$

We know $-\bar{\alpha}/\beta_4 < \mu_1$. By Lemma 2.3 (i),

$$\hat{\alpha}\beta_4 + \bar{\alpha}\beta_5 > 0,$$

which is equivalent to (2.11) by Lemma 2.3 (c). Thus, Theorem 2.5 (A) gives an interval for Q in this case.

Suppose now that Theorem 2.5 (D) provides an interval for Q and $\mathcal{B}_1 > 0$. It follows that

$$\max \left\{ \frac{\alpha}{\beta_{12}} ; \frac{-\alpha}{\beta_{45}} \right\} < \mu_2.$$

Hence, part (A) of Theorem 2.5 can be rewritten as (2.10). Let us prove that $\mu_1 < \mu_2$. We know that $\hat{\alpha}/\beta_{35} < \mu_2$, which is equivalent to $\hat{\alpha}\beta_{13} < \bar{\alpha}\beta_{35}$ by Lemma 2.3 (f). Then, by Lemma 2.3 (g), $\hat{\alpha}/\beta_{35} > \mu_1$. Thus,

$$\mu_2 > \frac{\hat{\alpha}}{\beta_{35}} > \mu_1.$$

Hence, it remains to prove that

$$\mu_2 < \frac{\hat{\alpha}\beta_{14} + \bar{\alpha}\beta_{25}}{\mathcal{B}_1}. \tag{2.12}$$

We know $-\hat{\alpha}/\beta_2 < \mu_2$. By Lemma 2.3 (j),

$$\hat{\alpha}\beta_1 + \bar{\alpha}\beta_2 > 0,$$

which is equivalent to (2.12) by Lemma 2.3 (d). Thus, Theorem 2.5 (A) provides an interval for Q in this case.

Finally, let us consider the situation when Theorem 2.5 (B) provides an interval for Q and $\mathcal{B}_1 > 0$. We have

$$\max\left\{ \frac{\alpha}{\beta_{12}}; \frac{-\alpha}{\beta_{45}}; \frac{\hat{\alpha}\beta_{45} + \bar{\alpha}\beta_{12}}{\mathcal{B}_2} \right\} < \min\left\{ \frac{\hat{\alpha}}{\beta_{35}}; \frac{\bar{\alpha}}{\beta_{13}} \right\},$$

and hence

$$\hat{\alpha} > 0 \quad \text{and} \quad \bar{\alpha} > 0.$$

Also, as discussed after Theorem 2.4, $\mathcal{B}_1 > 0$ implies $\mathcal{B}_3 > 0$ and $\mathcal{B}_4 > 0$. There are three cases to consider.

Case 1: Assume that $\hat{\alpha}/\beta_{35} < \bar{\alpha}/\beta_{13}$. We will show that this case implies that part (D) provides an interval, and hence part (A) gives an interval as shown above. We have

$$\max\left\{ \frac{\alpha}{\beta_{12}}; \frac{-\alpha}{\beta_{45}} \right\} < \frac{\hat{\alpha}}{\beta_{35}}$$

and $\hat{\alpha} > 0$. Hence, part (D) of Theorem 2.5 can be rewritten as follows:

$$\max\left\{ \frac{\hat{\alpha}}{\beta_{35}}; \frac{\hat{\alpha}\beta_2\beta + \alpha\beta_{235}\beta_{12}}{\mathcal{B}_4} \right\} < Q \leq \mu_2.$$

The first inequality $\hat{\alpha}/\beta_{35} < \mu_2$ follows from Lemma 2.3 (f). Let us prove that

$$\frac{\hat{\alpha}\beta_2\beta + \alpha\beta_{235}\beta_{12}}{\mathcal{B}_4} < \mu_2 = \frac{\bar{\alpha}\beta_{25} + \alpha\beta_3}{\beta_1\beta_{25} + \beta_3\beta_{12}}.$$

This inequality is equivalent to the following, taking into account that $\alpha = \bar{\alpha} - \hat{\alpha}$:

$$\hat{\alpha}(\beta_2 \beta M - \beta_{235}\beta_{12}M + \beta_3 \mathcal{B}_4) < \bar{\alpha}(\beta_{25}\mathcal{B}_4 + \beta_3 \mathcal{B}_4 - \beta_{235}\beta_{12}M),$$

where

$$M = \beta_1 \beta_{25} + \beta_3 \beta_{12}.$$

After a number of rearrangements and simplifications, we obtain an equivalent inequality:

$$\hat{\alpha}\beta_1 \beta_{235}(-\mathcal{B}_1) < \bar{\alpha}\beta_2 \beta_{235}\mathcal{B}_1,$$

which is true.

Case 2: Assume that $\hat{\alpha}/\beta_{35} > \bar{\alpha}/\beta_{13}$. We will show that this case implies that part (C) provides an interval, and hence part (A) gives an interval as shown above. We have

$$\max\left\{ \frac{\alpha}{\beta_{12}}; \frac{-\alpha}{\beta_{45}} \right\} < \frac{\bar{\alpha}}{\beta_{13}}$$

and $\bar{\alpha} > 0$. Hence, part (C) of Theorem 2.5 can be rewritten as follows:

$$\max\left\{ \frac{\bar{\alpha}}{\beta_{13}}; \frac{\bar{\alpha}\beta_4 \beta - \alpha\beta_{134}\beta_{45}}{\mathcal{B}_3} \right\} < Q \leq \mu_1.$$

The first inequality $\bar{\alpha}/\beta_{13} < \mu_1$ follows from Lemma 2.3 (e). Let us prove that

$$\frac{\bar{\alpha}\beta_4 \beta - \alpha\beta_{134}\beta_{45}}{\mathcal{B}_3} < \mu_1 = \frac{\hat{\alpha}\beta_{14} - \alpha\beta_3}{\beta_3 \beta_{45} + \beta_5 \beta_{14}}.$$

This inequality is equivalent to the following, taking into account that $\alpha = \bar{\alpha} - \hat{\alpha}$:

$$\bar{\alpha}(\beta_4 \beta L - \beta_{134}\beta_{45}L + \beta_3 \mathcal{B}_3) < \hat{\alpha}(\beta_{14}\mathcal{B}_3 + \beta_3 \mathcal{B}_3 - \beta_{134}\beta_{45}L),$$

where

$$L = \beta_3 \beta_{45} + \beta_5 \beta_{14}.$$

After a number of rearrangements and simplifications, we obtain an equivalent inequality:

$$\bar{\alpha}\beta_5\beta_{134}(-\mathcal{B}_1) < \hat{\alpha}\beta_4\beta_{134}\mathcal{B}_1,$$

which is true.

Case 3: Taking into account the previous two cases, we may assume that

$$\hat{\alpha}/\beta_{35} = \bar{\alpha}/\beta_{13} = \mu_1 = \mu_2.$$

We will show that this case implies that part (A) provides an interval for Q. We have

$$\max\left\{\frac{\alpha}{\beta_{12}}; \frac{-\alpha}{\beta_{45}}\right\} < \frac{\hat{\alpha}}{\beta_{35}} = \mu_1.$$

Hence, part (A) of Theorem 2.5 can be rewritten as follows:

$$\mu_1 < \frac{\hat{\alpha}\beta_{14} + \bar{\alpha}\beta_{25}}{\mathcal{B}_1}.$$

By Lemma 2.3 (c), this inequality is equivalent to $\hat{\alpha}\beta_4 + \bar{\alpha}\beta_5 > 0$, which is true. The proof is complete. ∎

In what follows, we shall assume that the parameters of the travel time functions in a network are random continuous variables. More precisely, free-flow travel times ($\alpha_i \geq 0$) for links follow specified probability distributions, and delay parameters ($\beta_i > 0$) have some general distribution. Let us define the random variables Ψ and Φ:

$$\Psi = \min\{\alpha_2 - \alpha_5; \alpha_4 - \alpha_1\} \quad \text{and} \quad \Phi = \max\{\alpha_2 - \alpha_5; \alpha_4 - \alpha_1\}.$$

Lemma 2.4 *The probability of Braess' paradox to occur in the classical network configuration N/N^+ satisfies the following bounds:*

$$0.5\,\mathbb{P}[\Psi - \alpha_3 > 0] \leq \mathbb{P}[\textit{Braess' paradox occurs}] \leq 0.5\,\mathbb{P}[\Phi - \alpha_3 > 0].$$

Proof: It is not difficult to see that

$$
\begin{aligned}
\mathbb{P}[\text{Braess' paradox occurs}] &\geq \mathbb{P}[((2.7) \text{ is true}) \text{ AND } (\hat{\alpha} > 0 \text{ AND } \bar{\alpha} > 0)] \\
&= \mathbb{P}[(2.7) \text{ is true}] \times \mathbb{P}[\hat{\alpha} > 0 \text{ AND } \bar{\alpha} > 0] \\
&= 0.5\,\mathbb{P}[\alpha_2 - \alpha_5 > \alpha_3 \text{ AND } \alpha_4 - \alpha_1 > \alpha_3] \\
&= 0.5\,\mathbb{P}[\min\{\alpha_2 - \alpha_5; \alpha_4 - \alpha_1\} > \alpha_3] \\
&= 0.5\,\mathbb{P}[\Psi - \alpha_3 > 0].
\end{aligned}
$$

Furthermore,

$$\mathbb{P}[\text{Braess' paradox does not occur}]$$
$$\geq \mathbb{P}[((2.7) \text{ is false}) \text{ OR } (\hat{\alpha} \leq 0 \text{ AND } \bar{\alpha} \leq 0)]$$
$$= \mathbb{P}[(2.7) \text{ is false}] + \mathbb{P}[\hat{\alpha} \leq 0 \text{ AND } \bar{\alpha} \leq 0]$$
$$\quad - \mathbb{P}[((2.7) \text{ is false}) \text{ AND } (\hat{\alpha} \leq 0 \text{ AND } \bar{\alpha} \leq 0)]$$
$$= 0.5 + 0.5\,\mathbb{P}[\hat{\alpha} \leq 0 \text{ AND } \bar{\alpha} \leq 0]$$
$$= 0.5 + 0.5\,(1 - \mathbb{P}[\hat{\alpha} > 0 \text{ OR } \bar{\alpha} > 0])$$
$$= 1 - 0.5\,\mathbb{P}[\alpha_2 - \alpha_5 > \alpha_3 \text{ OR } \alpha_4 - \alpha_1 > \alpha_3]$$
$$= 1 - 0.5\,\mathbb{P}[\max\{\alpha_2 - \alpha_5; \alpha_4 - \alpha_1\} > \alpha_3]$$
$$= 1 - 0.5\,\mathbb{P}[\Phi - \alpha_3 > 0].$$

Therefore,

$$\mathbb{P}[\text{Braess' paradox occurs}] = 1 - \mathbb{P}[\text{Braess' paradox does not occur}]$$
$$\leq 1 - (1 - 0.5\,\mathbb{P}[\Phi - \alpha_3 > 0])$$
$$= 0.5\,\mathbb{P}[\Phi - \alpha_3 > 0].$$

The proof is complete. ∎

2.2.1 Uniform and Exponential Distributions

We start with a relatively simple case when free-flow travel times for links are uniformly distributed. Without loss of generality, we assume that those random variables have support on the interval [0,1].

Theorem 2.7 *Let the free-flow travel times (α_i) for links in the classical network configuration N/N^+ follow the uniform distribution on [0,1], and let $\mathbb{P}_{\text{UN}}$ denote the probability of Braess' paradox to occur in such a network configuration. Then*
$$0.025 \leq \mathbb{P}_{\text{UN}} < 0.142.$$

Proof: The probability density function (pdf) of a uniform distribution for free-flow travel times is as follows:
$$f_{\alpha_i}(x) = \begin{cases} 1 & \text{if } 0 \leq x \leq 1, \\ 0 & \text{otherwise.} \end{cases}$$

It is easy to see that

$$f_{-\alpha_i}(x) = \begin{cases} 1 & \text{if } -1 \le x \le 0, \\ 0 & \text{otherwise.} \end{cases}$$

Let us consider the random variable $\alpha_i - \alpha_j$. The following convolution gives the pdf of the random variable $\alpha_i + (-\alpha_j)$:

$$\begin{aligned} f_{\alpha_i - \alpha_j}(z) &= (f_{\alpha_i} * f_{-\alpha_j})(z) \\ &= \int_{-\infty}^{+\infty} f_{\alpha_i}(x) f_{-\alpha_j}(z - x)\,dx \\ &= \int_0^1 f_{-\alpha_j}(z - x)\,dx. \end{aligned}$$

The integrand is 1 if and only if $-1 \le z - x \le 0$, i.e. $z \le x \le z + 1$. Taking into account that $0 \le x \le 1$, there are two cases to consider. If $-1 \le z < 0$, then

$$f_{\alpha_i - \alpha_j}(z) = \int_0^{z+1} 1\,dx = 1 + z.$$

If $0 \le z \le 1$, then

$$f_{\alpha_i - \alpha_j}(z) = \int_z^1 1\,dx = 1 - z.$$

Thus,

$$f_{\alpha_i - \alpha_j}(z) = \begin{cases} 1 + z & \text{if } -1 \le z < 0, \\ 1 - z & \text{if } 0 \le z \le 1, \\ 0 & \text{otherwise.} \end{cases}$$

Let Ψ be the random variable $\min\{\alpha_2 - \alpha_5; \alpha_4 - \alpha_1\}$, and let $r \in [0, 1]$. For the pdf $f_\Psi(x)$ of the random variable Ψ, we have

$$\begin{aligned} \int_r^1 f_\Psi(x)\,dx &= \mathbb{P}[\Psi > r] \\ &= \mathbb{P}[\alpha_2 - \alpha_5 > r \text{ AND } \alpha_4 - \alpha_1 > r] \\ &= \left(\int_r^1 (1 - z)\,dz \right)^2 \\ &= 0.25(1 - r)^4. \end{aligned}$$

Thus,

$$\int_r^1 f_\Psi(x)dx = 0.25(1-r)^4,$$

where $r \in [0, 1]$. Hence,

$$f_\Psi(x) = \begin{cases} (1-x)^3 & \text{if } 0 \le x \le 1, \\ 0 & \text{if } x > 1. \end{cases}$$

Now, let us consider the random variable $\Omega = \Psi - \alpha_3$. We obtain

$$\begin{aligned} f_\Omega(z) &= (f_\Psi * f_{-\alpha_3})(z) \\ &= \int_{-\infty}^{+\infty} f_\Psi(z-y)f_{-\alpha_3}(y)dy \\ &= \int_{-1}^0 f_\Psi(z-y)dy. \end{aligned}$$

Consider the case $0 \le z \le 1$. This implies $z - y \ge 0$ because $y \le 0$. If $z - y \le 1$ (i.e. $y \ge z-1$) then $f_\Psi(z-y) = (1-z+y)^3$; and $f_\Psi(z-y) = 0$ if $z-y > 1$ (i.e. $y < z - 1$). Therefore, for $0 \le z \le 1$,

$$f_\Omega(z) = \int_{z-1}^0 (1-z+y)^3 dy = 0.25(1-z)^4$$

and

$$\mathbb{P}[\Omega > 0] = \int_0^1 0.25(1-z)^4 dz = 0.05.$$

Thus, by Lemma 2.4,

$$\begin{aligned} \mathbb{P}[\text{Braess' paradox occurs}] &\ge 0.5\,\mathbb{P}[\Psi - \alpha_3 > 0] \\ &= 0.5\,\mathbb{P}[\Omega > 0] \\ &= 0.025. \end{aligned}$$

Let us find an upper bound for the probability of Braess' paradox to happen in the network configuration in question. Let Φ be the random variable $\max\{\alpha_2 - \alpha_5; \alpha_4 - \alpha_1\}$. For $r \in [0, 1]$, we obtain

$$\mathbb{P}[\Phi < r] = \mathbb{P}[\alpha_2 - \alpha_5 < r \text{ AND } \alpha_4 - \alpha_1 < r]$$

$$= \left(\int_{-1}^{r} f_{\alpha_2 - \alpha_5}(x) dx \right)^2$$

$$= \left(1 - \int_{r}^{1} f_{\alpha_2 - \alpha_5}(x) dx \right)^2$$

$$= \left(1 - 0.5(1 - r)^2 \right)^2$$

$$= 1 - (1 - r)^2 + 0.25(1 - r)^4.$$

Now, for $r \in [0, 1]$,

$$\int_{r}^{1} f_\Phi(x) dx = \int_{r}^{+\infty} f_\Phi(x) dx = \mathbb{P}[\Phi \geq r] = (1 - r)^2 - 0.25(1 - r)^4.$$

By differentiating, we obtain

$$f_\Phi(x) = (x - 1)(x^2 - 2x - 1) = x^3 - 3x^2 + x + 1, \quad 0 \leq x \leq 1.$$

Let us now consider the random variable $\Phi - \alpha_3$. Its pdf for $z \geq 0$ is calculated as follows:

$$f_{\Phi - \alpha_3}(z) = \int_{-\infty}^{+\infty} f_\Phi(z - x) f_{-\alpha_3}(x) dx = \int_{-1}^{0} f_\Phi(z - x) dx.$$

Consider the case $z \in [0, 1]$. Now, $z \geq 0$ and $x \leq 0$ imply $z - x \geq 0$. Also, if $z - x > 1$, then $f_\Phi(z - x) = 0$, so we may assume that $z - x \leq 1$, i.e. $x \geq z - 1$. We have

$$f_{\Phi - \alpha_3}(z) = \int_{z-1}^{0} \left((z - x)^3 - 3(z - x)^2 + (z - x) + 1 \right) dx$$

$$= \frac{1}{4} \left(-z^4 + 4z^3 - 2z^2 - 4z + 3 \right), \quad 0 \leq z \leq 1.$$

Finally, by Lemma 2.4, we obtain

$$\mathbb{P}[\text{Braess' paradox occurs}] \leq 0.5 \, \mathbb{P}[\Phi - \alpha_3 > 0]$$

$$= 0.5 \int_{0}^{1} \frac{1}{4} \left(-z^4 + 4z^3 - 2z^2 - 4z + 3 \right) dz$$

$$= 0.5 \times \frac{17}{60}$$

$$= 0.141(6)$$

$$< 0.142.$$

The proof is complete.   ∎

The following theorem is devoted to the situation when free-flow travel times for links are exponentially distributed.

Theorem 2.8 *Let the free-flow travel times (α_i) for links in the classical network configuration N/N^+ follow the exponential distribution, and let $\mathbb{P}_{EX}$ denote the probability of Braess' paradox to occur in such a network configuration. Then*

$$0.041 < \mathbb{P}_{EX} < 0.209.$$

Proof: The probability density function (pdf) of an exponential distribution for free-flow travel times is as follows:

$$f_{\alpha_i}(x) = \begin{cases} \frac{1}{\theta}e^{-x/\theta} & \text{if } x \geq 0, \\ 0 & \text{if } x < 0, \end{cases}$$

where θ is a scale parameter. It is easy to see that

$$f_{-\alpha_i}(x) = \begin{cases} \frac{1}{\theta}e^{x/\theta} & \text{if } x \leq 0, \\ 0 & \text{if } x > 0. \end{cases}$$

Let us consider the random variable $\alpha_i - \alpha_j$. The pdf of the random variable $\alpha_i + (-\alpha_j)$ is given by the following convolution:

$$\begin{aligned} f_{\alpha_i-\alpha_j}(z) &= (f_{\alpha_i} * f_{-\alpha_j})(z) \\ &= \int_{-\infty}^{+\infty} f_{\alpha_i}(x) f_{-\alpha_j}(z-x)dx \\ &= \frac{1}{\theta} \int_0^{+\infty} e^{-x/\theta} f_{-\alpha_j}(z-x)dx. \end{aligned}$$

Assume that $z \leq 0$. This implies $z - x \leq 0$ because $x \geq 0$. We obtain

$$\begin{aligned} f_{\alpha_i-\alpha_j}(z) &= \frac{1}{\theta^2} \int_0^{+\infty} e^{-x/\theta} e^{(z-x)/\theta} dx \\ &= \frac{1}{\theta^2} \int_0^{+\infty} e^{(z-2x)/\theta} dx \\ &= \frac{1}{\theta^2} \left[-\frac{\theta}{2} e^{(z-2x)/\theta} \right]_0^{+\infty} \\ &= \frac{e^{z/\theta}}{2\theta}. \end{aligned}$$

The function $f_{\alpha_i - \alpha_j}(z)$ is symmetric, and therefore

$$f_{\alpha_i - \alpha_j}(z) = \begin{cases} \frac{1}{2\theta} e^{-z/\theta} & \text{if } z \geq 0, \\ \frac{1}{2\theta} e^{z/\theta} & \text{if } z < 0. \end{cases}$$

Let Ψ be the random variable $\min\{\alpha_2 - \alpha_5; \alpha_4 - \alpha_1\}$, and let $r \geq 0$. For the pdf $f_\Psi(x)$ of the random variable Ψ, we have

$$\int_r^{+\infty} f_\Psi(x)dx = \mathbb{P}[\Psi > r]$$
$$= \mathbb{P}[\alpha_2 - \alpha_5 > r \text{ AND } \alpha_4 - \alpha_1 > r]$$
$$= \left(\int_r^{+\infty} \frac{e^{-z/\theta}}{2\theta} dz \right)^2$$
$$= 0.25\, e^{-2r/\theta}.$$

Thus,

$$\int_r^{+\infty} f_\Psi(x)dx = 0.25\, e^{-2r/\theta},$$

and hence

$$f_\Psi(x) = \frac{e^{-2x/\theta}}{2\theta}, \quad x \geq 0.$$

Now, let us consider the random variable $\Omega = \Psi - \alpha_3$ and determine its pdf:

$$f_\Omega(z) = (f_\Psi * f_{-\alpha_3})(z)$$
$$= \int_{-\infty}^{+\infty} f_\Psi(z - x)f_{-\alpha_3}(x)dx$$
$$= \frac{1}{\theta} \int_{-\infty}^{0} f_\Psi(z - x)e^{x/\theta}dx.$$

Consider the case $z \geq 0$. This implies $z - x \geq 0$ because $x \leq 0$. Therefore,

$$f_\Omega(z) = \frac{1}{2\theta^2} \int_{-\infty}^{0} e^{(3x - 2z)/\theta} dx = \frac{e^{-2z/\theta}}{6\theta}, \quad z \geq 0,$$

and

$$\mathbb{P}[\Omega > 0] = \int_0^{+\infty} \frac{e^{-2z/\theta}}{6\theta} dz = \frac{1}{12}.$$

Thus, by Lemma 2.4,

$$\mathbb{P}[\text{Braess' paradox occurs}] \geq 0.5\,\mathbb{P}[\Psi - \alpha_3 > 0]$$
$$= 0.5\,\mathbb{P}[\Omega > 0]$$
$$= 1/24$$
$$= 0.041(6)$$
$$> 0.041.$$

The second half of this proof for the upper bound is left to the reader as an exercise. ∎

2.2.2 Erlang-2, Erlang-3 and Erlang-4 Distributions

For traffic networks consisting of motorway sections, class A roads or a mixture of both, statistical tests showed that the distribution of parameters of travel time functions follow the Erlang-k distribution for small values of k as well as some other distributions, which will be discussed in the section devoted to simulation.

The technical result formulated in Lemma 2.5 is given without proof, and it will be used in the analysis of the Erlang distributions below.

Lemma 2.5 *The following equality holds:*

$$\int e^{v-x/a} \sum_{i=0}^{n} c_i x^i \, dx = e^{v-x/a} \sum_{i=0}^{n} \tau_i x^i + C,$$

where

$$\tau_n = -ac_n \quad \text{and} \quad \tau_i = a(i+1)\tau_{i+1} - ac_i, \ \ 0 \leq i \leq n-1.$$

Alternatively, the coefficients τ_m, $0 \leq m \leq n$, can be calculated directly:

$$\tau_m = -\sum_{j=0}^{n-m} \frac{(m+j)!}{m!} a^{j+1} c_{m+j}.$$

In particular,

$$\tau_0 = -\sum_{i=0}^{n} i!\, a^{i+1} c_i.$$

Our next theorem is devoted to the situation when free-flow travel times for links follow the Erlang-2 distribution. Similar to the results from the previous section, the formulated lower and upper bounds are true for any distribution of the delay parameters (β_i).

Theorem 2.9 *Let the free-flow travel times (α_i) for links in the classical network configuration N/N^+ follow the Erlang-2 distribution $\Gamma(2, \theta)$, and let $\mathbb{P}_{E2}$ denote the probability of Braess' paradox to occur in such a network configuration. Then*

$$0.025 < \mathbb{P}_{E2} < 0.163.$$

Proof: I. The probability density function of the Erlang-2 distribution for free-flow travel times is as follows:

$$f_{\alpha_i}(x) = \begin{cases} \epsilon\, x\, e^{-x/\theta} & \text{if } x \geq 0, \\ 0 & \text{if } x < 0, \end{cases}$$

where $\epsilon = \frac{1}{\theta^2}$ and θ is a scale parameter. It is easy to see that

$$f_{-\alpha_i}(x) = \begin{cases} -\epsilon\, x\, e^{x/\theta} & \text{if } x \leq 0, \\ 0 & \text{if } x > 0. \end{cases}$$

Let us consider the random variable $\alpha_i - \alpha_j$. The pdf of the random variable $\alpha_i + (-\alpha_j)$ is given by the following convolution:

$$\begin{aligned} f_{\alpha_i - \alpha_j}(z) &= (f_{\alpha_i} * f_{-\alpha_j})(z) \\ &= \int_{-\infty}^{+\infty} f_{\alpha_i}(x) f_{-\alpha_j}(z - x)\, dx \\ &= \int_{0}^{+\infty} \epsilon\, x\, e^{-x/\theta} f_{-\alpha_j}(z - x)\, dx. \end{aligned}$$

Assume that $z \leq 0$. This implies $z - x \leq 0$ because $x \geq 0$. We obtain

$$\begin{aligned} f_{\alpha_i - \alpha_j}(z) &= \epsilon^2 \int_{0}^{+\infty} x\, e^{-x/\theta}(x - z) e^{(z-x)/\theta}\, dx \\ &= \epsilon^2 \int_{0}^{+\infty} (x^2 - zx) e^{(z-2x)/\theta}\, dx. \end{aligned}$$

Let us apply Lemma 2.5. We have $n = 2, a = \theta/2, c_2 = 1, c_1 = -z, c_0 = 0$. Also,

$$\tau_0 = -\sum_{i=0}^{2} i!\, a^{i+1} c_i = \frac{1}{4}\theta^2(z - \theta).$$

By Lemma 2.5,

$$
\begin{aligned}
f_{\alpha_i - \alpha_j}(z) &= \epsilon^2 \left[e^{(z-2x)/\theta}(\tau_2 x^2 + \tau_1 x + \tau_0) \right]_0^{+\infty} \\
&= \epsilon^2 e^{z/\theta}(-\tau_0) \\
&= \frac{(\theta - z)e^{z/\theta}}{4\theta^2}, \quad z \le 0.
\end{aligned}
$$

The function $f_{\alpha_i - \alpha_j}(z)$ is symmetric, and therefore

$$
f_{\alpha_i - \alpha_j}(z) = \begin{cases}
\frac{1}{4\theta^2}(\theta + z)e^{-z/\theta} & \text{if } z \ge 0, \\
\frac{1}{4\theta^2}(\theta - z)e^{z/\theta} & \text{if } z < 0.
\end{cases}
$$

II. For the pdf $f_\Psi(x)$ of the random variable $\Psi = \min\{\alpha_2 - \alpha_5; \alpha_4 - \alpha_1\}$ and a real number $r \ge 0$, we have

$$
\begin{aligned}
\int_r^{+\infty} f_\Psi(x)dx &= \mathbb{P}[\Psi > r] \\
&= \mathbb{P}[\alpha_2 - \alpha_5 > r \text{ AND } \alpha_4 - \alpha_1 > r] \\
&= \left(\int_r^{+\infty} \frac{(z+\theta)e^{-z/\theta}}{4\theta^2}dz \right)^2 \\
&= \frac{1}{16\theta^4} \left(\int_r^{+\infty} (z+\theta)e^{-z/\theta}dz \right)^2.
\end{aligned}
$$

By Lemma 2.5, $n = 1, a = \theta, c_1 = 1, c_0 = 0, \tau_1 = -\theta, \tau_0 = -2\theta^2$. Therefore,

$$
\begin{aligned}
\int_r^{+\infty} f_\Psi(x)dx &= \frac{1}{16\theta^4} \left(\left[(-\theta z - 2\theta^2)e^{-z/\theta}\right]_r^{+\infty} \right)^2 \\
&= \frac{1}{16\theta^4} \left(\theta(r + 2\theta)e^{-r/\theta} \right)^2 \\
&= \frac{1}{16\theta^2}(r + 2\theta)^2 e^{-2r/\theta}.
\end{aligned}
$$

Thus,

$$\int_{r}^{+\infty} f_{\Psi}(x)dx = \frac{1}{16\theta^2}(r+2\theta)^2 e^{-2r/\theta}.$$

By differentiating, we obtain

$$f_{\Psi}(x) = \frac{1}{8\theta^3}(x^2 + 3\theta x + 2\theta^2)e^{-2x/\theta}, \quad x \geq 0.$$

III. Now, let us consider the random variable $\Omega = \Psi - \alpha_3$ and its pdf:

$$f_{\Omega}(z) = (f_{\Psi} * f_{-\alpha_3})(z)$$
$$= \int_{-\infty}^{+\infty} f_{\Psi}(z-x)f_{-\alpha_3}(x)dx$$
$$= \int_{-\infty}^{0} f_{\Psi}(z-x)(-\epsilon x)e^{x/\theta}dx.$$

Consider the case $z \geq 0$. This implies $z - x \geq 0$ because $x \leq 0$. Therefore,

$$f_{\Omega}(z) = \int_{-\infty}^{0} \frac{1}{8\theta^3} \left((z-x)^2 + 3\theta(z-x) + 2\theta^2\right) e^{2(x-z)/\theta}(-\epsilon x)e^{x/\theta}dx$$
$$= -\frac{1}{8\theta^5}\int_{-\infty}^{0} e^{(3x-2z)/\theta}\left(x^3 - (2z+3\theta)x^2 + (z^2 + 3\theta z + 2\theta^2)x\right)dx.$$

Let us apply Lemma 2.5. We have $n = 3, a = -\theta/3, c_3 = 1, c_2 = -2z - 3\theta,$
$c_1 = z^2 + 3\theta z + 2\theta^2, c_0 = 0.$ Also,

$$\tau_0 = -\sum_{i=0}^{3} i!\, a^{i+1}c_i = -\frac{\theta^2}{27}(3z^2 + 13\theta z + 14\theta^2).$$

By Lemma 2.5,

$$f_{\Omega}(z) = -\frac{1}{8\theta^5}e^{-2z/\theta}\tau_0$$
$$= \frac{e^{-2z/\theta}}{6^3\theta^3}(3z^2 + 13\theta z + 14\theta^2), \quad z \geq 0.$$

We obtain

$$\mathbb{P}[\Omega > 0] = \int_{0}^{+\infty} \frac{e^{-2z/\theta}}{6^3\theta^3}(3z^2 + 13\theta z + 14\theta^2)dz.$$

Again, by Lemma 2.5, $n = 2$, $a = \theta/2$, $c_2 = 3$, $c_1 = 13\theta$, $c_0 = 14\theta^2$ and

$$\tau_0 = -\sum_{i=0}^{2} i!\, a^{i+1} c_i = -11\theta^3.$$

Hence, by Lemma 2.5,

$$\mathbb{P}[\Omega > 0] = \frac{1}{6^3 \theta^3} \int_0^{+\infty} e^{-2z/\theta}(3z^2 + 13\theta z + 14\theta^2)\,dz = \frac{1}{6^3\theta^3}(-\tau_0) = \frac{11}{6^3}.$$

Thus, by Lemma 2.4,

$$\mathbb{P}[\text{Braess' paradox occurs}] \geq 0.5\, \mathbb{P}[\Psi - \alpha_3 > 0]$$
$$= 0.5\, \mathbb{P}[\Omega > 0]$$
$$> 0.025.$$

IV. In the second half of this proof, we will find an upper bound for the probability of Braess' paradox to happen in the network in question. Let us start with the random variable $\Phi = \max\{\alpha_2 - \alpha_5; \alpha_4 - \alpha_1\}$. For $r \geq 0$, we obtain

$$\mathbb{P}[\Phi < r] = \mathbb{P}[\alpha_2 - \alpha_5 < r \text{ AND } \alpha_4 - \alpha_1 < r]$$
$$= \left(\int_{-\infty}^{r} f_{\alpha_2 - \alpha_5}(x)\,dx\right)^2$$
$$= \left(1 - \int_{r}^{+\infty} f_{\alpha_2 - \alpha_5}(x)\,dx\right)^2$$
$$= \left(1 - \int_{r}^{+\infty} \frac{1}{4\theta^2}(x + \theta)e^{-x/\theta}\,dx\right)^2 \quad \text{\{Using part II\}}$$
$$= \left(1 - \frac{1}{4\theta}(r + 2\theta)e^{-r/\theta}\right)^2$$
$$= 1 - \frac{1}{2\theta}(r + 2\theta)e^{-r/\theta} + \frac{1}{16\theta^2}(r + 2\theta)^2 e^{-2r/\theta}.$$

Now, for $r \geq 0$,

$$\int_{r}^{+\infty} f_\Phi(x)\,dx = \mathbb{P}[\Phi \geq r] = \frac{1}{2\theta}(r + 2\theta)e^{-r/\theta} - \frac{1}{16\theta^2}(r + 2\theta)^2 e^{-2r/\theta}.$$

Therefore,

$$f_\Phi(x) = \frac{e^{-x/\theta}}{2\theta^2}(x+\theta) - \frac{e^{-2x/\theta}}{8\theta^3}(x^2 + 3\theta x + 2\theta^2), \quad x \geq 0.$$

V. Let us now consider the random variable $\Phi - \alpha_3$. Its pdf for $z \geq 0$ is calculated as follows:

$$f_{\Phi-\alpha_3}(z) = \int_{-\infty}^{+\infty} f_\Phi(z-x) f_{-\alpha_3}(x) dx = \int_{-\infty}^{0} f_\Phi(z-x)(-\epsilon x) e^{x/\theta} dx.$$

Taking into account that $z \geq 0$ and $x \leq 0$ imply $z - x \geq 0$, we have

$$f_{\Phi-\alpha_3}(z) = \frac{1_{\bullet}}{2\theta^4} \int_{-\infty}^{0} e^{(2x-z)/\theta}(x^2 - (z+\theta)x) dx$$

$$+ \frac{1}{8\theta^5} \int_{-\infty}^{0} e^{(3x-2z)/\theta} \left(x^3 - (2z+3\theta)x^2 + (z^2 + 3\theta z + 2\theta^2)x \right) dx.$$

By Lemma 2.5 for the first integral, we have: $n = 2$, $a = -\theta/2$, $c_2 = 1$, $c_1 = -z - \theta$, $c_0 = 0$ and

$$\tau_0 = -\sum_{i=0}^{2} i! \, a^{i+1} c_i = \frac{\theta^2}{4}(z + 2\theta).$$

Hence, evaluating the first integral by Lemma 2.5 and using part III for the second one, we obtain

$$f_{\Phi-\alpha_3}(z) = \frac{1}{8\theta^2} e^{-z/\theta}(z + 2\theta) - f_\Omega(z), \quad z \geq 0.$$

Now, let us calculate

$$\frac{1}{8\theta^2} \int_{0}^{+\infty} e^{-z/\theta}(z + 2\theta) dz.$$

Using Lemma 2.5, we obtain: $n = 1$, $a = \theta$, $c_1 = 1$, $c_0 = 2\theta$ and

$$\tau_0 = -\sum_{i=0}^{1} i! \, a^{i+1} c_i = -3\theta^2.$$

Therefore,

$$\frac{1}{8\theta^2} \int_{0}^{+\infty} e^{-z/\theta}(z + 2\theta) dz = \frac{1}{8\theta^2}(-\tau_0) = \frac{3}{8}.$$

Finally, using Lemma 2.4 and part III for $\int_0^{+\infty} f_\Omega(z)dz$, we deduce

$$\mathbb{P}[\text{Braess' paradox occurs}] \leq 0.5\,\mathbb{P}[\Phi - \alpha_3 > 0]$$
$$= 0.5 \int_0^{+\infty} \left(\frac{1}{8\theta^2} e^{-z/\theta}(z + 2\theta) - f_\Omega(z) \right) dz$$
$$= 0.5\,(3/8 - 11/216)$$
$$< 0.163.$$

The proof is complete. ∎

In the following theorem, we analyse the Erlang-3 distribution for free-flow travel times. Because the proof of this result is rather long, we only present a sketch of the five steps outlined in the proof of Theorem 2.9.

Theorem 2.10 *Let the free-flow travel times (α_i) for links in the classical network configuration N/N^+ follow the Erlang-3 distribution $\Gamma(3, \theta)$, and let $\mathbb{P}_{E3}$ denote the probability of Braess' paradox to occur in such a network configuration. Then*

$$0.016 < \mathbb{P}_{E3} < 0.129.$$

Proof: I. The pdf of the Erlang-3 distribution for free-flow travel times is

$$f_{\alpha_i}(x) = \begin{cases} \epsilon\,x^2\,e^{-x/\theta} & \text{if } x \geq 0, \\ 0 & \text{if } x < 0, \end{cases}$$

where $\epsilon = \frac{1}{2\theta^3}$ and θ is a scale parameter. Hence,

$$f_{-\alpha_i}(x) = \begin{cases} \epsilon\,x^2\,e^{x/\theta} & \text{if } x \leq 0, \\ 0 & \text{if } x > 0, \end{cases}$$

and, assuming $z \leq 0$,

$$f_{\alpha_i - \alpha_j}(z) = \int_{-\infty}^{+\infty} f_{\alpha_i}(x) f_{-\alpha_j}(z - x)dx$$
$$= \int_0^{+\infty} \epsilon\,x^2\,e^{-x/\theta} f_{-\alpha_j}(z - x)dx \quad \{z \leq 0\ \&\ x \geq 0 \Rightarrow z - x \leq 0\}$$
$$= \epsilon^2 \int_0^{+\infty} (x^4 - 2zx^3 + z^2x^2)e^{(z-2x)/\theta}dx$$
$$= \frac{1}{16\theta^3}(z^2 - 3z\theta + 3\theta^2)e^{z/\theta}.$$

By symmetry,

$$f_{\alpha_i - \alpha_j}(z) = \begin{cases} \frac{1}{16\theta^3}(z^2 + 3z\theta + 3\theta^2)e^{-z/\theta} & \text{if } z \geq 0, \\ \frac{1}{16\theta^3}(z^2 - 3z\theta + 3\theta^2)e^{z/\theta} & \text{if } z < 0. \end{cases}$$

II. For the pdf $f_\Psi(x)$ of the random variable $\Psi = \min\{\alpha_2 - \alpha_5; \alpha_4 - \alpha_1\}$ and a real number $r \geq 0$, we have

$$\int_r^{+\infty} f_\Psi(x)dx = \mathbb{P}[\Psi > r]$$

$$= \mathbb{P}[\alpha_2 - \alpha_5 > r \text{ AND } \alpha_4 - \alpha_1 > r]$$

$$= \left(\int_r^{+\infty} \frac{1}{16\theta^3}(z^2 + 3z\theta + 3\theta^2)e^{-z/\theta}dz \right)^2$$

$$= \frac{1}{2^8\theta^4} \left(r^2 + 5\theta r + 8\theta^2 \right)^2 e^{-2r/\theta}.$$

Therefore,

$$f_\Psi(x) = \frac{1}{2^7\theta^5}(x^4 + 8\theta x^3 + 26\theta^2 x^2 + 39\theta^3 x + 24\theta^4) e^{-2x/\theta}, \quad x \geq 0.$$

III. Consider the random variable $\Omega = \Psi - \alpha_3$ and assume that $z \geq 0$:

$$f_\Omega(z) = \int_{-\infty}^{+\infty} f_\Psi(z - x)f_{-\alpha_3}(x)dx$$

$$= \int_{-\infty}^0 f_\Psi(z - x)\epsilon x^2 e^{x/\theta}dx \quad \{z \geq 0 \text{ \& } x \leq 0 \Rightarrow z - x \geq 0\}$$

$$= \int_{-\infty}^0 \frac{1}{2^7\theta^5} \left((z - x)^4 + 8\theta(z - x)^3 + 26\theta^2(z - x)^2 \right.$$

$$\left. + 39\theta^3(z - x) + 24\theta^4 \right) e^{2(x-z)/\theta}\epsilon x^2 e^{x/\theta}dx$$

$$= \frac{1}{2^8\theta^8} \int_{-\infty}^0 e^{(3x-2z)/\theta} \left(x^6 - (4z + 8\theta)x^5 + (6z^2 + 24\theta z + 26\theta^2)x^4 \right.$$

$$- (4z^3 + 24\theta z^2 + 52\theta^2 z + 39\theta^3)x^3$$

$$\left. + (z^4 + 8\theta z^3 + 26\theta^2 z^2 + 39\theta^3 z + 24\theta^4)x^2 \right) dx$$

$$= \frac{e^{-2z/\theta}}{2^7 3^5\theta^5} \left(9z^4 + 108\theta z^3 + 522\theta^2 z^2 + 1187\theta^3 z + 1079\theta^4 \right), \quad z \geq 0.$$

We obtain

$$\mathbb{P}[\Omega > 0] = \int_0^{+\infty} \frac{e^{-2z/\theta}}{2^7 3^5 \theta^5} \left(9z^4 + 108\theta z^3 + 522\theta^2 z^2 + 1187\theta^3 z + 1079\theta^4\right) dz$$

$$= \frac{169}{2^6 3^4}.$$

By Lemma 2.4,

$$\mathbb{P}_{E3} \geq 0.5\,\mathbb{P}[\Psi - \alpha_3 > 0]$$
$$= 0.5\,\mathbb{P}[\Omega > 0]$$
$$> 0.016.$$

IV. For the random variable $\Phi = \max\{\alpha_2 - \alpha_5; \alpha_4 - \alpha_1\}$ and a real number $r \geq 0$, we obtain

$$\mathbb{P}[\Phi < r] = \mathbb{P}[\alpha_2 - \alpha_5 < r \text{ AND } \alpha_4 - \alpha_1 < r]$$

$$= \left(\int_{-\infty}^r f_{\alpha_2 - \alpha_5}(x)dx\right)^2$$

$$= \left(1 - \int_r^{+\infty} f_{\alpha_2 - \alpha_5}(x)dx\right)^2$$

$$= \left(1 - \int_r^{+\infty} \frac{1}{16\theta^3}(x^2 + 3x\theta + 3\theta^2)e^{-x/\theta}dx\right)^2$$

$$= \left(1 - \frac{1}{16\theta^2}(r^2 + 5\theta r + 8\theta^2)e^{-r/\theta}\right)^2$$

$$= 1 - \frac{1}{8\theta^2}(r^2 + 5\theta r + 8\theta^2)e^{-r/\theta}$$

$$+ \frac{1}{2^8\theta^4}\left(r^2 + 5\theta r + 8\theta^2\right)^2 e^{-2r/\theta}.$$

Now, for $r \geq 0$,

$$\int_r^{+\infty} f_\Phi(x)dx = \mathbb{P}[\Phi \geq r]$$

$$= \frac{1}{8\theta^2}(r^2 + 5\theta r + 8\theta^2)e^{-r/\theta}$$

$$- \frac{1}{2^8\theta^4}\left(r^2 + 5\theta r + 8\theta^2\right)^2 e^{-2r/\theta}.$$

Therefore, using part II,

$$f_\Phi(x) = \frac{e^{-x/\theta}}{8\theta^3}(x^2 + 3\theta x + 3\theta^2) - f_\Psi(x), \quad x \geq 0.$$

V. Consider the random variable $\Phi - \alpha_3$ and its pdf for $z \geq 0$:

$$f_{\Phi - \alpha_3}(z) = \int_{-\infty}^{+\infty} f_\Phi(z - x) f_{-\alpha_3}(x) dx$$

$$= \int_{-\infty}^{0} f_\Phi(z - x) \epsilon x^2 e^{x/\theta} dx \qquad \{z \geq 0 \ \& \ x \leq 0 \Rightarrow z - x \geq 0\}$$

$$= \frac{1}{2^4 \theta^6} \int_{-\infty}^{0} e^{(2x-z)/\theta} \left(x^4 - (2z+3\theta)x^3 + (z^2+3\theta z+3\theta^2)x^2 \right) dx$$

$$- \int_{-\infty}^{0} f_\Psi(z - x) \epsilon x^2 e^{x/\theta} dx \qquad \{\text{Using part III}\}$$

$$= \frac{e^{-z/\theta}}{2^7 \theta^3} (2z^2 + 12\theta z + 21\theta^2) - f_\Omega(z), \quad z \geq 0.$$

By Lemma 2.4, we obtain

$$\mathbb{P}_{E3} \leq 0.5 \, \mathbb{P}[\Phi - \alpha_3 > 0]$$

$$= 0.5 \int_0^{+\infty} \left(\frac{e^{-z/\theta}}{2^7 \theta^3} (2z^2 + 12\theta z + 21\theta^2) - f_\Omega(z) \right) dz$$

$$= 0.5 \left(\frac{37}{2^7} - \frac{169}{2^6 3^4} \right)$$

$$< 0.129.$$

■

Our next theorem considers the Erlang-4 distribution for free-flow travel times. Because the proof of this result is rather long, we only present a sketch of the five steps outlined in the proof of Theorem 2.9.

Theorem 2.11 *Let the free-flow travel times* (α_i) *for links in the classical network configuration* N/N^+ *follow the Erlang-4 distribution* $\Gamma(4, \theta)$, *and let* $\mathbb{P}_{E4}$ *denote the probability of Braess' paradox to occur in such a network configuration. Then*

$$0.01 < \mathbb{P}_{E4} < 0.103.$$

Proof: I. The pdf of the Erlang-4 distribution for free-flow travel times is

$$f_{\alpha_i}(x) = \begin{cases} \epsilon \, x^3 \, e^{-x/\theta} & \text{if } x \geq 0, \\ 0 & \text{if } x < 0, \end{cases}$$

where $\epsilon = 1/(6\theta^4)$ and θ is a scale parameter. Hence,

$$f_{-\alpha_i}(x) = \begin{cases} -\epsilon\, x^3\, e^{x/\theta} & \text{if } x \leq 0, \\ 0 & \text{if } x > 0, \end{cases}$$

and, assuming $z \leq 0$,

$$
\begin{aligned}
f_{\alpha_i - \alpha_j}(z) &= \int_{-\infty}^{+\infty} f_{\alpha_i}(x) f_{-\alpha_j}(z-x)\, dx \\
&= \int_{0}^{+\infty} \epsilon\, x^3\, e^{-x/\theta} f_{-\alpha_j}(z-x)\, dx \quad \{z \leq 0\ \&\ x \geq 0 \Rightarrow z - x \leq 0\} \\
&= \epsilon^2 \int_{0}^{+\infty} (x^6 - 3zx^5 + 3z^2 x^4 - z^3 x^3) e^{(z-2x)/\theta}\, dx \\
&= \frac{1}{2^5 3\theta^4}(15\theta^3 - 15\theta^2 z + 6\theta z^2 - z^3) e^{z/\theta}.
\end{aligned}
$$

By symmetry,

$$f_{\alpha_i - \alpha_j}(z) = \begin{cases} \frac{1}{2^5 3\theta^4}(15\theta^3 + 15\theta^2 z + 6\theta z^2 + z^3) e^{-z/\theta} & \text{if } z \geq 0, \\ \frac{1}{2^5 3\theta^4}(15\theta^3 - 15\theta^2 z + 6\theta z^2 - z^3) e^{z/\theta} & \text{if } z < 0. \end{cases}$$

II. For the pdf $f_\Psi(x)$ of the random variable $\Psi = \min\{\alpha_2 - \alpha_5; \alpha_4 - \alpha_1\}$ and a real number $r \geq 0$, we have

$$
\begin{aligned}
\int_{r}^{+\infty} f_\Psi(x)\, dx &= \mathbb{P}[\Psi > r] \\
&= \mathbb{P}[\alpha_2 - \alpha_5 > r \text{ AND } \alpha_4 - \alpha_1 > r] \\
&= \left(\int_{r}^{+\infty} \frac{1}{2^5 3\theta^4}(z^3 + 6\theta z^2 + 15\theta^2 z + 15\theta^3) e^{-z/\theta}\, dz \right)^2 \\
&= \frac{1}{2^{10} 9\theta^6} \left(r^3 + 9\theta r^2 + 33\theta^2 r + 48\theta^3 \right)^2 e^{-2r/\theta}.
\end{aligned}
$$

Therefore, for $x \geq 0$,

$$f_\Psi(x) = \frac{e^{-2x/\theta}}{2^9 9\theta^7}(x^6 + 15\theta x^5 + 102\theta^2 x^4 + 396\theta^3 x^3 + 918\theta^4 x^2 + 1215\theta^5 x + 720\theta^6).$$

III. Consider the random variable $\Omega = \Psi - \alpha_3$ and assume that $z \geq 0$:

$$f_\Omega(z) = \int_{-\infty}^{+\infty} f_\Psi(z-x)f_{-\alpha_3}(x)dx$$

$$= \int_{-\infty}^{0} f_\Psi(z-x)(-\epsilon x^3)e^{x/\theta}dx \quad \{z \geq 0 \,\&\, x \leq 0 \Rightarrow z - x \geq 0\}$$

$$= \int_{-\infty}^{0} \frac{1}{2^9 9 \theta^7}\Big((z-x)^6 + 15\theta(z-x)^5 + 102\theta^2(z-x)^4$$

$$+ 396\theta^3(z-x)^3 + 918\theta^4(z-x)^2 + 1215\theta^5(z-x) + 720\theta^6\Big)$$

$$\times e^{2(x-z)/\theta}(-\epsilon x^3)e^{x/\theta}dx$$

$$= -\frac{1}{2^{10}3^3\theta^{11}}\int_{-\infty}^{0} e^{(3x-2z)/\theta}\Big(x^9 - (6z+15\theta)x^8$$

$$+ (15z^2 + 75\theta z + 102\theta^2)x^7$$

$$- (20z^3 + 150\theta z^2 + 408\theta^2 z + 396\theta^3)x^6$$

$$+ (15z^4 + 150\theta z^3 + 612\theta^2 z^2 + 1188\theta^3 z + 918\theta^4)x^5$$

$$- (6z^5 + 75\theta z^4 + 408\theta^2 z^3 + 1188\theta^3 z^2 + 1836\theta^4 z + 1215\theta^5)x^4$$

$$+ (z^6 + 15\theta z^5 + 102\theta^2 z^4 + 396\theta^3 z^3 + 918\theta^4 z^2$$

$$+ 1215\theta^5 z + 720\theta^6)x^3\Big)dx$$

$$= \frac{e^{-2z/\theta}}{6^9\theta^7}\Big(27z^6 + 621\theta z^5 + 6,354\theta^2 z^4 + 36,780\theta^3 z^3$$

$$+ 126,474\theta^4 z^2 + 244,621\theta^5 z + 207,780\theta^6\Big), \quad z \geq 0.$$

We obtain

$$\mathbb{P}[\Omega > 0] = \int_{0}^{+\infty} \frac{e^{-2z/\theta}}{6^9\theta^7}\Big(27z^6 + 621\theta z^5 + 6,354\theta^2 z^4 + 36,780\theta^3 z^3$$

$$+ 126,474\theta^4 z^2 + 244,621\theta^5 z + 207,780\theta^6\Big)\,dz = \frac{216,538}{6^9}.$$

By Lemma 2.4,

$$\mathbb{P}_{E4} \geq 0.5\,\mathbb{P}[\Psi - \alpha_3 > 0]$$

$$= 0.5\,\mathbb{P}[\Omega > 0]$$

$$> 0.010.$$

IV. For the random variable $\Phi = \max\{\alpha_2 - \alpha_5; \alpha_4 - \alpha_1\}$ and a real number $r \geq 0$, we obtain

$$\mathbb{P}[\Phi < r] = \mathbb{P}[\alpha_2 - \alpha_5 < r \text{ AND } \alpha_4 - \alpha_1 < r]$$

$$= \left(\int_{-\infty}^{r} f_{\alpha_2 - \alpha_5}(x) dx \right)^2$$

$$= \left(1 - \int_{r}^{+\infty} f_{\alpha_2 - \alpha_5}(x) dx \right)^2$$

$$= \left(1 - \int_{r}^{+\infty} \frac{1}{2^5 3\theta^4} (15\theta^3 + 15\theta^2 x + 6\theta x^2 + x^3) e^{-x/\theta} dx \right)^2$$

$$= \left(1 - \frac{1}{2^5 3\theta^3} (r^3 + 9\theta r^2 + 33\theta^2 r + 48\theta^3) e^{-r/\theta} \right)^2$$

$$= 1 - \frac{1}{2^4 3\theta^3} (r^3 + 9\theta r^2 + 33\theta^2 r + 48\theta^3) e^{-r/\theta}$$

$$\quad + \frac{1}{2^{10} 9\theta^6} (r^3 + 9\theta r^2 + 33\theta^2 r + 48\theta^3)^2 e^{-2r/\theta}.$$

Now, for $r \geq 0$,

$$\int_{r}^{+\infty} f_{\Phi}(x) dx = \mathbb{P}[\Phi \geq r]$$

$$= \frac{1}{2^4 3\theta^3} (r^3 + 9\theta r^2 + 33\theta^2 r + 48\theta^3) e^{-r/\theta}$$

$$\quad - \frac{1}{2^{10} 9\theta^6} (r^3 + 9\theta r^2 + 33\theta^2 r + 48\theta^3)^2 e^{-2r/\theta}.$$

Therefore, using part II,

$$f_{\Phi}(x) = \frac{e^{-x/\theta}}{2^4 3\theta^4} (x^3 + 6\theta x^2 + 15\theta^2 x + 15\theta^3) - f_{\Psi}(x), \quad x \geq 0.$$

V. Consider the random variable $\Phi - \alpha_3$ and its pdf for $z \geq 0$:

$$f_{\Phi - \alpha_3}(z) = \int_{-\infty}^{+\infty} f_{\Phi}(z - x) f_{-\alpha_3}(x) dx$$

$$= \int_{-\infty}^{0} f_{\Phi}(z - x)(-\epsilon x^3) e^{x/\theta} dx \quad \{z \geq 0 \ \& \ x \leq 0 \Rightarrow z - x \geq 0\}$$

$$= \frac{1}{2^5 9\theta^8} \int_{-\infty}^{0} e^{(2x-z)/\theta} \left(x^6 - (3z + 6\theta)x^5 \right.$$

$$\quad \left. + (3z^2 + 12\theta z + 15\theta^2)x^4 - (z^3 + 6\theta z^2 + 15\theta^2 z + 15\theta^3)x^3 \right) dx$$

$$\quad - \int_{-\infty}^{0} f_{\Psi}(z - x)(-\epsilon x^3) e^{x/\theta} dx \quad \{\text{Using part III}\}$$

$$= \frac{e^{-z/\theta}}{2^8 3\theta^4} (z^3 + 12\theta z^2 + 54\theta^2 z + 90\theta^3) - f_{\Omega}(z), \quad z \geq 0.$$

By Lemma 2.4, we obtain

$$\mathbb{P}_{E4} \le 0.5\,\mathbb{P}[\Phi - \alpha_3 > 0]$$

$$= 0.5 \int_0^{+\infty} \left(\frac{e^{-z/\theta}}{2^8 3\theta^4}(z^3 + 12\theta z^2 + 54\theta^2 z + 90\theta^3) - f_\Omega(z) \right) dz$$

$$= 0.5 \left(\frac{29}{27} - \frac{216,538}{6^9} \right)$$

$$< 0.103.$$

$\blacksquare$

It may be pointed out that the theorems devoted to the exponential and Erlang distributions are relevant to the χ^2-distribution, which is a particular case of those distributions if we set $\theta = 2$. In the next section, we will look at a generalization of the previous results for the Erlang-k distribution, which converges to the normal distribution for large values of k.

2.2.3 Generalization for the Erlang-k Distribution

Let us generalize the previous results and consider the Erlang-k distribution for free-flow travel times. As illustrated in the next section, this distribution might be important in the context of road networks because a number of consecutive motorway sections can be modelled by a single link. The parameters of such a link could be described by the Erlang-k distribution, where k might have large values.

We will only consider the upper bound because the lower bound becomes too small (< 0.01) for $k \ge 5$. Also, for the sake of simplicity, we sacrifice a non-dominant term ($-0.5 \int_0^{+\infty} f_\Omega(z)dz$) in the upper bound. The absolute value of this term actually constitutes the aforementioned lower bound.

In contrast to previous theorems, the generalization will not directly provide an explicit numerical upper bound for Braess' paradox to occur. Instead, we will deduce some formula that can be used for calculation of the numerical upper bound. Note that in the third sum of this formula, the upper limit $(k - 2 - f - l)$ may be equal to -1, in which case the entire sum is equal to 0.

Theorem 2.12 *Let the free-flow travel times* (α_i) *for links in the classical network configuration* N/N^+ *follow the Erlang-k distribution* $\Gamma(k, \theta)$, *and let* $\mathbb{P}_{Ek}$ *denote the probability of Braess' paradox to occur in such a network configuration. Then*

$$\mathbb{P}_{Ek} < \sum_{l=0}^{k-1} \sum_{f=0}^{k-1-l} \frac{\binom{k-1+f}{f}}{2^{2k+f}} \left\{ 1 + \sum_{j=0}^{k-2-f-l} 2^{s_j+1-k} \binom{2k-3-s_j}{k-1} \frac{s_j}{k-1-s_j} \right\},$$

where $s_j = f + l + j$.

Proof: The pdf of the Erlang-k distribution for free-flow travel times is

$$f_{\alpha_i}(x) = \begin{cases} \epsilon\, x^{k-1}\, e^{-x/\theta} & \text{if } x \geq 0, \\ 0 & \text{if } x < 0, \end{cases}$$

where

$$\epsilon = \frac{1}{(k-1)!\,\theta^k}.$$

Hence,

$$f_{-\alpha_i}(x) = \begin{cases} \epsilon\, (-x)^{k-1}\, e^{x/\theta} & \text{if } x \leq 0, \\ 0 & \text{if } x > 0, \end{cases}$$

and, assuming $z \leq 0$,

$$\begin{aligned}
f_{\alpha_i - \alpha_j}(z) &= \int_{-\infty}^{+\infty} f_{\alpha_i}(x) f_{-\alpha_j}(z-x)\,dx \\
&= \int_0^{+\infty} \epsilon\, x^{k-1}\, e^{-x/\theta} f_{-\alpha_j}(z-x)\,dx \quad \{z \leq 0 \ \& \ x \geq 0 \Rightarrow z-x \leq 0\} \\
&= \epsilon^2 \int_0^{+\infty} e^{(z-2x)/\theta} (x^2 - zx)^{k-1}\,dx \\
&= \epsilon^2 \int_0^{+\infty} e^{(z-2x)/\theta} \sum_{i=0}^{k-1} \binom{k-1}{i} (-z)^i x^{2k-2-i}\,dx.
\end{aligned}$$

For evaluating the last integral, we apply Lemma 2.5: $n = 2k - 2$, $a = \theta/2$, $c_j = 0$ if $0 \leq j \leq k-2$ and

$$c_{2k-2-i} = \binom{k-1}{i}(-z)^i \quad \text{if } 0 \leq i \leq k-1.$$

Also,

$$\tau_0 = -\sum_{j=k-1}^{2k-2} j!(\theta/2)^{j+1} c_j = -\sum_{i=0}^{k-1} (\theta/2)^{2k-1-i}(2k-2-i)!\,c_{2k-2-i}.$$

We obtain

$$f_{\alpha_i - \alpha_j}(z) = \epsilon^2 e^{z/\theta}(-\tau_0)$$

$$= \epsilon^2 e^{z/\theta} \sum_{i=0}^{k-1} (\theta/2)^{2k-1-i} \binom{k-1}{i} (2k-2-i)!(-z)^i, \quad z \leq 0.$$

By symmetry,

$$f_{\alpha_i - \alpha_j}(z) = \begin{cases} \epsilon^2 e^{-z/\theta} \sum_{i=0}^{k-1} c_i' z^i & \text{if } z \geq 0, \\ \epsilon^2 e^{z/\theta} \sum_{i=0}^{k-1} c_i'(-z)^i & \text{if } z < 0, \end{cases}$$

where

$$c_i' = (\theta/2)^{2k-1-i} \binom{k-1}{i} (2k-2-i)!.$$

For the pdf $f_\Psi(x)$ of the random variable $\Psi = \min\{\alpha_2 - \alpha_5; \alpha_4 - \alpha_1\}$ and a real number $r \geq 0$, we have

$$\int_r^{+\infty} f_\Psi(x)dx = \mathbb{P}[\Psi > r] \tag{2.13}$$

$$= \mathbb{P}[\alpha_2 - \alpha_5 > r \text{ AND } \alpha_4 - \alpha_1 > r] \tag{2.14}$$

$$= \left(\int_r^{+\infty} \epsilon^2 e^{-z/\theta} \sum_{i=0}^{k-1} c_i' z^i dz \right)^2 \quad \text{\{By Lemma 2.5\}} \tag{2.15}$$

$$= \epsilon^4 e^{-2r/\theta} \left(\sum_{i=0}^{k-1} \tau_i' r^i \right)^2, \tag{2.16}$$

where

$$\tau_i' = - \sum_{t=0}^{k-1-i} \frac{(i+t)!}{i!} \theta^{t+1} c_{i+t}'$$

$$= -\frac{1}{i!} \theta^{2k-i} ((k-1)!)^2 \sum_{t=0}^{k-1-i} 2^{i+t+1-2k} \binom{2k-2-i-t}{k-1}.$$

For the random variable $\Phi = \max\{\alpha_2 - \alpha_5; \alpha_4 - \alpha_1\}$ and a real number $r \geq 0$, we obtain

$$\mathbb{P}[\Phi < r] = \mathbb{P}[\alpha_2 - \alpha_5 < r \text{ AND } \alpha_4 - \alpha_1 < r]$$

$$= \left(\int_{-\infty}^{r} f_{\alpha_2-\alpha_5}(x)dx \right)^2$$

$$= \left(1 - \int_{r}^{+\infty} f_{\alpha_2-\alpha_5}(x)dx \right)^2$$

$$= \left(1 - \int_{r}^{+\infty} \epsilon^2 e^{-x/\theta} \sum_{i=0}^{k-1} c_i' x^i dx \right)^2 \qquad \text{\{Using (2.15), (2.16)\}}$$

$$= \left(1 - \epsilon^2 e^{-r/\theta} \sum_{i=0}^{k-1} \tau_i' r^i \right)^2$$

$$= 1 - 2\epsilon^2 e^{-r/\theta} \sum_{i=0}^{k-1} \tau_i' r^i + \epsilon^4 e^{-2r/\theta} \left(\sum_{i=0}^{k-1} \tau_i' r^i \right)^2.$$

Now, for $r \geq 0$, we have

$$\int_{r}^{+\infty} f_\Phi(x)dx = \mathbb{P}[\Phi > r]$$

$$= 2\epsilon^2 e^{-r/\theta} \sum_{i=0}^{k-1} \tau_i' r^i - \epsilon^4 e^{-2r/\theta} \left(\sum_{i=0}^{k-1} \tau_i' r^i \right)^2 \qquad \text{\{Using (2.13), (2.16)\}}$$

$$= 2\epsilon^2 e^{-r/\theta} \sum_{i=0}^{k-1} \tau_i' r^i - \int_{r}^{+\infty} f_\Psi(x)dx.$$

By differentiating, we obtain for $x \geq 0$:

$$f_\Phi(x) + f_\Psi(x) = 2\epsilon^2 e^{-x/\theta} \sum_{i=0}^{k-1} (i\tau_i' x^{i-1} - \tau_i' x^i/\theta)$$

$$= 2\epsilon^2 e^{-x/\theta} \sum_{i=0}^{k-1} \left((i+1)\tau_{i+1}' - \tau_i'/\theta \right) x^i$$

$$= g(x).$$

Note that in the last sum, we set $\tau_k' = 0$.

Let us now consider the random variable $\Phi - \alpha_3$ and its pdf for $z \geq 0$:

$$
\begin{aligned}
f_{\Phi-\alpha_3}(z) &= \int_{-\infty}^{+\infty} f_\Phi(z-x) f_{-\alpha_3}(x)\, dx \\
&= \int_{-\infty}^{+\infty} g(z-x) f_{-\alpha_3}(x)\, dx - \int_{-\infty}^{+\infty} f_\Psi(z-x) f_{-\alpha_3}(x)\, dx \\
&= \int_{-\infty}^{0} g(z-x)\epsilon(-x)^{k-1} e^{x/\theta}\, dx - f_\Omega(z),
\end{aligned}
$$

where $\Omega = \Psi - \alpha_3$. Taking into account that $z \geq 0$ and $x \leq 0$ imply $z - x \geq 0$, we have

$$
\begin{aligned}
f_{\Phi-\alpha_3}(z) + f_\Omega(z) &= \int_{-\infty}^{0} 2\epsilon^2 e^{(x-z)/\theta} \left(\sum_{i=0}^{k-1} \left((i+1)\tau'_{i+1} - \tau'_i/\theta\right)(z-x)^i \right) \\
&\qquad\qquad \times \epsilon(-1)^{k-1} x^{k-1} e^{x/\theta}\, dx \\
&= \epsilon' \int_{-\infty}^{0} e^{(2x-z)/\theta} x^{k-1} \sum_{i=0}^{k-1} \tau''_i (z-x)^i\, dx.
\end{aligned}
$$

In the last expression, we denote

$$
\epsilon' = (-1)^{k-1} 2\epsilon^3 \quad \text{and} \quad \tau''_i = (i+1)\tau'_{i+1} - \tau'_i/\theta.
$$

By rearranging,

$$
\begin{aligned}
x^{k-1} \sum_{i=0}^{k-1} \tau''_i (z-x)^i &= x^{k-1} \sum_{i=0}^{k-1} \tau''_i \sum_{j=0}^{i} \binom{i}{j} z^j (-x)^{i-j} \\
&= \sum_{i=0}^{k-1} \sum_{j=0}^{i} \tau''_i \binom{i}{j} z^j (-1)^{i-j} x^{k-1+i-j} \\
&= \sum_{t=0}^{k-1} x^{k-1+t} \left(\sum_{i=t}^{k-1} \tau''_i \binom{i}{i-t}(-1)^t z^{i-t} \right) \\
&= \sum_{t=0}^{k-1} c_{k-1+t}\, x^{k-1+t},
\end{aligned}
$$

where

$$c_{k-1+t} = \sum_{i=t}^{k-1} \tau_i'' \binom{i}{i-t} (-1)^t z^{i-t}, \quad 0 \le t \le k-1.$$

We obtain

$$f_{\Phi-\alpha_3}(z) + f_\Omega(z) = \epsilon' \int_{-\infty}^0 e^{(2x-z)/\theta} \sum_{t=0}^{k-1} c_{k-1+t} \, x^{k-1+t} dx.$$

By Lemma 2.5, $n = 2k - 2$, $a = -\theta/2$, $c_j = 0$ if $0 \le j \le k - 2$ and c_{k-1+t} are defined above for $0 \le t \le k - 1$. Also,

$$\hat{\tau}_0 = -\sum_{j=0}^{n} j! \, a^{j+1} c_j$$

$$= -\sum_{j=k-1}^{2k-2} j! \left(-\frac{\theta}{2}\right)^{j+1} c_j$$

$$= -\sum_{t=0}^{k-1} (k-1+t)! \left(-\frac{\theta}{2}\right)^{k+t} c_{k-1+t}$$

$$= -\sum_{t=0}^{k-1} \sum_{i=t}^{k-1} (k-1+t)! \left(-\frac{\theta}{2}\right)^{k+t} \tau_i'' \binom{i}{i-t}(-1)^t z^{i-t}$$

$$= -\sum_{t=0}^{k-1} \sum_{i=t}^{k-1} \hat{\tau}_{t,i} z^{i-t},$$

where

$$\hat{\tau}_{t,i} = (k-1+t)! \left(-\frac{\theta}{2}\right)^{k+t} \tau_i'' \binom{i}{i-t}(-1)^t.$$

By rearranging,

$$\hat{\tau}_0 = -\sum_{l=0}^{k-1} \left(\sum_{f=0}^{k-1-l} \hat{\tau}_{f,f+l} \right) z^l,$$

where

$$\hat{\tau}_{f,f+l} = (k-1+f)! \left(-\frac{\theta}{2}\right)^{k+f} \binom{f+l}{l} (-1)^f \left\{ (f+l+1)\tau_{f+l+1}' - \tau_{f+l}'/\theta \right\}.$$

By Lemma 2.5, we obtain

$$f_{\Phi-\alpha_3}(z) + f_\Omega(z) = \epsilon' e^{-z/\theta} \hat{\tau}_0, \quad z \ge 0.$$

Now, by Lemma 2.4,

$$\mathbb{P}_{Ek} \leq 0.5\,\mathbb{P}[\Phi - \alpha_3 > 0]$$

$$= 0.5 \int_0^{+\infty} f_{\Phi-\alpha_3}(z)dz$$

$$= 0.5 \int_0^{+\infty} \epsilon' e^{-z/\theta} \hat{\tau}_0 \, dz - 0.5 \int_0^{+\infty} f_\Omega(z)dz$$

$$< 0.5\epsilon' \int_0^{+\infty} e^{-z/\theta} \hat{\tau}_0 \, dz$$

$$= 0.5\epsilon' \int_0^{+\infty} e^{-z/\theta} \sum_{l=0}^{k-1} \left(\sum_{f=0}^{k-1-l} (-\hat{\tau}_{f,f+l}) \right) z^l dz.$$

By Lemma 2.5, $n = k - 1$, $a = \theta$, $c_l = \sum_{f=0}^{k-1-l}(-\hat{\tau}_{f,f+l})$ for $0 \leq l \leq k-1$ and

$$\tau_0 = -\sum_{l=0}^{k-1} l!\,\theta^{l+1} c_l.$$

We obtain by Lemma 2.5:

$$\mathbb{P}_{Ek} < 0.5\epsilon'(-\tau_0) = 0.5\epsilon' \sum_{l=0}^{k-1} l!\,\theta^{l+1} \sum_{f=0}^{k-1-l} (-\hat{\tau}_{f,f+l}).$$

Let us rearrange the last term in the expression for $\hat{\tau}_{f,f+l}$ taking into account that $s_j = f + l + j$:

$$(f+l+1)\tau'_{f+l+1} - \tau'_{f+l}/\theta = \frac{\theta^{2k-f-l-1}\,((k-1)!)^2}{(f+l)!\,2^k} \left\{ \sum_{j=0}^{k-1-f-l} 2^{s_j+1-k} \right.$$

$$\times \binom{2k-2-s_j}{k-1} - \sum_{j=0}^{k-2-f-l} 2^{s_j+2-k} \binom{2k-3-s_j}{k-1} \left. \right\}$$

$$= \frac{\theta^{2k-f-l-1}\,((k-1)!)^2}{(f+l)!\,2^k} \left\{ 1 + \sum_{j=0}^{k-2-f-l} 2^{s_j+1-k} \binom{2k-3-s_j}{k-1} \frac{s_j}{k-1-s_j} \right\}.$$

Note that the last sum is equal to 0 if the upper limit is -1.
   Finally, we obtain

$$\mathbb{P}_{Ek} < 0.5\epsilon' \sum_{l=0}^{k-1} l!\,\theta^{l+1} \sum_{f=0}^{k-1-l} (-\hat{\tau}_{f,f+l})$$

$$= (-1)^{k-1}\epsilon^3 \sum_{l=0}^{k-1} l!\,\theta^{l+1} \sum_{f=0}^{k-1-l} (k-1+f)! \left(-\frac{\theta}{2}\right)^{k+f} \binom{f+l}{l}(-1)^{f+1}$$

$$\times \frac{\theta^{2k-f-l-1}\,((k-1)!)^2}{(f+l)!\,2^k}\left\{1+\sum_{j=0}^{k-2-f-l} 2^{s_j+1-k}\binom{2k-3-s_j}{k-1}\frac{s_j}{k-1-s_j}\right\}$$

$$=\sum_{l=0}^{k-1}\sum_{f=0}^{k-1-l}\frac{\binom{k-1+f}{f}}{2^{2k+f}}\left\{1+\sum_{j=0}^{k-2-f-l} 2^{s_j+1-k}\binom{2k-3-s_j}{k-1}\frac{s_j}{k-1-s_j}\right\},$$

as required. ∎

It is an interesting conjecture that the bound of Theorem 2.12 tends to 0 as $k \to +\infty$. We will only present numerical values of this bound for some values of the shape parameter k:

$$\mathbb{P}_{E5} < 0.090, \quad \mathbb{P}_{E6} < 0.072, \quad \mathbb{P}_{E7} < 0.058, \quad \mathbb{P}_{E8} < 0.047,$$
$$\mathbb{P}_{E9} < 0.038, \quad \mathbb{P}_{E10} < 0.031, \quad \mathbb{P}_{E11} < 0.026, \quad \mathbb{P}_{E12} < 0.021,$$
$$\mathbb{P}_{E13} < 0.017, \quad \mathbb{P}_{E14} < 0.014, \quad \mathbb{P}_{E15} < 0.012, \quad \mathbb{P}_{E16} < 0.010,$$
$$\mathbb{P}_{E28} < 0.001.$$

2.2.4 Simulation Results

In this section, we present the results of a computer simulation for the probability of Braess' paradox to occur in the classical network configuration N/N^+. The simulation is based on Theorem 2.6 and a random generation of the parameters of the travel time functions from specified probability distributions. More precisely, for a given probability distribution, the inverse of its cumulative probability density function is used to generate instances of the free-flow travel times (α_i), and a similar procedure is used to generate instances of the delay parameters (β_i). Up to 1 million instances were generated to calculate each probability presented in the following tables; typically, more instances were needed for small probabilities to reduce simulation errors. The probabilities are given to one significant figure if they are less than 0.1 and to two significant figures otherwise.

The probabilities of Braess' paradox to occur for different distributions of free-flow travel times and delay parameters are shown in Table 2.5. An important observation is that for a given distribution of free-flow travel times, the choice of the distribution of delay parameters practically does not affect the likelihood of Braess' paradox to occur. Some influence of the delay parameters can be observed when free-flow travel times follow the Erlang-28 distribution, however the probabilities themselves are very small.

Table 2.5 Probabilities for Braess' paradox to occur for different distributions of free-flow travel times and delay parameters.

Delay Parameter β_i:	Free-Flow Travel Times α_i:					
	Unif.	*Exp.*	*Erlang-2*	*Erlang-4*	*Erlang-16*	*Erlang-28*
Uniform	0.05	0.09	0.06	0.03	0.0006	0.00002
Exponential	0.05	0.09	0.06	0.03	0.0007	0.00003
Erlang-2	0.05	0.09	0.06	0.03	0.0007	0.00002
Erlang-4	0.05	0.10	0.06	0.03	0.0006	0.00002
Erlang-16	0.05	0.10	0.06	0.03	0.0006	0.00002
Erlang-28	0.05	0.10	0.07	0.03	0.0005	0.00001
Weibull(8,1)	0.06	0.11	0.07	0.03	0.0006	0.00002
Lognormal(0,1)	0.05	0.09	0.06	0.03	0.0006	0.00003
Beta(6,2)	0.05	0.10	0.07	0.03	0.0006	0.00001

Taking into account the observation made above, it seems reasonable to analyse situations when probability distributions for free-flow travel times and delay parameters are the same. Our statistical tests showed that free-flow travel times of links in road networks can be modelled by the Erlang-k distribution for small values of k, as well as the Weibull and lognormal distributions. This is of no surprise because those distribution can be very similar for some sets of their parameters as can be seen in Figure 2.7. This figure also illustrates the known fact that the Erlang-k distribution converges to the normal distribution for large values of k. On the other hand, with different parameters, the lognormal and Erlang-k distributions can be extremely right-skewed, which make them very different from the symmetric normal distribution, whereas the Weibull distribution can be skewed to any side depending on its parameters.

Tables 2.6 and 2.7 show the probabilities of Braess' paradox occurring when the parameters of travel time functions follow the Weibull distribution or the lognormal distribution. As can be seen from Table 2.6, the probabilities practically do not depend on the scale parameter λ if the shape parameter is fixed. Similarly, the probabilities in Table 2.7 do not depend on the M-parameter for a given S-parameter.

It is interesting to note that the highest probability of 14% in all simulations is achieved for the lognormal distribution when the S-parameter is at least 5; that is, when the distribution is extremely right-skewed. Such a distribution, however, is not common for modelling the parameters of travel time functions of roads and it may be of rather theoretical interest.

The beta distribution may be important in the context of Braess' paradox because of the variety of shapes provided by this distribution, taking into

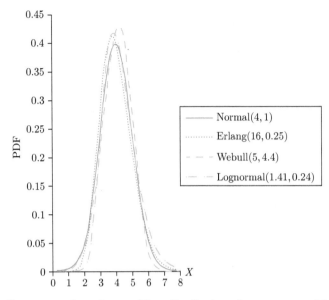

Fig. 2.7 Illustration of similarity of four distributions for some sets of their parameters.

Table 2.6 Probabilities for the Weibull distribution Weibull(k,λ).

	Scale Parameter λ:		
Shape Parameter k:	1	10	50
2	0.03	0.03	0.03
3	0.01	0.01	0.01
4	0.003	0.003	0.003
5	0.0005	0.0006	0.0005
6	0.00009	0.0001	0.00009
7	0.00002	0.00002	0.00003
10	$< 10^{-5}$	$< 10^{-5}$	$< 10^{-5}$

Table 2.7 Probabilities for the lognormal distribution Lognormal(M, S).

	S-parameter:				
M-parameter:	0.5	1	3	5	50
0	0.03	0.09	0.13	0.14	0.14
1	0.03	0.09	0.13	0.14	0.14
5	0.03	0.09	0.13	0.14	0.14

Table 2.8 Probabilities for the beta distribution Beta(A, B).

A-parameter:	B-parameter: 0.5	1	3	5	50
0.5	0.07	0.09	0.10	0.11	0.11
1	0.03	0.05	0.07	0.08	0.09
3	0.0006	0.003	0.01	0.02	0.04
5	0.00001	0.0001	0.001	0.004	0.02

Table 2.9 Probabilities for the normal distribution $\mathcal{N}(\mu, \sigma^2)$.

Mean μ:	Standard Deviation σ: 20	24	30	40	48
120	$< 10^{-5}$	0.0001	0.001	0.007	-
240	$< 10^{-5}$	$< 10^{-5}$	$< 10^{-5}$	$< 10^{-5}$	0.00009
480	$< 10^{-5}$	$< 10^{-5}$	$< 10^{-5}$	$< 10^{-5}$	$< 10^{-5}$

Table 2.10 (Cont.) Probabilities for the normal distribution $\mathcal{N}(\mu, \sigma^2)$.

Mean μ:	Standard Deviation σ: 60	80	96	120	160
120	-	-	-	-	-
240	0.001	0.007	-	-	-
480	$< 10^{-5}$	$< 10^{-5}$	0.00008	0.001	0.007

account that Braess' paradox can be observed in various applied contexts. Table 2.8 confirms another general observation that if a distribution is very left-skewed (e.g. $A = 5$, $B = 0.5$ for the beta distribution), then the corresponding probability of Braess' paradox to occur is very small; and the highest values of the probability are achieved for very right-skewed distributions.

The normal distribution with a 'large' negative tail may not be appropriate to model the random behaviour of free-flow travel times. However, because of its importance, some normal distributions can be used in the context of Braess' paradox as a first approximation, and hence we will only consider one family of the normal distribution with a rather 'small' negative tail, that is, when $\mu/\sigma \geq 3$. Note that in our simulations all generated instances with at least one negative parameter were rejected.

As can be seen in Tables 2.9 and 2.10, the probability of Braess' paradox to occur for the normal distribution of the parameters is approximately the same if the ratio μ/σ is fixed. A further observation for the normal distribution (and perhaps any distribution) with a fixed positive mean is that the smaller the standard deviation, the smaller the probability of Braess' paradox to occur. This is easy to understand for a very small standard deviation because in this case, with very high probability, the parameters α_i would be very close to the positive mean, and hence $\hat{\alpha}$ and $\bar{\alpha}$ would be negative and there would be no Braess' paradox by Theorem 2.6; that is, the probability of Braess' paradox happening would be very small.

A marvellous open problem would be to investigate the probability of Braess' paradox to occur in large traffic networks when a single link is added. Our hypothesis is that such probabilities are rather small. Some insight could be gained from the generalized traffic network discussed in Section 2.1.1. For this network, suppose that the (a, b)-path consists of eight motorway sections, and the same is true for the (b, d)-, (a, c)-, (c, d)- and (b, c)-paths. Assume also that the parameters of travel time functions for each section are modelled by the Erlang-2 distribution. Using the addition property of this distribution, such a generalized network can be reduced to the classical network configuration, where the parameters of travel time functions for each link are modelled by the Erlang-16 distribution. From Table 2.5, the probability of Braess' paradox to occur in the generalized network when a link on the (b, c)-path is removed is 0.0006. Another interesting addition to the aforementioned open problem would be to consider non-linear BPR functions.

Based on the results of this section, we can conclude that typical probabilities for Braess' paradox to occur in the classical network configuration do not exceed 10%, and they are very low for some distributions of the parameters of travel time functions. If the classical network configuration consists of motorway sections and class A roads and the parameters of the travel time functions are modelled by the Erlang-2 distribution, then the probability of Braess' paradox to occur is 6%. Also, let us summarize three observations made in this section:

1. For a given distribution of free-flow travel times, the choice of the distribution of delay parameters practically does not affect the likelihood of Braess' paradox to occur.
2. If the distribution of the parameters of travel time functions is very left-skewed, then the corresponding probability of Braess' paradox to occur is very small. The highest values of the probability are achieved for very right-skewed distributions.

3. If the distribution of the parameters of travel time functions has a fixed positive mean, then the smaller the standard deviation, the smaller the probability that Braess' paradox occurs.

2.3 Exercises

2.3.1

Let us consider the classical network configuration N/N^+ with the following parameters:

$$\alpha_1 = \alpha_3 = \alpha_5 = 1, \quad \alpha_2 = \alpha_4 = 4$$

and

$$\beta_1 = \beta_5 = 1, \quad \beta_2 = \beta_3 = \beta_4 = 0.$$

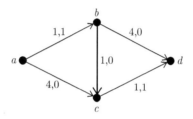

Fig. 2.8 Classical network configuration N/N^+ with specified parameters.

(i) Apply Theorem 2.6 to decide if Braess' paradox occurs in N/N^+ for some value of the total flow Q.

(ii) Find the equilibrium time and the equilibrium flow in N for $Q = 2$.

(iii) Using Lemma 2.2 or otherwise, find the equilibrium time and the equilibrium flow in N^+ for $Q = 2$. Hence decide whether Braess' paradox happens in the network configuration N/N^+ for $Q = 2$.

(iv) Apply Theorems 2.1–2.4 and find all values of Q for which Braess' paradox occurs in N/N^+. Verify your result using Corollary 2.2.

(v) Find the equilibrium time in N, and then in N^+, for all values of Q in the form of travel time functions T_{Eq} and T_{Eq}^+. Interpret the relationship between these two functions. (You might find helpful to plot them.)

(vi) A number of modifications have been planned for some links in the aforementioned network configuration N/N^+. The updated free-flow travel times and delay parameters are as follows:

$$\alpha_1 = \alpha_3 = \alpha_5 = 1, \quad \alpha_2 = \alpha_4 = 3$$

and

$$\beta_1 = \beta_2 = \beta_3 = \beta_4 = \beta_5 = 1.$$

Analyse this modified network configuration from the viewpoint of Braess' paradox and the pseudo-paradox.

2.3.2

Explain, without using Proposition 2.1, why the following statement is true: if Braess' paradox occurs in the classical network configuration N/N^+, then, at equilibrium, the path P_3 has a non-vanishing flow in N^+.

2.3.3

Prove Pas–Principio's result (Corollary 2.1) using Theorems 2.1–2.4 in Section 2.1.2.

2.3.4

Suppose that in the classical network configuration, α_i are uniformly distributed on $[0, m], m > 0$, and β_i are uniformly distributed on $[0, n], n > 0$. Show that Braess' paradox occurs in such a network configuration if and only if it happens in the same networks where α_i and β_i are uniformly distributed on $[0, 1]$. What does this imply about the probabilities of Braess' paradox to occur in those two network configurations?

2.3.5*

Let the free-flow travel times (α_i) for links in the classical network configuration N/N^+ follow the exponential distribution, and let $\mathbb{P}_{\mathrm{EX}}$ denote the probability of Braess' paradox to occur in such a network configuration. Show that

$$\mathbb{P}_{\mathrm{EX}} < 0.209,$$

which is the second half of Theorem 2.8. Hint: first, find $\mathbb{P}[\Phi < r]$, then $\mathbb{P}[\Phi \geq r]$ and the pdf $f_\Phi(x)$ for $x \geq 0$. Next, determine the pdf $f_{\Phi - \alpha_3}(z)$ for $z \geq 0$. Finally, apply Lemma 2.4 and find the required upper bound.

2.3.6

Prove Lemma 2.5.

2.3.7*

Find the pdf of the random variable $\Omega = \Psi - \alpha_3$ in part III of the proof of Theorem 2.11 by applying the alternative convolution:

$$f_\Omega(z) = \int_{-\infty}^{+\infty} f_{-\alpha_3}(z - x) f_\Psi(x) dx,$$

where $z \geq 0$.

2.3.8

Calculate the upper bound of Theorem 2.12 for the Erlang-4 distribution (to 3 dp).

2.3.9

Prove the following identity from the final part of the proof of Theorem 2.12:

$$\sum_{j=0}^{k-1-f-l} 2^{s_j+1-k} \binom{2k - 2 - s_j}{k - 1} - \sum_{j=0}^{k-2-f-l} 2^{s_j+2-k} \binom{2k - 3 - s_j}{k - 1}$$

$$= 1 + \sum_{j=0}^{k-2-f-l} 2^{s_j+1-k} \binom{2k - 3 - s_j}{k - 1} \frac{s_j}{k - 1 - s_j},$$

where $s_j = f + l + j$.

$$2.3.10^*$$

Develop a spreadsheet or a computer code for evaluating the upper bound of Theorem 2.12, and hence find the numerical upper bound for the Erlang-18 distribution (to 3 dp).

2.4 Solutions

2.3.1

(i) We have

$$\hat{\alpha} = \alpha_2 - \alpha_3 - \alpha_5 = 2, \quad \bar{\alpha} = \alpha_4 - \alpha_1 - \alpha_3 = 2.$$

Hence, (2.7)–(2.9) are satisfied and Braess' paradox occurs in N/N^+ by Theorem 2.6. ∎

(ii) The travel time for the path P_1 is

$$T_1 = \alpha_1 + \beta_1 f_{ab} + \alpha_2 + \beta_2 f_{bd} = 5 + f_{ab}.$$

The travel time for the path P_2 is

$$T_2 = \alpha_4 + \beta_4 f_{ac} + \alpha_5 + \beta_5 f_{cd} = 5 + f_{cd} = 5 + Q - f_{ab} = 7 - f_{ab}.$$

If one of the paths has a vanishing flow, then $f_{ab} = 0$ or 2 and there is no equilibrium because an unused path would have a shorter travel time. Hence, the both paths have non-vanishing flows and their travel times must be equal, that is, $5 + f_{ab} = 7 - f_{ab}$ or $f_{ab} = 1$.

Thus, the equilibrium time in N for $Q = 2$ is

$$T_{Eq} = 6,$$

and the equilibrium flow is as follows:

$$f_{ab} = 1, \quad f_{bd} = 1, \quad f_{ac} = 1, \quad f_{cd} = 1.$$

Alternatively, Lemma 2.1 can be used for this part. ∎

(iii) By Lemma 2.2 (a), we have $\beta_{35} = \beta_{13} = 1$, and hence part (a) of the lemma covers the case $0 < Q \leq 2$. For $Q = 2$, we obtain

$$T_{Eq}^+ = \alpha_{135} + 2\beta_{135} = 3 + 2 \times 2 = 7.$$

Part (a) of Lemma 2.2 corresponds to the situation when, at equilibrium, the only path with non-vanishing flow is P_3. Hence, the equilibrium flow in N^+ is as follows:

$$f_{ab} = f_{bc} = f_{cd} = 2, \quad f_{ac} = f_{bd} = 0.$$

Because $T_{Eq}^+ > T_{Eq}$, Braess' paradox happens in the network configuration N/N^+ for $Q = 2$. ∎

(iv) We have $\mu_1 = \mu_2 = 2$, $\mathcal{B}_1 = 1$ and $\mathcal{B}_2 = 3$. Now, by Theorem 2.1, Braess' paradox occurs if $2 < Q < 4$, and by Theorem 2.2, Braess' paradox happens if $4/3 < Q \leq 2$. Theorems 2.3 and 2.4 provide no intervals for Braess' paradox. Thus, Braess' paradox occurs in N/N^+ if and only if

$$4/3 < Q < 4.$$

Corollary 2.2 gives the same interval. ∎

(v) Because $\alpha = 0$, cases (a) and (b) of Lemma 2.1 are impossible, and using case (c), we obtain

$$T_{Eq} = 5 + Q/2, \quad Q > 0.$$

Now, by Lemma 2.2 (a), $T_{Eq}^+ = 3 + 2Q$ if $0 < Q \leq 2$. The cases (b)–(e) of this lemma are impossible. Using case (f), we obtain $T_{Eq}^+ = 5 + Q/2$ if $Q \geq 4$. Finally, by Lemma 2.2 (g), $g = 4 - Q$ and

$$T_{Eq}^+ = 5 + Q/2 + g/2 = 7 \quad \text{if} \quad 2 < Q < 4.$$

Therefore, the travel time functions T_{Eq}^+ is as follows:

$$T_{Eq}^+ = \begin{cases} 3 + 2Q & \text{if } 0 < Q \leq 2, \\ 7 & \text{if } 2 < Q < 4, \\ 5 + Q/2 & \text{if } Q \geq 4. \end{cases}$$

Thus, some improvement of travel times in the network N when adding the link (b, c) only occurs for small values of Q ($Q < 4/3$). The extent of this improvement can be seen from the equilibrium functions. Braess' paradox occurs in N/N^+ if $4/3 < Q < 4$. The pseudo-paradox happens if $Q \geq 4$; that is, under this condition the travel times at equilibria in N and N^+ are the same and there is no improvement in the network.   ∎

(vi) By Theorem 2.5, there is no Braess' paradox because $\mathcal{B}_1 = 0$. Now, by Corollary 2.4 (d), the pseudo-paradox occurs in this network configuration if $Q > 0.5$.

Alternatively, using Lemmas 2.1 and 2.2, we obtain

$$T_{Eq} = 4 + Q \quad \text{if} \quad Q > 0,$$

$$T_{Eq}^+ = \begin{cases} 3 + 3Q & \text{if} \ 0 < Q \leq 0.5, \\ 4 + Q & \text{if} \ Q > 0.5. \end{cases}$$

Thus, there is no Braess' paradox, and the pseudo-paradox occurs in this network configuration if $Q > 0.5$.   ∎

2.3.2

Suppose that, at equilibrium, the path P_3 has a vanishing flow in N^+; that is, the flow at equilibrium in N^+ may only use P_1 and P_2, or just one of these paths. It is easy to see that the same flow is an equilibrium flow in N, and therefore the travel times at equilibria are equal. Thus, Braess' paradox cannot happen if at equilibrium the path P_3 has a vanishing flow in N^+. Because Braess' paradox does occur, we conclude that, at equilibrium, the path P_3 has a non-vanishing flow in N^+.   ∎

2.3.3

For the network configuration M/M^+, we have

$$\alpha = 0, \ \ \bar{\alpha} = \hat{\alpha} = \alpha_2 - \alpha_3, \ \ \mu_1 = \mu_2 = (\alpha_2 - \alpha_3)/\beta_{12}, \ \ \mathcal{B}_1 = \beta_1^2 - \beta_2^2.$$

The Braess number $\mathcal{B}_1$ is positive because $\beta_1 > \beta_2$ in M/M^+. Under the conditions of Theorem 2.1, Braess' paradox occurs if

$$\frac{\alpha_2 - \alpha_3}{\beta_{12}} < Q < \frac{2\beta_{12}(\alpha_2 - \alpha_3)}{\beta_1^2 - \beta_2^2} = \frac{2(\alpha_2 - \alpha_3)}{\beta_1 - \beta_2}.$$

Now,

$$\mathcal{B}_2 = \beta_1^2 + 2\beta_{12}^2 - \beta_2^2 = \beta_{12}(3\beta_1 + \beta_2),$$

which is a positive number. Therefore, by Theorem 2.2, Braess' paradox occurs if

$$\frac{2(\alpha_2 - \alpha_3)}{3\beta_1 + \beta_2} < Q \le \frac{\alpha_2 - \alpha_3}{\beta_{12}}.$$

Note that the lower bound is less than the upper bound because $\beta_1 > \beta_2$. Thus, the above inequalities can be written together as

$$\frac{2(\alpha_2 - \alpha_3)}{3\beta_1 + \beta_2} < Q < \frac{2(\alpha_2 - \alpha_3)}{\beta_1 - \beta_2}.$$

The upper and lower bounds of Theorems 2.3 and 2.4 provide no intervals for Q.   ∎

2.3.4

Let α_i' and β_i' be random variables, which are uniformly distributed on [0,1]. Then, $\alpha_i = m\alpha_i'$, $\beta_i = n\beta_i'$ and the inequalities (2.7)–(2.9) of Theorem 2.6 can be rewritten as follows:

$$n\beta_1' n\beta_5' > n\beta_2' n\beta_4',$$
$$n\beta_1'(m\alpha_2' - m\alpha_3' - m\alpha_5') + n\beta_2'(m\alpha_4' - m\alpha_1' - m\alpha_3') > 0,$$
$$n\beta_4'(m\alpha_2' - m\alpha_3' - m\alpha_5') + n\beta_5'(m\alpha_4' - m\alpha_1' - m\alpha_3') > 0.$$

Taking into account that $n > 0$ and $m > 0$, these inequalities can be simplified:

$$\beta_1' \beta_5' > \beta_2' \beta_4',$$

$$\beta_1'(\alpha_2' - \alpha_3' - \alpha_5') + \beta_2'(\alpha_4' - \alpha_1' - \alpha_3') > 0,$$
$$\beta_4'(\alpha_2' - \alpha_3' - \alpha_5') + \beta_5'(\alpha_4' - \alpha_1' - \alpha_3') > 0.$$

Thus, the inequalities (2.7)–(2.9) of Theorem 2.6 are satisfied if and only if the same inequalities hold for the variables α_i' and β_i'. It follows that the probabilities of Braess' paradox to occur in the two network configurations are equal. ∎

2.3.5

We will find an upper bound for the probability of Braess' paradox to happen in the network configuration of Theorem 2.8. Let us start with the random variable $\Phi = \max\{\alpha_2 - \alpha_5; \alpha_4 - \alpha_1\}$. For $r \geq 0$, we obtain

$$\mathbb{P}[\Phi < r] = \mathbb{P}[\alpha_2 - \alpha_5 < r \text{ AND } \alpha_4 - \alpha_1 < r]$$
$$= \left(\int_{-\infty}^{r} f_{\alpha_2 - \alpha_5}(x)dx \right)^2$$
$$= \left(1 - \int_{r}^{+\infty} f_{\alpha_2 - \alpha_5}(x)dx \right)^2$$
$$= \left(1 - 0.5\, e^{-r/\theta} \right)^2$$
$$= 1 - e^{-r/\theta} + 0.25\, e^{-2r/\theta}.$$

Now, for $r \geq 0$,

$$\int_{r}^{+\infty} f_\Phi(x)dx = \mathbb{P}[\Phi \geq r] = e^{-r/\theta} - 0.25\, e^{-2r/\theta}.$$

Therefore,

$$f_\Phi(x) = \frac{e^{-x/\theta}}{\theta} - \frac{e^{-2x/\theta}}{2\theta}, \quad x \geq 0.$$

Let us now consider the random variable $\Phi - \alpha_3$. Its pdf for $z \geq 0$ is calculated as follows:

$$f_{\Phi - \alpha_3}(z) = \int_{-\infty}^{+\infty} f_\Phi(z - x) f_{-\alpha_3}(x)dx = \int_{-\infty}^{0} f_\Phi(z - x) \frac{1}{\theta} e^{x/\theta} dx.$$

Taking into account that $z \geq 0$ and $x \leq 0$ imply $z - x \geq 0$, we have

$$f_{\Phi - \alpha_3}(z) = \frac{1}{\theta^2} \int_{-\infty}^{0} e^{(2x-z)/\theta} dx - \frac{1}{2\theta^2} \int_{-\infty}^{0} e^{(3x-2z)/\theta} dx$$
$$= \frac{1}{2\theta} e^{-z/\theta} - f_{\Omega}(z), \quad z \geq 0.$$

In the last equality, we used a formula for $f_{\Omega}(z)$ from the proof of Theorem 2.8. Finally, by Lemma 2.4, we obtain

$$\mathbb{P}[\text{Braess' paradox occurs}] \leq 0.5 \, \mathbb{P}[\Phi - \alpha_3 > 0]$$
$$= 0.5 \int_{0}^{+\infty} \left(\frac{1}{2\theta} e^{-z/\theta} - f_{\Omega}(z) \right) dz$$
$$= 0.5 \, (1/2 - 1/12)$$
$$= 0.208(3)$$
$$< 0.209.$$

The proof is complete.   ∎

2.3.6

By differentiating, we obtain

$$e^{v-x/a} \sum_{i=0}^{n} c_i x^i = -\frac{1}{a} e^{v-x/a} \sum_{i=0}^{n} \tau_i x^i + e^{v-x/a} \sum_{i=1}^{n} i\tau_i x^{i-1}.$$

Therefore,

$$\sum_{i=0}^{n} c_i x^i = -\frac{1}{a} \sum_{i=0}^{n} \tau_i x^i + \sum_{i=0}^{n-1} (i+1)\tau_{i+1} x^i$$
$$= -\frac{\tau_n}{a} x^n + \sum_{i=0}^{n-1} \left((i+1)\tau_{i+1} - \frac{\tau_i}{a} \right) x^i.$$

This implies

$$\tau_n = -ac_n \quad \text{and} \quad \tau_i = a(i+1)\tau_{i+1} - ac_i, \quad 0 \leq i \leq n-1.$$

Using these formulae, we have

$$\tau_{n-t} = -\left[a^{t+1}n(n-1) \times ... \times (n-t+1)c_n\right.$$

$$\left. + a^t(n-1) \times ... \times (n-t+1)c_{n-1} + ... + ac_{n-t}\right].$$

Thus,

$$\tau_m = -\sum_{j=0}^{n-m} \frac{(m+j)!}{m!}a^{j+1}c_{m+j}, \quad 0 \le m \le n.$$

∎

2.3.7

Assuming that $z \ge 0$, we obtain

$$f_\Omega(z) = \int_{-\infty}^{+\infty} f_{-\alpha_3}(z-x)f_\Psi(x)dx \qquad \{f_{-\alpha_3}(z-x) = 0 \text{ if } x < z\}$$

$$= \int_{z}^{+\infty} \epsilon(x-z)^3 e^{(z-x)/\theta} f_\Psi(x)dx \qquad \{x \ge z \ge 0 \Rightarrow x \ge 0\}$$

$$= \int_{z}^{+\infty} \epsilon(x-z)^3 e^{(z-x)/\theta} \frac{e^{-2x/\theta}}{2^9 9 \theta^7}(x^6 + 15\theta x^5 + 102\theta^2 x^4 + 396\theta^3 x^3$$

$$+ 918\theta^4 x^2 + 1215\theta^5 x + 720\theta^6)dx$$

$$= \frac{1}{2^{10}3^3\theta^{11}} \int_{z}^{+\infty} e^{(z-3x)/\theta} \left(x^9 + (15\theta - 3z)x^8 \right.$$

$$+ (102\theta^2 - 450\theta z + 3z^2)x^7$$

$$+ (396\theta^3 - 306\theta^2 z + 450\theta z^2 - z^3)x^6$$

$$+ (918\theta^4 - 1188\theta^3 z + 306\theta^2 z^2 - 15\theta z^3)x^5$$

$$+ (1215\theta^5 - 2754\theta^4 z + 1188\theta^3 z^2 - 102\theta^2 z^3)x^4$$

$$+ (720\theta^6 - 3645\theta^5 z + 2754\theta^4 z^2 - 396\theta^3 z^3)x^3$$

$$+ (-2160\theta^6 z + 3645\theta^5 z^2 - 918\theta^4 z^3)x^2$$

$$\left. + (2160\theta^6 z^2 - 1215\theta^5 z^3)x - 720\theta^6 z^3\right) dx.$$

By Lemma 2.5, $n = 9$, $a = \theta/3$ and

$$\tau_9 = -\frac{\theta}{3}$$
$$\tau_8 = \theta(z - 6\theta)$$

$$\tau_7 = \frac{\theta}{3}(-3z^2 + 53\theta z - 150\theta^2)$$

$$\tau_6 = \frac{\theta}{9}(3z^3 - 156\theta z^2 + 1289\theta^2 z - 2238\theta^3)$$

$$\tau_5 = \frac{\theta^2}{9}(51z^3 - 1230\theta z^2 + 6142\theta^2 z - 7230\theta^3)$$

$$\tau_4 = \frac{\theta^3}{3^3}(1173z^3 - 16,842\theta z^2 + 55,496\theta^2 z - 47,085\theta^3)$$

$$\tau_3 = \frac{\theta^4}{3^4}(15,384z^3 - 141,726\theta z^2 + 320,399\theta^2 z - 207,780\theta^3)$$

$$\tau_2 = \frac{\theta^5}{3^4}(40,170z^3 - 240,141\theta z^2 + 378,719\theta^2 z - 207,780\theta^3)$$

$$\tau_1 = \frac{\theta^6}{3^5}(178,755z^3 - 655,242\theta z^2 + 757,438\theta^2 z - 415,560\theta^3)$$

$$\tau_0 = \frac{\theta^7}{3^6}(353,715z^3 - 655,242\theta z^2 + 757,438\theta^2 z - 415,560\theta^3).$$

Therefore,

$$\begin{aligned}
f_\Omega(z) &= \frac{-e^{-2z/\theta}}{2^{10}3^3\theta^{11}}(\tau_9 z^9 + \ldots + \tau_1 z + \tau_0)\\
&= \frac{e^{-2z/\theta}}{6^9\theta^7}\left(27z^6 + 621\theta z^5 + 6,354\theta^2 z^4 + 36,780\theta^3 z^3\right.\\
&\quad \left. + 126,474\theta^4 z^2 + 244,621\theta^5 z + 207,780\theta^6\right), \quad z \geq 0.
\end{aligned}$$

∎

2.3.8

For $k = 4$, let us denote

$$S(l, f) = \frac{\binom{3+f}{f}}{2^{8+f}}\left\{1 + \sum_{j=0}^{2-f-l} 2^{s_j-3}\binom{5-s_j}{3}\frac{s_j}{3-s_j}\right\}.$$

Hence,

$$\mathbb{P}_{E4} < \sum_{l=0}^{3}\sum_{f=0}^{3-l} S(l, f).$$

We obtain

$$S(0,0) = \frac{5}{2^9}, \; S(0,1) = \frac{10}{2^9}, \; S(0,2) = \frac{10}{2^9}, \; S(0,3) = \frac{5}{2^9}, \; S(1,0) = \frac{5}{2^9},$$

$$S(1,1) = \frac{8}{2^9}, \; S(1,2) = \frac{5}{2^9}, \; S(2,0) = \frac{4}{2^9}, \; S(2,1) = \frac{4}{2^9}, \; S(3,0) = \frac{2}{2^9}.$$

Thus,

$$\mathbb{P}_{E4} < \frac{29}{2^8} < 0.114. \qquad \blacksquare$$

2.3.9

$$\sum_{j=0}^{k-1-f-l} 2^{s_j+1-k} \binom{2k-2-s_j}{k-1} - \sum_{j=0}^{k-2-f-l} 2^{s_j+2-k} \binom{2k-3-s_j}{k-1}$$

$$= 2^{f+l+(k-1-f-l)+1-k} \binom{2k-2-f-l-(k-1-f-l)}{k-1}$$

$$+ \sum_{j=0}^{k-2-f-l} 2^{s_j+1-k} \left\{ \binom{2k-2-s_j}{k-1} - 2\binom{2k-3-s_j}{k-1} \right\}$$

$$= \binom{k-1}{k-1} + \sum_{j=0}^{k-2-f-l} 2^{s_j+1-k} \frac{(2k-2-s_j)! - 2(2k-3-s_j)!\,(k-1-s_j)}{(k-1)!(k-1-s_j)!}$$

$$= 1 + \sum_{j=0}^{k-2-f-l} 2^{s_j+1-k} \frac{(2k-3-s_j)!}{(k-1)!\,(k-2-s_j)!} \times \frac{2k-2-s_j-2(k-1-s_j)}{k-1-s_j}$$

$$= 1 + \sum_{j=0}^{k-2-f-l} 2^{s_j+1-k} \binom{2k-3-s_j}{k-1} \frac{s_j}{k-1-s_j}. \qquad \blacksquare$$

2.3.10

$$\mathbb{P}_{E18} < 0.007.$$

Acknowledgements

The material of Section 2.1 is based on the article [20], released under the Creative Commons Attribution License.

References

[1] R. Arnott and K. Small, The economics of traffic congestion, *American Scientist*, **82** (1994), 446–455.

[2] M. J. Beckmann, C. B. McGuire and C. B. Winsten, *Studies in the Economics of Transportation*, New Haven, CT: Yale University Press, 1956.

[3] D. Braess, A. Nagurney and T. Wakolbinger, On a paradox of traffic planning, *Transportation Science*, **39** (4)(2005), 446–450. (The article was originally published as: D. Braess, Über ein Paradoxon aus der Verkehrsplanung, *Unternehmensforschung*, **12** (1968), 258–268).

[4] D. Branston, Link capacity functions: a review, *Transportation Research*, **10** (1976), 223–236.

[5] Bureau of Public Roads, *Traffic Assignment Manual,* US Department of Commerce, Urban Planning Division, Washington, DC, 1964.

[6] W. Byung-Wook, A differential game model of Nash equilibrium on a congested traffic network, *Networks*, **23** (6)(1993), 557–565.

[7] S. Dafermos and A. Nagurney, On some traffic equilibrium theory paradoxes, *Transportation Research*, **B 18** (2)(1984), 101–110.

[8] A. Haurie and P. Marcotte, On the relationship between Nash-Cournot and Wardrop equilibria, *Networks*, **15** (3)(1985), 295–308.

[9] Y. A. Korilis, A. A. Lazar and A. Orda, Avoiding the Braess paradox in non-cooperative networks, *Journal of Applied Probability*, **36** (1)(1999), 211–222.

[10] W.-H. Lin and H. K. Lo, Investigating Braess' paradox with time-dependent queues, *Transportation Science*, **43** (1)(2009), 117–126.

[11] A. Nagurney, The negation of the Braess paradox as demand increases: The wisdom of crowds in transportation networks, *Europhysics Letters*, **91** (4)(2010), 1–5.

[12] A. Nagurney, D. C. Parkes and P. Daniele, The Internet, evolutionary variational inequalities, and the time-dependent Braess paradox, *Computational Management Science*, **4** (4)(2007), 355–375.

[13] E. I. Pas and S. L. Principio, Braess' paradox: some new insights, *Transportation Research*, **B 31** (3)(1997), 265–276.

[14] T. Roughgarden and É. Tardos, How bad is selfish routing? *Journal of the ACM*, **49** (2)(2002), 236–259.

[15] A. S. Schulz and N. S. Moses, On the performance of user equilibria in traffic networks, MIT Sloan School of Management, 2002, *Proceedings of the 14th ACM-SIAM Symposium on Discrete Algorithms*, Baltimore, MD, 2003.

[16] G. Valiant and T. Roughgarden, Braess' paradox in large random graphs, *Proceedings of the 7th Annual ACM Conference on Electronic Commerce (EC)*, Ann Arbor, MI, 2006, 296–305.

[17] A. A. Walters, The theory and measurement of private and social cost of highway congestion, *Econometrica*, **29** (1971), 676–697.

[18] J. G. Wardrop, Some theoretical aspects of road traffic research, *Proceedings of the Institution of Civil Engineers*, PART II, **1** (1952), 325–378.

[19] H. Yang and M. G. H. Bell, A capacity paradox in network design and how to avoid it, *Transportation Research*, **A 32** (7)(1998), 539–545.

[20] V. Zverovich and E. Avineri, Braess' paradox in a generalised traffic network, *Journal of Advanced Transportation*, **49** (1)(2015), 114–138.

3

Emergency Response: Navigable Networks and Optimal Routing in Hazardous Indoor Environments

V. Zverovich, P. Boguslawski, L. Mahdjoubi and F. Fadli

The extreme importance of emergency response in complex buildings during natural and human-induced disasters has been widely acknowledged. Preparedness and rapid response are crucial issues for saving human lives, and the ability to identify the paths of egress during an emergency is critical for rescue and emergency services. Applications to prioritize indoor routes for emergency situations in a complex built facility have been restricted to building simulations and network approaches. These types of applications often failed to account for the complexity and trade-offs needed to select the optimal indoor path during an emergency situation. There is a need for efficient algorithms for finding safest indoor routes, which would take into account the three-dimensional (3D) structure of buildings, their relevant semantics and the nature and shape of hazards. In Section 3.1, we discuss algorithms for safest routes and balanced routes in buildings, where an extreme event with many epicentres is occurring. In a balanced route, a trade-off between route length and hazard proximity is made. The algorithms are based on the integration of a multi-attribute decision-making technique, Dijkstra's classical algorithm and the introduced hazard proximity numbers, hazard propagation coefficient and proximity index for a route. In Section 3.2, a step change is proposed for finding the optimal indoor routes for search and rescue teams. We present an algorithm that is based on a novel approach integrating the Analytic Hierarchy Process (AHP), the propagation of hazard and other techniques, and where three criteria are used: hazard proximity, distance/travel time and route complexity. The important feature of the algorithm is its ability to generate an optimal route depending on user's needs. The test results demonstrate the robustness of the algorithm with respect to different parameters, and its insensitivity to different scenarios of uncontrolled evacuation. Quality models supporting real-time decision-making and allowing the implementation of

Modern Applications of Graph Theory. Vadim Zverovich, Oxford University Press (2021). © Vadim Zverovich.
DOI: 10.1093/oso/9780198856740.003.0003

automated methods are very important. In Section 3.3, we discuss a novel automated construction of the Variable Density Network (VDN) for determining egress paths in dangerous environments. VDN is used for deriving a navigable network in an indoor building environment, including a full 3D topological model. The accuracy of the proposed approach is compared with key methods for navigable network generation using the actual floor plan of the Doha World Trade Center. The main benefit of the VDN approach is an increased prediction accuracy of egress route planning.

3.1 Automated Selection of Safest and Balanced Routes

Emergency response in the built environment is now being widely studied, with a significant surge of interest in this area after 9/11. The focus of such studies is on rescue and evacuation, which are based on route finding and indoor navigation ([15, 35, 37, 39, 44, 45, 66]). Despite many publications in this field in recent times, there is still a lack of appropriate evacuation algorithms and their implementations ([39, 41, 47, 65]).

In this section, we will present an algorithm for finding the safest route in a building, where an extreme event with many epicentres is occurring. Another algorithm produces a balanced route, which is achieved by a trade-off between route length and hazard proximity. The algorithms are based on a novel approach that integrates the recently developed 3D building model (summarized in Section 3.1.1), a multi-attribute decision-making technique, Dijkstra's classical algorithm (see Section 1.3.4) and the introduced hazard proximity numbers, hazard propagation coefficient and proximity index for a route. This study proposes the enhanced concept of hazard proximity with two criteria: distance and the number of obstructions (i.e. walls and floors) between the epicentre of an extreme event and points in the building. The algorithms are validated by testing them on different buildings and discussing the results in Section 3.1.4. Note that the underlying 3D model is constructed automatically and it includes all the semantics (3D or others) necessary for the aforementioned algorithms.

3.1.1 The Building Information Modelling – Geography Information Science (BIM-GIS) Model

The research in this chapter is based on the 3D building model recently developed by Boguslawski et al. [12] and Barki et al. [4]. This model is an integration of BIM technology and GIS analysis. Ideally, the Industry Foundation Classes (IFC) format could be used to exchange model between the design environment and GIS analysis tool. However, the information content in IFC is very high for GIS applications such as emergency management. The original model needs to be simplified in the process of generalization to reduce the storage cost and extract building elements and geometry essential for reconstruction of indoor spatial relations. This can be achieved through development of generalization procedures or using simplified models, which can be automatically generated in commercial software packages, for example Green Building XML (gbXML).

We consider the gbXML format as a simplified BIM model, which is the main data format used as an input in our research: a detailed geometry is simplified to a level sufficient to preserve the adjacency relationship between rooms, which is essential. The information about room volumes and wall surfaces, including openings (i.e. doors and windows), is accompanied by attributes such as room names etc. The openings are used for conventional navigation and egress routes computation. However, other alternative routes can also be considered in case of direct hazard; for example, walls or partitions can be drilled to get access to adjacent rooms. This requires additional information about construction materials, which can be obtained from the original model. The external links in gbXML allow one to look up the required parameters if the original model is available (e.g. the IFC model). For example, suppose that there are two adjacent rooms (without a door between them) modelled by nodes u and v, and we know that the wall between them is thin, which is an attribute of the link uv. Then, the link uv is made available for navigation in the network and the user is informed if this link happens to be in the egress route.

Our models are Boundary Representations (B-Rep) of the concerned buildings, where volumes (cells) are enclosed by faces, faces by edges, and edges by points. A B-Rep may be modelled and implemented by various data structures, for example the radial edge [71], G-Maps [43] or recently developed dual half-edges (DHE) [6]. The DHE structure has been adopted to encode the model geometry as the underlying navigable network can be simultaneously and automatically constructed together with the 3D model.

A simple model of a building stored in the Autodesk Revit format was exported to the gbXML model and reconstructed using the primal/dual DHE data structure (see Figure 3.1). Surfaces stored in the gbXML model represent boundaries of spaces (i.e. rooms). Information about adjacent spaces and incorporated openings, for example doors and windows, is attached to each surface. Original surfaces are represented with DHE as double-sided faces. Adjacency information is used to group surfaces into sets representing

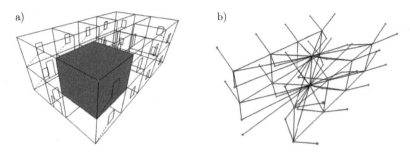

Fig. 3.1 Simple model reconstructed using the DHE data structure: a) structure of a building includes one selected room (grey cell); windows are connected to wall boundaries by bridge edges (dotted lines); b) graph of connections between rooms.

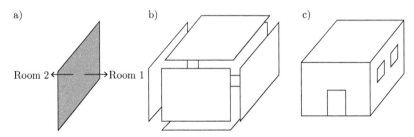

Fig. 3.2 DHE model reconstruction from gbXML: a) gbXML surfaces include information about adjacent rooms; b) surfaces bounding a room are merged and form a cell; c) room with incorporated openings.

boundaries of separate spaces. They are merged together (see Figure 3.2) along adjacent edges into closed cells representing rooms using the cardboard and tape method [6]. The resulting model is a cell complex, where rooms are represented as cells in the primal structure with an associated dual node unambiguously representing this cell. Adjacent cells are connected by dual edges bounded by dual nodes (see Figure 3.3).

A door between two rooms is represented as a zero-volume cell with an associated unique node. Thus, there are two dual edges connecting the first room node to the door node, and the door node to the second room node. The same idea is applied in case of a door between an internal room and an external space. The latter is represented as a cell or a set of connected cells if it was partitioned. The complete graph of indoor connections is shown in Figure 3.1b. Some openings, which are not directly connected to the boundary of the enclosing surface, for example windows, are connected to the surface boundary by bridge edges (dotted lines in Figure 3.1a). A bridge edge is not

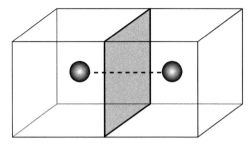

Fig. 3.3 Two rooms represented by dual nodes are connected by a dual link penetrating a shared face.

part of the original model but it is introduced to preserve a valid topology of the B-Rep model.

The structure of the model presented in Figure 3.1 is simple and was reconstructed without additional improvement or validation. However, models with more complex structure exported to the gbXML format must be processed first to reconstruct a valid navigable network. Some common issues are unclosed and overlapping cells. Such models are valid for most of engineering analyses but not for GIS, which requires a complete topology with a proper representation of spatial relations among objects. For validated models, the graph of connections reflecting spatial relationships among cells in the complex is created automatically using the DHE construction operators. For further details, see the article by Boguslawski et al. [12].

Summarizing the BIM-GIS model, a building in general consists of several connected rooms that have volumes (corridors, offices, storage spaces etc. are considered as rooms too), so they are represented by primal cells. The geometry of a room is modelled by the links and nodes of a cell, and relations between adjacent rooms are represented with dual links connecting the corresponding cells. Those relations are described in terms of access level from one room to the adjacent one: access to the next room is by a door; the next room is not accessible because of a wall, but if the wall is thin a hole can be made. It may not be possible to get directly to the next room if the wall is made of concrete. This is an example of a basic set of attributes that are assigned to connections between rooms and then used as weights in graph traversal algorithms, for example Dijkstra's algorithm. Rooms are not the only objects in a building that are important. Walls, doors, and windows are essential and included in the model. They are represented as cells with geometry, and some attributes are assigned to them.

3.1.2 Test Building and Tessellation of Corridors

Figure 3.4a shows a hypothetical building created in Revit. The building is then exported to gbXML and reconstructed with the DHE structure (Figure 3.4b). The latter is the primal model, which includes the geometry of the building. The graph of connections, that is, the dual model, is then automatically created (Figure 3.4c). The building has three exits and three stairwells, and all floors have a similar structure, except for the exits at the ground floor. The plan of the ground floor is shown in Figure 3.5.

For navigation purposes, all corridors and large open spaces in a building should be partitioned automatically to generate a navigable network reflecting real navigation routes. Without a proper partition, an incorrect distance between nodes (cells) might be calculated and hence wrong connections

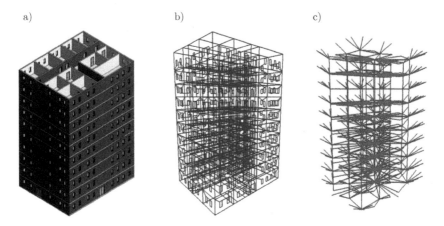

Fig. 3.4 a) Test building; b) DHE reconstruction; c) graph of connections.

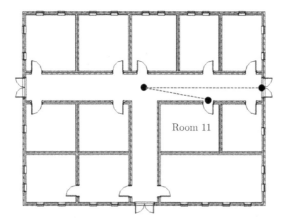

Room 11

Fig. 3.5 Plan of the ground floor of the test building.

(links) might be selected for evacuation. This is illustrated in Figure 3.5, where a person has to go from Room 11 to the nearest exit. According to the current model with the corridor modelled by just one node, the person would follow a dashed line, while the actual walking pattern should be different; one possible path is shown in Figure 3.6 (the bold line). To achieve an appropriate tessellation, we use an approach based on the Voronoi Diagrams (VD). VD partition a space into a set of adjacent cells represented as a graph. The dual graph to VD consists of links connecting adjacent dual nodes, which represent primal cells. These links form a network we use for navigation. It should be noted that for some concave shapes, a Voronoi tessellation may produce cells that are split into several unconnected parts; for example, if a boundary of a corridor overlaps with the cell and some parts of the divided cell are not connected to the cell enclosing the dual node. However, in our implementation this situation does not exist. All dual nodes are enclosed by exactly one Voronoi cell.

For the corridors of the building in Figure 3.4a, this is illustrated in Figure 3.6, where one possible tessellation of the corridor is shown. Note that some cells in this tessellation have a node (for example the node u in Figure 3.6), which is either a door or a concave corner. Hence, such a cell is already associated with one primal node in the 3D model. However, for other cells, new tessellation nodes are introduced with the corresponding links; an example is the node v in Figure 3.6. Note that different tessellations are possible and the density of tessellation may be higher for larger areas. The choice of the tessellation density depends on the precision, which is required in the navigation route.

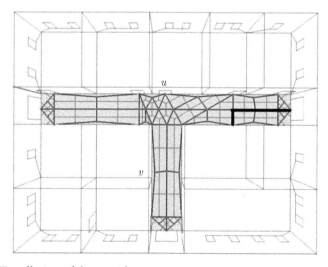

Fig. 3.6 Tessellation of the corridor.

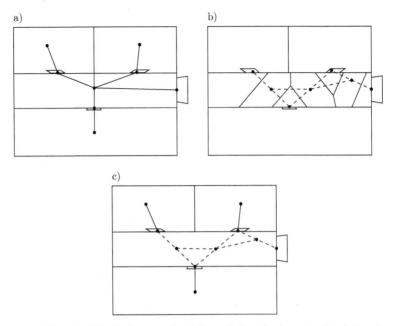

Fig. 3.7 a) Graph G: logical network of the building (only navigable links via doors are shown); b) navigable network generated for a corridor: solid lines represent a Voronoi tessellation, dotted lines represent the navigable network; c) graph G^+: combined networks.

Let us denote by G the original 3D dual model of the building without tessellation, that is, the graph of connections between cells, and by G^+ the 3D model of the building with all the necessary tessellations. Thus, G^+ has some additional nodes such as corner points and new nodes added during the tessellation process of corridors and large open spaces. Examples of the graphs G and G^+ are given in Figure 3.7, where only navigable links are shown. Note that non-navigable links, which are not shown in Figure 3.7, represent physical obstructions, for example the link between the nodes in the left and right top corners. Now, the shortest route from a selected room to the nearest exit can easily be calculated using Dijkstra's algorithm.

The automated construction of variable density navigable networks will be explored in Section 3.3.

3.1.3 Algorithms for Safest and Balanced Routes

In this section, two algorithms for finding safest and balanced routes in buildings are presented. The algorithms are designed for complex buildings, for example high-rise buildings and shopping malls, where an extreme event is occurring. Typically, for people it is safer to stay further away from the epicentres of an extreme event in terms of the distance and the number of obstructions (e.g. walls). In this study, we consider extreme events having this natural property. Examples of such events are fire, and terrorist activities, such as bomb attacks or hostage-taking situations. For the latter, the location of a bomb or a terrorist is an *epicentre*, whereas for fire the *epicentres* can be defined as points (nodes) in the building, where the temperature exceeds a certain threshold. Walls, floors and ceilings are called *obstructions*. For a particular point p in the building and an extreme event with the epicentre z, the hazard for that point is a function of the direct distance from p to the epicentre z and the minimum number of obstructions between p and z. Note that an extreme event may have many epicentres.

For convenience, we summarize the notation, which will be used in this section:

$AS(P_\rho)$	Aggregate score for P_ρ
AS^*	Maximal aggregate score for all routes in R
$D(e)$	Length of link e in metres
$D(P_\rho)$	Length of P_ρ in metres
G	Graph of connections
G^+	Graph G extended by tessellation nodes
GM	Geometric mean
$H(v)$	Hazard proximity number for node v
$H_i(v)$	Hazard proximity number for node v with respect to (w.r.t.) z_i
$HD(e)$	Hazard proximity number for link e
P_ρ	(p, q)-route for the propagation coefficient ρ
$PI(P_\rho)$	Proximity index for P_ρ
R	Set of (p, q)-routes
$S_D(P_\rho)$	Distance score for P_ρ
$S_{PI}(P_\rho)$	Proximity index score for P_ρ
WGM	Weighted geometric mean
$b(v, z_i)$	Minimum number of obstructions between v and z_i
$d(v, z_i)$	Direct distance from v to z_i in metres
$d^+ (d^-)$	Maximal (minimal) length of routes in R
$l(P_\rho)$	Number of links in P_ρ
$p^+ (p^-)$	Maximal (minimal) proximity index of routes in R
$r(e)$	Proximity ratio for link e
$r_i(e)$	Proximity ratio for link e w.r.t. z_i
t	Hazard tolerance, $t = 0, 0.5$ or 1
z_i	i-th epicentre
ρ	Propagation coefficient
ρ_{max}	Maximal propagation coefficient
τ	Adapted propagation coefficient, $\tau = 1 + \rho/100$

Safest Routes

The first objective is to find the safest route in a building; that is, the total hazard proximity from a route to all the epicentres of the extreme event should be minimized. As explained above, the hazard has two criteria: distance and the number of obstructions. These criteria are very natural and should be considered together. For example, the distance of 40 m between a person and an epicentre of one of the aforementioned extreme events in a direct visibility might be considered not as safe as the distance of 30 m with two concrete walls between the person and the epicentre. It may be pointed out that we consider a generic extreme event. For a particular event, for example fire, our algorithms can be further developed by taking into account more precise models for heat propagation, smoke spread and structural collapse.

Thus, we consider an extreme event with k epicentres and two criteria for safety: distance and the number of obstructions. Based on these criteria, Algorithm 3.1 finds the safest available (p, q)-route in a building. Note that the application of this algorithm is threefold: it can be used by a rescue team to get from one of the entrances to a particular place in the building; as an evacuation algorithm from room p to one of the exits; or for navigation from point p to point q in the building.

Let $d(v, z_i)$ denote the direct distance from node v to the epicentre z_i in metres, and $b(v, z_i)$ the minimum number of obstructions between v and z_i. In the first step of Algorithm 3.1, the direct distances and the number of obstructions are calculated in DB_Procedure. More precisely, using the standard geometric technique, the direct distance $d(v, z_i)$ from node v to the epicentre z_i in metres is calculated for all nodes $v \in V(G^+)$ and all epicentres. Then, the Breadth-First Search Algorithm (see Section 1.3.1) is run in the graph G from the node z_i. It finds the number of links in shortest paths from z_i to all other nodes in G. Because each link in those paths represents a physical obstruction, the number of links in such a (z_i, v)-path is the minimum number of obstructions (i.e. walls and floors) between z_i and v. This number is denoted by $b(v, z_i)$ and it should be further adjusted for doors and exit doors. For example, if v represents a door with two adjacent rooms r_1 and r_2, then we set

$$b(v, z_i) = \min\{b(r_1, z_i), b(r_2, z_i)\}$$

because the links vr_1 and vr_2 do not represent a physical obstruction. Note that for some nodes, such as corner nodes and the 'new' nodes in the tessellation, the numbers $b(v, z_i)$ have not been calculated because those nodes are not in the dual graph G. Hence, for each such node $v \in V(G^+) - V(G)$ we set

$b(v, z_i) = b(w, z_i)$, where the node $w \in V(G)$ represents the cell whose tessellation contains v.

It may be pointed out that the calculation of the parameters $b(v, z_i)$ is only possible because the dual graph incorporates all the necessary 3D information about the building.

Algorithm 3.1: Safest (p, q)-route in a building, where an extreme event is occurring.

Input:	The graphs G and G^+, which constitute the 3D model of the building.
	The epicentres of an extreme event (nodes z_i, $i = 1$, $2, ..., k$).
	Node p; node q (optional; q is one of the exits by default).
	The maximal propagation coefficient $\rho_{\max}$.
Output:	Safest (p, q)-route P.

(1) Run DB_Procedure to produce $d(v, z_i)$ and $b(v, z_i)$ for all $v \in V(G^+)$ and $i = 1, 2, ..., k$.
(2) Run HP_Procedure with $\rho = \rho_{\max}$ to produce $HD(e)$ for all links in G^+.
(3) Run Dijkstra's algorithm in the graph G^+ from node p with link weights $HD(e)$. It produces the safest available (p, q)-route P in the building with two criteria: distance and the number of obstructions.
(4) Report P. Algorithm stops.

DB_Procedure: Calculation of direct distances and the minimum number of obstructions.

Input:	The graphs G and G^+. Nodes z_i, $i = 1, 2, ..., k$.
Output:	$d(v, z_i)$ and $b(v, z_i)$ for all $v \in V(G^+)$ and $i = 1, 2, ..., k$.

(1) Repeat Steps 2 to 5 for each node z_i.
(2) For each node $v \in V(G^+)$, calculate $d(v, z_i)$, the direct distance from v to z_i in metres.

(3) Run the Breadth-First Search Algorithm in the graph G from node z_i. It returns the number of links in shortest (v, z_i)-paths in G for all nodes v; that is, the minimum number of obstructions $b(v, z_i)$ between v and z_i for all nodes $v \in V(G)$.

(4) For each node $v \in V(G)$ representing a door, update $b(v, z_i)$ as follows:

   (a) If v is an exit door, then we set $b(v, z_i) = b(r, z_i)$, where the node r represents a room adjacent to v.

   (b) If v is a door, but not an exit door, then

$$b(v, z_i) = \min\{b(r_1, z_i), b(r_2, z_i)\},$$

where r_1 and r_2 represent two rooms adjacent to v.

(5) For each node $v \in V(G^+) - V(G)$, set $b(v, z_i) = b(w, z_i)$, where $w \in V(G)$ represents the cell whose tessellation contains v.

(6) Report $d(v, z_i)$ and $b(v, z_i)$ for all $v \in V(G^+)$ and $i = 1, 2, ..., k$. Algorithm stops.

HP_Procedure: Calculation of hazard proximity numbers.

Input: The graphs G and G^+. Nodes $z_i, i = 1, 2, ..., k$.
            The propagation coefficient $\rho \geq 0$.
            $d(v, z_i)$ and $b(v, z_i)$ for all $v \in V(G^+)$ and $i = 1, 2, ..., k$.

Output: Hazard proximity numbers $HD(e)$ for all links in G^+.

(1) Calculate $\tau = 1 + \frac{\rho}{100}$.

(2) Compute the hazard proximity numbers:

$$H_i(v) = \frac{100}{\tau^{\sqrt{d(v,z_i) \times b(v,z_i)}}}$$

for each node $v \in V(G^+)$ and each $i = 1, 2, ..., k$.

(3) Calculate $H(v) = \max_{1 \leq i \leq k} H_i(v)$ for each node $v \in V(G^+)$.

(4) For each link $e = uv \in E(G^+)$, compute the hazard proximity numbers for links:

$$HD(e) = 0.5[H(u) + H(v)] \times D(e),$$

where $D(e)$ is the length of e in metres.

(5) Report $HD(e)$ for all links in G^+. Algorithm stops.

The second step of Algorithm 3.1 is to compute the hazard proximity numbers for all links in G^+ for the given propagation coefficient $\rho = \rho_{\max}$, where ρ represents the degree of hazard propagation—it is explained in more detail below. This is done in the HP_Procedure. Initially, for each epicentre z_i, the hazard proximity numbers $H_i(v)$ are calculated for all nodes in G^+. These numbers go from small positive values, which mean 'very far from the epicentre', to 100, which stands for 'inside the epicentre z_i'. The hazard proximity numbers are based on the parameters $d(v, z_i)$ and $b(v, z_i)$, which should first be replaced by one variable representing their average. For two variables with different numerical ranges it is appropriate to use a (weighted) geometric mean. Since $d(v, z_i)$ and $b(v, z_i)$ have different ranges, the weighted geometric mean is applied:

$$WGM = \left(d(v, z_i)^{w_1} b(v, z_i)^{w_2} \right)^{\frac{1}{w_1 + w_2}}$$

where w_1 and w_2 are the relative weights for the direct distance and the number of obstructions. Because the direct distance is as important as the number of obstructions, we can assume that the corresponding weights for the two variables are in proportion 50:50; that is, $w_1 = 0.5$ and $w_2 = 0.5$. However, these weights can be adjusted if necessary, for example for buildings with many large open spaces and few obstructions. Thus, the above formula is simplified to the standard geometric mean:

$$GM = \sqrt{d(v, z_i) \times b(v, z_i)}.$$

Next, the values of geometric means should be transformed to the scale going from 100 to 0 taking into account the propagation coefficient $\rho \geq 0$. This is achieved by using the following formula:

$$\frac{100}{\left(1 + \frac{\rho}{100}\right)^{\sqrt{d(v,z_i) \times b(v,z_i)}}}.$$

Now, if we denote $\tau = 1 + \frac{\rho}{100}$, then a well-justified formula for the *hazard proximity numbers with respect to* z_i is obtained:

$$H_i(v) = \frac{100}{\tau^{\sqrt{d(v,z_i) \times b(v,z_i)}}}.$$

For instance, if $\rho = 100$, then hazard proximity numbers for nodes propagate quickly from 100 (in the epicentres) to small positive numbers (far from the epicentres). This puts a strong emphasis on the epicentre and the rooms in its close proximity. In contrast, if ρ is a small positive number, then the propagation is slow, thus putting less emphasis on the epicentre and the nearby rooms. In the extreme case $\rho = 0$ there is no propagation; that is, all hazard proximity numbers for nodes are equal to 100.

Having calculated $H_i(v)$ for all the nodes in G^+ and all values of $i = 1, 2, ..., k$, the following formula is used to compute the *hazard proximity numbers for nodes*:

$$H(v) = \max_{1 \leq i \leq k} H_i(v).$$

Here we assume that the hazard at a particular node is equal to the maximal hazard proximity number at this node for all the epicentres; this approach is justified in many cases. A different formula can be easily incorporated in the algorithm if it is necessary, for example, to take into account the cumulative effect of all the hazard proximity numbers $H_i(v)$. Further, for each link $e = uv$ in the graph G^+, the *hazard proximity number* $HD(e)$ *for link* e is determined by calculating the arithmetic average of the hazard proximity numbers of its end-nodes, and then by multiplying the resulting number by the length of e in metres:

$$HD(e) = 0.5[H(u) + H(v)] \times D(e).$$

The last operation is important because G^+ is not a homogeneous network. For example, let us suppose that one link is 2 metres long and another is 10 metres long, and they both have the same hazard proximity number, say 10. If they both are used in a navigation route, then it is natural to assume that a travel time for a longer link would be approximately five times longer, so the hazard proximity number for the longer link should be 50. In other words, if we subdivided the longer link into five 2-metre-long links to make the network more homogeneous, then those five links would approximately contribute 50 to the total hazard proximity of the route.

The final stage of Algorithm 3.1 is to run Dijkstra's algorithm (see Section 1.3.4) in the graph G^+ from node p with link weights $HD(e)$. It produces the safest available (p, q)-route P in the building with respect to two criteria: distance and the number of obstructions. Note that the formula for the hazard proximity numbers is based on the weights $w_1 = 0.5$ and $w_2 = 0.5$ for $d(v, z_i)$ and $b(v, z_i)$; however, different weights can be used in the formula if necessary.

Balanced Routes

The second objective is to produce a balanced (p, q)-route in a building, where an extreme event is occurring. This is achieved in Algorithm 3.2, where one of the input parameters is the *hazard tolerance coefficient* t. The hazard tolerance is a trade-off between distance and safety and it can be equal to 0, 0.5 or 1. If $t = 1$, then the shortest route will be generated. If $t = 0$, then the algorithm finds the safest available route. If t is not specified and there are enough routes in the set R, then a route with the 50/50 balance of distance/safety will be reported as the balanced route.

Typically, there are many (p, q)-routes. The shortest route might go through the epicentre and be dangerous, whereas the safest route might be the longest one. A member of the rescue team, who is fully protected from the hazard, may wish to use the shortest route even if it is the most dangerous one; that is, their hazard tolerance t is equal to 1. In contrast, an unprotected person with a respiratory disease may want to use the safest evacuation route, whatever its length, in which case the hazard tolerance t is 0. The hazard tolerance is an optional parameter, and at the moment there are only two values. If it is not given, then by default $t = 0.5$. The default value of 0.5 simply means that the hazard tolerance has not been specified, and this number will be used as a relative weight for the distance attribute, thus the relative weight for the hazard proximity will be 0.5 too. We use one variable t in this context because the single formula for the aggregate score $AS(P_\rho)$ will be applied for all values of t. Thus, if t is not specified, then a route with a right balance of distance/proximity will be chosen.

The first part of Algorithm 3.2 runs DB_Procedure to determine all parameters $d(v, z_i)$ and $b(v, z_i)$. Then, the binary search is carried out with respect to ρ. The first run is for $\rho = 0$, producing the shortest (p, q)-route P_0 because all hazard proximity numbers for links are $100D(e)$. The route P_0 is included in the set R. The next run is for $\rho = \rho_{\max}$. If the resulting route coincides with P_0, then there is no interval for the binary search and it is terminated. Otherwise, the route is different from P_0 and it is included in R. The next run is for $\rho = 0.5\rho_{\max}$. There are three possibilities here. If the resulting route is a new one, then it is included in R and the binary search continues for two intervals $(0; 0.5\rho_{\max})$ and $(0.5\rho_{\max}; \rho_{\max})$. If the resulting route coincides with one of the routes in the set R, then one of the intervals is removed from the search and the other interval is used in the binary search. For example, if the route coincides with P_0, then the binary search continues for the interval $(0.5\rho_{\max}; \rho_{\max})$, whereas the interval $(0; 0.5\rho_{\max})$ is removed. This procedure is terminated if at least one of the stopping criteria is satisfied: a specified size of R, a specified length of the widest interval and a running time. For our test buildings, the procedure goes on until seven routes are found or the length of the widest interval is less than 0.1.

Algorithm 3.2: Balanced (p, q)-route in a building, where an extreme event is occurring.

Input:	The graphs G and G^+, which constitute the 3D model of the building.
	The epicentres of an extreme event (nodes $z_i, i = 1, 2, ..., k$).
	Node p; node q (optional; q is one of the exits by default).
	The maximal propagation coefficient $\rho_{\max}$.
	Hazard tolerance t: $t = 0$ or 1 (optional; $t = 0.5$ by default).
Output:	Balanced (p, q)-route (t is not specified); safest route ($t = 0$); shortest route ($t = 1$).

(1) Run DB_Procedure to produce $d(v, z_i)$ and $b(v, z_i)$ for all $v \in V(G^+)$ and $i = 1, 2, ..., k$.

(2) Set $R = \emptyset$, where R is a set of (p, q)-routes.

(3) Carry out the binary search with respect to $\rho, 0 \le \rho \le \rho_{\max}$, starting with $\rho = 0, \rho_{\max}, 0.5\rho_{\max}$ etc. For each value of ρ, implement the following:

   (a) Run HP_Procedure for the specified value of ρ to produce $HD(e)$ for all links in G^+.

   (b) Run Dijkstra's algorithm in the graph G^+ from node p with link weights $HD(e)$. It produces the (p, q)-route P_ρ corresponding to the propagation coefficient ρ.

   (c) Set $R = R \cup \{P_\rho\}$ if $P_\rho \notin R$. (The shortest route P_0 belongs to R.)

   Go to Step 4 if at least one of the stopping criteria is satisfied.

(4) For each route $P_\rho \in R$:

   (a) Calculate the length of P_ρ: $D(P_\rho) = \sum_{e \in P_\rho} D(e)$.

   (b) Compute the proximity ratios

$$r_i(e) = \frac{\sqrt{d(u, z_i) \times b(u, z_i)} + \sqrt{d(v, z_i) \times b(v, z_i)}}{2D(e)}$$

   for each $i = 1, 2, ..., k$ and each link $e = uv$ in P_ρ.

   (c) Calculate $r(e) = \min_{1 \le i \le k} r_i(e)$ for each link $e = uv$ in P_ρ.

(d) Calculate the proximity index

$$PI\,(P_\rho) = \frac{l\,(P_\rho)}{\sum_{e \in P_\rho} \frac{1}{r(e)}}.$$

PI is the harmonic mean of proximity ratios $r(e)$, and $l\,(P_\rho)$ is the number of links in P_ρ.

(5) If R consists of one route, then report P_0 and stop the algorithm.

(6) Compute the following:

$$d^- = \min_{P_\rho \in R} D\,(P_\rho); \qquad d^+ = \max_{P_\rho \in R} D\,(P_\rho);$$

$$p^- = \min_{P_\rho \in R} PI\,(P_\rho); \qquad p^+ = \max_{P_\rho \in R} PI\,(P_\rho).$$

(7) For each route $P_\rho \in R$, calculate the aggregate score:

$$AS\,(P_\rho) = 100\,t \left(1 - \left(\frac{D\,(P_\rho) - d^-}{d^+ - d^-} \right)^2 \right)$$
$$+ 100(1 - t) \left(1 - \left(\frac{p^+ - PI\,(P_\rho)}{p^+ - p^-} \right)^2 \right).$$

(8) Compute $AS^* = \max_{P_\rho \in R} AS\,(P_\rho)$.

(9) Report P_ρ for which $AS(P_\rho) = AS^*$. Algorithm stops.

In the next block of Algorithm 3.2, the total length of each route in R is calculated. Then, the *proximity ratios* are computed for each link in a route. They are based on the geometric averages of parameters $d(v, z_i)$ and $b(v, z_i)$ for end-nodes of the link and its length. In contrast to hazard proximity numbers, the proximity ratios do not depend on the propagation coefficient ρ, and a small proximity ratio means a close proximity to one of the epicentres. For a given link, the final proximity ratio $r(e)$ is the smallest proximity ratio for that link: $r(e) = \min_{1 \leq i \leq k} r_i(e)$. The *proximity index* for a route P is the harmonic mean of $r(e)$ for all links in the route:

$$PI\,(P) = \frac{l\,(P)}{\sum_{e \in P} \frac{1}{r(e)}}$$

where $l(P)$ is the number of links in P. Note that the proximity index is an average of *rates*. Also, the proximity index should not be dominated by sections of a route with large proximity ratios, and the impact of small proximity ratios is important. Therefore, the harmonic mean is an appropriate measure for the proximity index. Since the proximity index is independent on the propagation coefficient, it can be used for comparison of the routes from the set R.

In the final part of the algorithm, a multi-attribute decision-making technique is used to rank the routes in R and choose a balanced (p, q)-route. First of all, the maximal and minimal values of the lengths and proximity indices are calculated for all routes in R:

$$d^- = \min_{P \in R} D(P); \qquad d^+ = \max_{P \in R} D(P);$$
$$p^- = \min_{P \in R} PI(P); \qquad p^+ = \max_{P \in R} PI(P).$$

Then, quadratic value functions are applied for rating the routes with respect to two attributes, the distance and the proximity index. Different value functions were tested, and it turned out that the most appropriate one is quadratic. For each route in R, the scores for these attributes are given by the following formulae, respectively:

$$S_D(P) = 100 \left(1 - \left(\frac{D(P) - d^-}{d^+ - d^-} \right)^2 \right)$$

and

$$S_{PI}(P) = 100 \left(1 - \left(\frac{p^+ - PI(P)}{p^+ - p^-} \right)^2 \right).$$

Finally, the *aggregate score* is calculated as a weighted average of the routes' scores where the weights depend on the tolerance coefficient t. More precisely, the weights for the scores S_D and S_{PI} are t and $1 - t$, respectively. The balanced (p, q)-route is one with the highest aggregate score. For example, if $t = 1$, then the shortest route is chosen. If $t = 0$, then the algorithm returns the route with the highest proximity index, that is, the safest available route. If t is not specified and there are enough routes in the set R, then a route with the 50/50 balance of distance/proximity will be reported as the balanced route.

3.1.4 Testing the Algorithms

We start testing Algorithms 3.1 and 3.2 with a virtual building shown in Figure 3.8. This 10-floor building has 5 stairwells and 5 exits, and at each level there are 5 rooms connected by a long corridor as can be seen in Figure 3.9. The relatively large number of stairwells is needed to illustrate the behaviour of the algorithms.

In what follows, we set $\rho_{\max} = 100$ for Algorithms 3.1 and 3.2. However, further testing is needed to decide which values of the maximal propagation coefficient are appropriate for different buildings. The epicentre of an extreme event (labelled by a star) is on the fourth floor, as can be seen in Figure 3.10. The starting point (the node p) is located on the top floor, above the epicentre, and the node q is not specified. Thus, we are looking for a route from p to one of the exits. It is not difficult to see that Algorithm 3.1 is a particular case of Algorithm 3.2 if we set $t = 0$ in the latter; that is, the former finds a route with the highest proximity index. Thus, it is enough to test Algorithm 3.2 for different values of the hazard tolerance t.

For the above scenario, the binary search of Algorithm 3.2 produces five different routes with the following propagation coefficients: $\rho = 0, 7, 10, 11, 12$. The lengths and the proximity indices for those routes are summarized in

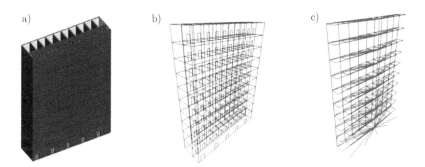

Fig. 3.8 a) Virtual building; b) DHE reconstruction; c) graph of connections.

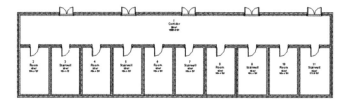

Fig. 3.9 Plan of the ground floor.

Table 3.1 Test results for Floor 9.

Route	P_0	P_7	P_{10}	P_{11}	P_{12}
Distance (m)	49.7	56.3	62.9	69.5	76.2
Proximity index	1.73	2.87	3.62	4.28	4.87
Status	Shortest	-	Balanced	-	Safest

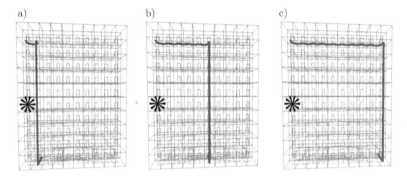

a)   b)   c)

Fig. 3.10 a) The shortest route P_0; b) the balanced route P_{10}; c) the safest route P_{12}.

Table 3.1. As can be seen in the table, Algorithm 3.2 returns the shortest route P_0 if $t = 1$, and the safest route P_{12} if $t = 0$. If t is not specified, then the balanced route P_{10} is returned by the algorithm.

The route P_0 is shown in Figure 3.10a. It goes through the first stairwell, which is very close to the epicentre, so it is the most dangerous route, but the shortest one. The second route P_7 goes through the second stairwell; it is safer but longer. The routes P_{10}, P_{11} and P_{12} go through the third, fourth and fifth stairwells, respectively. The routes P_{10} and P_{12} are shown in Figures 3.10b and c. The scatter plot for the two parameters of the five routes is given in Figure 3.11. It is not surprising that there is a very strong positive correlation (at 1% significance level) between distance and proximity index. Also, the routes form a so-called efficient frontier in the sense that no route is 'dominated' by another one; that is, for any two routes one of them is safer but longer.

Different floors for the starting room have been tested. In general, Algorithm 3.2 produces good results; however, in some cases there is an unexpected behaviour. For example, if the starting point is located on the sixth floor, then the binary search finds five routes, see Table 3.2. The first four routes are similar to the results for the top floor; they go through the first four

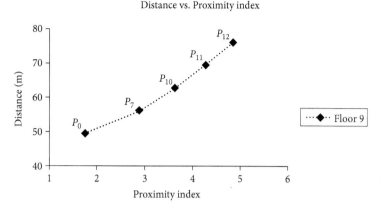

Fig. 3.11 The scatter plot for routes $P_0, P_7, P_{10}, P_{11}, P_{12}$.

Table 3.2 Test results for Floor 6.

Route	P_0	P_{12}	P_{22}	P_{25}	P_{26}
Distance (m)	37.7	44.3	50.9	57.5	93.1
Proximity index	1.50	2.36	2.85	3.25	3.87
Status	Shortest	-	-	Balanced	Safest

stairwells, respectively. However, the fifth route P_{26} does not go directly to the fifth stairwell. According to the algorithm, it is safer first to go upstairs and then use the fifth stairwell, as can be seen in Figure 3.12. This is reflected in the corresponding proximity indices: 3.87 for P_{26} and 3.59 for the route that goes directly to the fifth stairwell. Note that P_{26} is much longer compared to other routes, so the balanced route P_{25} might be considered as a more reasonable one.

The aforementioned behaviour is not necessarily unreasonable; it depends on the type of hazard. For some extreme events it might be deemed as safe, for others (e.g. fire) as unsafe. In the latter case, this problem can be rectified differently. The first approach is to include another criterion, route complexity, which will be investigated in the next section. The complexity of the route P_{26} would be rather high because it goes upstairs and uses two staircases, thus decreasing the likelihood that it will be eventually chosen. Another approach is to use a better model for hazard propagation in a building if the nature of hazard is known; however, such models are out of scope of this chapter where we consider a generic extreme event.

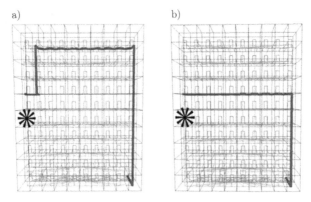

Fig. 3.12 Starting point p on Floor 6: a) the safest route P_{26} ($D = 93.1$ m, $PI = 3.87$); b) the safest route P_{29} with an additional penalty for going upstairs ($D = 64.2$ m, $PI = 3.59$).

For the time being, we can use a simple approach based on a binary input variable x. It is equal to 1 if going upstairs is undesirable, and 0 otherwise. By default, $x = 1$, in which case we add an additional 'penalty' for going upstairs in terms of distance. This penalty increases the hazard proximity numbers for links representing sections of a staircase that go up, thus making them undesirable in the routes. Note that such links are not forbidden completely because in some cases going upstairs is unavoidable. This adjustment of Algorithm 3.2 produces the route P_{29} instead of P_{26}. The new route has a slightly worse proximity index (3.59 versus 3.87), but it is much shorter (64.2 m versus 93.1 m) and does not go upstairs. The adjusted algorithm returns P_{22} as a balanced route because the 'outlier' P_{26} was replaced by a much shorter route P_{29}.

Let us now consider the building of Figure 3.4, which has three stairwells. Instead of looking at simple situations with one or two epicentres, we simulate an extreme event with four epicentres as illustrated in Figure 3.13. This is an extremely tight situation because the west stairwell is blocked by two epicentres at the third and seventh floors, whereas another two epicentres are located in the east and north stairwells at the fifth floor. Thus, to avoid the epicentres on the fifth floor, one has to use the west stairwell, which is not safe either. This means that a very safe route from the top floor to one of the exits does not exist; that is, any route is very close to the hazard in such an extreme configuration of epicentres. Note that the hazard proximity numbers for rooms, which are shown in red/orange/yellow hues, correspond to the hazard proximity numbers for nodes in the dual model.

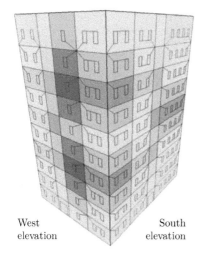 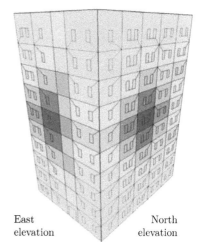

West elevation   South elevation   East elevation   North elevation

Fig. 3.13 Simulation of an extreme event with four epicentres.

Table 3.3 Test results.

Route	P_0	P_{11}	P_{165}
Distance (m)	39.6	57.3	111.5
Proximity index	0.73	1.32	1.64
Status	Shortest	Balanced	Safest

The shortest, balanced and safest routes produced by Algorithm 3.2 are shown in Figure 3.14, where stars represent the epicentres, and the information about the routes is summarized in Table 3.3. Note that ρ_{max} is now 200 because there are many epicentres. As can be seen in Figure 3.14, the shortest route goes through two epicentres in the west stairwell, the balanced route uses the north stairwell with one epicentre, and the safest available route tries to stay further away from the epicentres by first using the east stairwell down to the sixth floor, then the west stairwell down to the fourth floor and finally the north stairwell down to the ground floor.

The above extreme example makes it obvious that not only the global hazard proximity of a route is important but also a local proximity of route's nodes/links to the hazard should be taken into account. Indeed, the proximity index is a global parameter, which can be used to compare different routes. However, it does not tell us that the route is going through or very close to one of the epicentres. Some threshold values for the proximity ratios $r(e)$ could be used for this purpose if an epicentre is not located in a corridor or a large open

a)   b)   c)

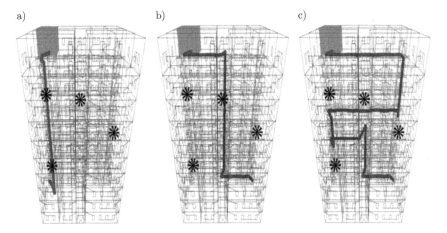

Fig. 3.14 a) The shortest route P_0; b) the balanced route P_{11}; c) the safest route P_{165}.

space. Otherwise, the direct distance $d(v, z_i)$ from v to the epicentre could be a better measure of local hazard proximity of the node v. The assessment of a route's local hazard proximity is out of scope of this chapter. However, in the context of the above example, the proximity ratio less than 0.5 means an extreme proximity to an epicentre, whereas the proximity ratio between 0.5 and 1 represents a close proximity. Thus, if a user opts for a balanced route in our example, then (s)he should be given the information that it is extremely close to the epicentre together with the option to choose the safest available route.

Finally, let us consider a more realistic example, which is based on the Doha World Trade Center. A typical actual floor of this building is shown in Figure 3.15, and its DHE reconstruction and tessellation are illustrated in Figure 3.16. The left and right pictures of Figure 3.17 demonstrate the shortest and safest routes between two rooms, respectively. Further, the 37-floor building of Figure 3.18 was generated using the aforementioned floor plan. To add even more complexity, the locations and the number of staircases were modified: there are three staircases between the ground floor and the third floor; three staircases at different locations between Floors 5 and 36; and six staircases on Floor 4 (three going up and three going down). The resulting 37-floor building is not exactly the Doha World Trade Center but it is based on the actual floor plan of this building. Figure 3.18 illustrates the shortest and safest routes in this building, where stars represent three hazard epicentres.

As mentioned above, the algorithms can be used for navigation from point p to point q in the building; this is illustrated in Figure 3.17. Also, they can

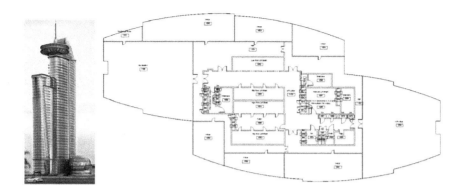

Fig. 3.15 The Doha World Trade Center and its typical floor.

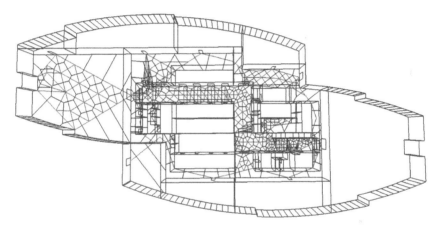

Fig. 3.16 DHE reconstruction and tessellation of the Doha World Trade Center.

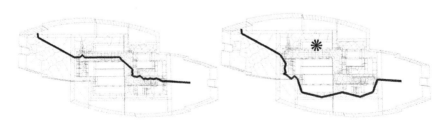

Fig. 3.17 The shortest and safest routes between two rooms (the star represents a hazard epicentre).

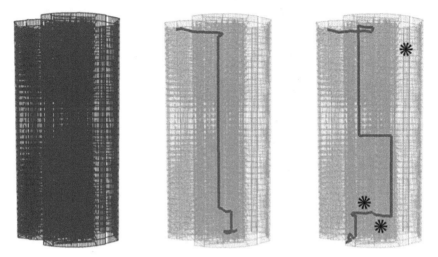

Fig. 3.18 The 37-floor building reconstructed with DHE, shortest and safest routes.

be used for evacuation from room p to one of the exits or by a rescue team to get from one of the entrances to a particular place in the building. This is illustrated in Figure 3.18. From the practical viewpoint, the 3D model of the building can be created in advance and kept in the cloud. On arrival, the rescue team should detect the event epicentres. For example, in case of fire, the team can scan the building with appropriate equipment or use temperature sensors in the building if they are available. Then, the model in the cloud is updated with epicentres and the safest available route is found. Notice that people inside the building will also have access to the updated model in the cloud, and there is a potential for using an indoor navigator.

In the next section, we will consider three criteria: distance, hazard proximity and route complexity. The multi-attribute rating technique will be applied for finding the user-optimal route; that is, a route which is reasonably short, safe and simple. If distribution of people in a building is known or can be predicted, then the distance criterion may be replaced by the time criterion, thus taking into account multiple agents in the building. Another interesting extension would be to take into account the dynamics of the situation, which is particularly important if a hazardous event develops rapidly.

3.2 Analytic Prioritization of Indoor Routes for Search and Rescue Operations

Managing efficiently disaster scenes in built facilities is critical for Search and Rescue (SAR) personnel to provide a time-critical response, and thus eliminate the potential for the loss of life, injury and damage. Prevailing approaches have used either heuristic/knowledge rules encapsulated in a prototype computer model or GIS-based emergency management systems. The major limitation of many systems developed so far is the lack of practical applicability to an emergency response that provides decision-makers with means to assess several alternatives based on multiple, conflicting criteria [27].

To deal with the complexity of variables in an emergency situation, the Analytic Hierarchy Process (AHP) has offered one of the best ways for providing consistent comparisons and selecting the best alternative among several conflicting criteria. As an approach designed to handle decisions with multiple attributes, the AHP has been applied in a wide range of areas. For instance, Wang et al. [68] recently applied the AHP for developing a comprehensive evaluation index system used to weight railway emergency plans. The AHP approach can be summarized into four main steps and the calculation procedure as follows: (1) structuring a problem into a decision hierarchy consisting of criteria and alternatives; (2) establishing pairwise comparisons between decision elements at each hierarchy level; (3) transforming comparison matrices into sets of weights; (4) aggregating the weights to rank the alternatives.

Despite the potential of the AHP technique, there is currently little research that applied this approach in the context of emergency response. In addition, there has been no attempt to integrate the powerful analytic approach with 3D indoor environments.

In this section, we propose an algorithm for finding the optimal routes for SAR teams in a building, where a multi-epicentre extreme event is occurring. The important feature of the algorithm is its ability to generate an optimal route depending on user's needs; that is, the user has an option to specify the preferential ranking of three basic criteria. The algorithm is based on a novel approach that integrates a new version of the AHP for multi-attribute decision-making, the direct iteration algorithm for eigenvectors/eigenvalues, statistical characteristics, the propagation of hazard, Duckham–Kulik's adapted algorithm for simplest paths, Dijkstra's classical algorithm for shortest paths and the binary search with three criteria: hazard proximity, distance and route complexity.

Furthermore, we discuss and validate subcriteria for the route complexity and further develop hazard proximity numbers and the proximity index for large open spaces in a building. The algorithm is validated and tested on a realistic complex building. We illustrate how the distance criterion can be replaced by travel time using Nelson–MacLennan–Pauls' formula for people's indoor speed and extend this formula for an SAR team moving in a counter-flow. Also, our algorithm is compared with an existing method.

The AHP Approach for Ranking Alternatives

An important integrated part of the presented algorithms is the AHP, which is one of the best multi-attribute rating techniques developed by Thomas Saaty [55, 56, 57] for the US government. The AHP is an eigenvalue approach designed to handle decisions with multiple attributes and it has been applied in a wide range of areas. A survey of the AHP, its applications and interesting facts were given by Saaty [57], Zahedi [74] and Foreman [23]. The most recent application of the AHP was developed by Turskis et al. [64] for selecting the foundation type for a single-storey dwelling house. Another interesting application of the fuzzy AHP to tunnel health evaluation was proposed by Zhang et al. [76].

The first stage in the AHP is to set up a decision hierarchy, which is called a *hierarchy tree*. This means that the decision problem is broken down into hierarchies of its decision elements. At the root of the tree is the most general objective of the problem, then all the relevant attributes are arranged at the first hierarchy level. At the next level of the hierarchy tree, some of the attributes might be broken down into more detail, and so forth. Finally, at the lowest level in the hierarchy tree, the available alternatives are set out.

At the second stage, each attribute is compared in turn with every other attribute at the same level of the hierarchy. For each comparison, a decision-maker has to determine to what extent one attribute is more (or less) important than the other attribute. When making such a decision, the following ratings are used ([55]): 1 (equally important), 3 (weakly more important), 5 (strongly more important), 7 (very strongly more important), 9 (extremely more important). For example, if attribute X is weakly more important than attribute Y, then X is 3 times more important than Y. Therefore, the latter attribute is only 1/3 as important as the former attribute, that is, 3 times less important. Ratings between the above numbers are allowed; hence, one may use direct numerical inputs on the scale from 1/9 (extremely less important) to 9 (extremely more important). Thus, for a given hierarchy of size n, one can construct a *reciprocal comparison $n \times n$ matrix A*, where its elements have the property

$$A_{i,j} = 1/A_{j,i}$$

and the main diagonal consists of 1s. The process of pairwise comparisons should be carried out for each hierarchy. Finally, the available alternatives are pairwisely compared with respect to the attributes in the levels above.

The third stage is to transform the comparison matrices into sets of weights representing the relative importance of all the attributes (or alternatives) at the same hierarchy level. For an $n \times n$ comparison matrix A of a particular hierarchy, let $\lambda_{\max}$ denote the largest eigenvalue and w be the corresponding normalized eigenvector; that is,

$$Aw = \lambda_{\max} w.$$

The components of the vector

$$w = (w_1, w_2, ..., w_n)$$

are the weights of the attributes (or alternatives) of that hierarchy, so that the first attribute (alternative) has weight w_1, the second w_2 and so on. Next, the Consistency Index (CI) of A is calculated as follows:

$$CI = \frac{\lambda_{\max} - n}{n - 1}.$$

The CI is then normalized by the Random Index (RI), which is the average CI over random entries of reciprocal matrices of the same order. The accurate list of RIs for $n \leq 100$ was given by Donegan and Dodd [19]. Thus, the Consistency Ratio (CR) is calculated as follows:

$$CR = \frac{CI}{RI}.$$

The comparison matrix is considered as consistent if CR < 0.05 for $n = 3$, CR < 0.09 for $n = 4$ and CR < 0.1 for $n > 4$. An inconsistent matrix should be reconsidered.

The final stage of the AHP is to aggregate the weights and compare the alternative options. To find the aggregate score of an alternative, all paths going from the root of the tree to this particular alternative are first identified. For each path, all weights along that path are multiplied together. Then, the resulting numbers are summed for all the above paths. Thus, each of the alternatives will be given an aggregate score between 0 and 1, to signify the priority of that alternative with respect to other alternatives.

3.2.1 Algorithm for Prioritization of Indoor Routes (PIR-Algorithm)

We start this subsection with problem definition. Suppose that there is a building where an extreme event with many epicentres is occurring. As defined in [77], the location of a bomb or a terrorist is an *epicentre*, whereas for fire the *epicentres* can be defined as points (nodes) in the building where the temperature exceeds a certain threshold. Note that a large extreme event would be typically represented by several epicentres, which are denoted by

$$z_1, z_2, ..., z_k.$$

Further, assume that an agent has to go from point p to point q in the building. For example, using one of the exits as an entrance point, an SAR team must find the 'best' route to trapped people in the building and then find a route back to one of the exits, which might be different to the first route. Taking into account that there is hazard in the building, such routes must be reasonably safe, simple and short/fast; they will be called *optimal routes*. Thus, three criteria should be considered for a route: hazard proximity, route complexity and distance. Note that the distance criterion may be replaced by travel time if information about the distribution of people in the building is available. The agent must be given an option to choose an appropriate balance of the aforementioned criteria. For instance, the agent might have personal protective equipment so that the hazard proximity becomes least important; another user might want to minimize the route complexity, thus making this criterion most important. The algorithm for finding the optimal routes must possess the aforementioned properties. Moreover, it must be robust and efficient and produce a reliable result.

In what follows, we will exploit the 3D building model recently developed by Boguslawski et al. [12]. An original BIM model was created in Autodesk Revit and exported to a surface representation stored in the gbXML format. Subsequently, a graph representation was reconstructed based on the model geometry, topology and semantical information. Topological relationships and information about doors between adjacent rooms were used to generate a logical network. In this network, denoted by G, nodes represent spaces (e.g. rooms), whereas links represent adjacency relationship among those spaces. Some of the links are marked as navigable if two adjacent spaces have a door between them. In addition, links between staircase nodes are considered for navigation, which automatically allow for navigation between the building floors. Afterwards, more detailed navigable networks are generated for specific spaces with a complex geometry and several doors, for example corridors. The navigable networks together with navigable links from the logical network G are used to form the unified navigable network G^*, where special links representing doors and staircases are introduced. A movement from one space to another can be detected when one of the special links is used. In other words, nodes belonging to an individual space are linked with nodes from another space by special links. The difference between the unified network G^* and the network G^+ used in Section 3.1 is the following. In the unified network G^*, the navigable sub-networks generated for complex spaces are consistently combined with the parts of the logical network G representing spaces without navigable sub-networks. Also, doors in G^* are represented by links. An example of the unified network G^* will be given in Section 3.2.2.

The input of PIR-Algorithm consists of the aforementioned graphs G and G^*, and the epicentres of an extreme event $z_1, z_2, ..., z_k$. The input also includes the start node p and the destination node q of the required (p, q)-route. Note that p may be the artificial node outside the building connected to all exits for finding a route going from one of the exits to the specified location q. Alternatively, the node p may be one of the locations in the building, in which case the destination node q is optional. If q is not specified in this case then by default it is one of the exits.

The optional maximal propagation coefficient ρ_{max} is also part of the input, by default $\rho_{max} = 100$. The propagation coefficient ρ represents the degree of hazard spread through a building, and this spread is reflected in hazard proximity numbers determined in Step 3(a) of PIR-Algorithm. For instance, if ρ is high, then hazard proximity numbers for nodes propagate quickly from 100 (in the epicentre) to small positive numbers (far from the epicentre). This puts a strong emphasis on the epicentre and the rooms in its close proximity. In contrast, if ρ is a small positive number, then the propagation is slow, thus putting less emphasis on the epicentre and the nearby rooms. In the extreme case $\rho = 0$ there is no propagation of hazard; that is, all hazard proximity numbers for nodes are equal.

Finally, a user can optionally choose the preferential ranking of the three criteria: distance (D), hazard proximity (HP) and route complexity (RC). By default,

$$HP > D > RC;$$

that is, HP is the most important criterion, D is the second most important and RC is the least important. For an advanced user, who might have personal protective equipment and/or wish to minimize the route complexity, seven options are available:

$$D > HP > RC,$$
$$D > RC > HP,$$

$$HP > D > RC,$$
$$HP > RC > D,$$
$$RC > D > HP,$$
$$RC > HP > D,$$
$$D = HP = RC.$$

In general, PIR-Algorithm has two parts. In the first module (Steps 1–4), a set R of feasible (p, q)-routes is generated. This set includes three 'extreme' routes: the shortest, the safest and the simplest routes, as well as a number of routes where the aforementioned criteria (D, HP, RC) are taken into account with different degrees of importance. It may be pointed out that the set R typically consists of all 'reasonable' (p, q)-routes, which is achieved by two binary searches. The second module of PIR-Algorithm (Steps 5–12) constitutes a stochastic version of the AHP. Using statistical and quantitative characteristics of the routes from the set R, the AHP prioritizes the routes with respect to the specified preferential ranking of the criteria. Thus, the algorithm finds the best (p, q)-route, the second-best (p, q)-route etc.

For convenience, we summarize the notation used in PIR-Algorithm:

$AS(P_j)$	Aggregate score for route P_j
C_j	Complexity of route P_j
$C(e)$	Complexity of link e
$D(e)$	Length of link e in metres
D_j	Length of P_j in metres
$E(G)$	Set of links in G
G	Graph of connections
G^*	Unified network
$H(v)$	Hazard proximity number for node v
$H_i(v)$	Hazard proximity number for node v w.r.t. z_i
$H(e)$	Pure hazard proximity number for link e
$HD(e)$	Hazard proximity number for link e
M_i	Comparison matrix for i-th hierarchy
P_ρ	(p, q)-Route for the propagation coefficient ρ based on distance/hazard proximity
P'_ρ	(p, q)-Route for the propagation coefficient ρ based on complexity/hazard proximity
PI_j	Proximity index for P_j
R	Set of (p, q)-routes
$b(v, z_i)$	Minimum number of obstructions (i.e. walls, floors, ceilings) between v and z_i
$d(v, z_i)$	Direct distance from v to z_i in metres

$l(P_j)$	Number of links in P_j
n	Number of routes in the set R
$r(e)$	Proximity ratio for link e
$r_i(e)$	Proximity ratio for link e w.r.t. z_i
$\boldsymbol{w}_i$	Normalized eigenvector corresponding to λ_i
z_i	i-th epicentre
λ_i	Largest eigenvalue of M_i
μ_X	Mean of parameter X
ρ	Propagation coefficient
σ_X	Sample standard deviation of parameter X
$\rho_{\max}$	Maximal propagation coefficient

PIR-Algorithm: Prioritization of (p, q)-routes in a building, where an extreme event is occurring.

MODULE 1

Input: The graphs G and G^*, which constitute the 3D model of the building.

The epicentres of an extreme event (nodes z_i, $i = 1, 2, ..., k$).

Node p; node q (optional; q is one of the exits by default).

The maximal propagation coefficient $\rho_{\max}$ (optional, by default $\rho_{\max} = 100$).

Preferential ranking of distance (D), hazard proximity (HP) and route complexity (RC) (optional, by default $HP > D > RC$; that is, HP is the most important criterion and D is the second most important one).

Output: Optimal (p, q)-route and information about its parameters.

(1) Calculate $d(v, z_i)$ and $b(v, z_i)$ for all $v \in V(G^*)$ and $i = 1, 2, ..., k$ by repeating the following steps for each node z_i:

    (a) For each node $v \in V(G^*)$, calculate $d(v, z_i)$, the direct distance from v to z_i in metres.

    (b) Run the Breadth-First Search Algorithm in the graph G from node z_i. It returns the number of links in shortest (v, z_i)-paths in G for all nodes v; that is, the minimum number of obstructions $b(v, z_i)$ between v and z_i for all nodes $v \in V(G)$.

    (c) For each node $v \in V(G^*) - V(G)$, set $b(v, z_i) = b(w, z_i)$, where $w \in V(G)$ represents the cell whose tessellation contains v.

(2) Set $R = \emptyset$, where R is a set of (p, q)-routes.

(3) Carry out the binary search with respect to ρ, $0 \leq \rho \leq \rho_{max}$, starting with $\rho = 0$, ρ_{max}, $0.5\rho_{max}$ etc. For each value of ρ, implement the following:

(a) Compute the hazard proximity numbers for nodes:

$$H_i(v) = \frac{100}{\left(1 + \frac{\rho}{100}\right)\sqrt{d(v,z_i) \times [1 + b(v,z_i)]}}$$

for each node $v \in V(G^*)$ and each $i = 1, 2, ..., k$. Then calculate $H(v) = \max_{1 \leq i \leq k} H_i(v)$ for each node $v \in V(G^*)$.

(b) For each link $e = uv \in E(G^*)$, compute the hazard proximity numbers for links:

$$HD(e) = 0.5[H(u) + H(v)] \times D(e),$$

where $D(e)$ is the length of e in metres.

(c) Run Dijkstra's algorithm in the graph G^* from node p with link weights $HD(e)$. It produces the (p, q)-route P_ρ corresponding to the propagation coefficient ρ.

(d) Set $R = R \cup \{P_\rho\}$ if $P_\rho \notin R$.

Go to Step 4 if at least one of the stopping criteria is satisfied (a specified number of generated routes, length of the widest interval and a running time).

(4) Carry out the binary search with respect to ρ, $0 \leq \rho \leq \rho_{max}$, starting with $\rho = 0$, ρ_{max}, $0.5\rho_{max}$ etc. For each value of ρ, implement the following:

(a) Compute the hazard proximity numbers for nodes:

$$H_i(v) = \frac{100}{\left(1 + \frac{\rho}{100}\right)\sqrt{d(v,z_i) \times [1 + b(v,z_i)]}}$$

for each node $v \in V(G^*)$ and each $i = 1, 2, ..., k$. Then calculate $H(v) = \max_{1 \leq i \leq k} H_i(v)$ for each node $v \in V(G^*)$.

(b) For each link $e = uv \in E(G^*)$, compute the pure hazard proximity numbers for links:

$$H(e) = 0.5[H(u) + H(v)].$$

(c) Run Duckham–Kulik's adapted algorithm in the graph G^* from node p with link weights $H(e) \times C(e)$. Here $C(e)$ is the complexity of link e; that is,

$C(e) = 0.3922 + 0.0049D(e)$ if e is a staircase link going up,

$C(e) = 0.3137 + 0.0049D(e)$ if e is a staircase link going down,

$C(e) = 0.1961 + 0.0049D(e)$ if e goes through a door, otherwise

$C(e) = 0.0490\alpha + 0.0049D(e)$ where α is the angle (in radians) between e and the 'previous' link c if e and c are considered as vectors.

The algorithm produces the (p, q)-route P'_ρ corresponding to the propagation coefficient ρ.

(d) Set $R = R \cup \{P'_\rho\}$ if $P'_\rho \notin R$.

Go to Step 5 (Module 2) if at least one of the stopping criteria is satisfied (a specified number of generated routes, length of the widest interval and a running time).

The First Part of PIR-Algorithm—Binary Searches

In the first step of PIR-Algorithm, the direct distances $d(v, z_i)$ and the number of obstructions $b(v, z_i)$ are calculated for all nodes v in G^* and all epicentres z_i. The set R of feasible (p, q)-routes is initialized in Step 2. Then, in Step 3, the binary search is carried out with respect to the propagation coefficient ρ, taking into account distance and hazard proximity, which are reflected in the hazard proximity numbers $HD(e)$. The first run is for $\rho = 0$, producing the shortest (p, q)-route P_0 because all hazard proximity numbers for links are $100D(e)$. The route P_0 is included in the set R. The next run is for $\rho = \rho_{\max}$. If the resulting route coincides with P_0, then there is no interval for the binary search and it is terminated. Otherwise, the route is different from P_0 and it is included in R. The next run is for $\rho = 0.5\rho_{\max}$. There are three possibilities here. If the resulting route is a new one, then it is included in R and the binary search continues for two intervals $(0; 0.5\rho_{\max})$ and $(0.5\rho_{\max}; \rho_{\max})$. If the resulting route coincides with one of the routes in the set R, then one of the intervals is removed from the search and the other interval is used in the binary search. For example, if the route coincides with P_0, then the binary search continues for the interval $(0.5\rho_{\max}; \rho_{\max})$, whereas the interval $(0; 0.5\rho_{\max})$ is removed. This procedure is terminated if at least one of the stopping criteria is satisfied: a specified number of generated routes included in R, a specified length of the widest interval and a running time.

The hazard proximity numbers for nodes and links in Steps 3(a) and 3(b) were introduced in [77] (see also Section 3.1), where a slightly different formula for $H_i(v)$ was used, which is not applicable for large open spaces:

$$H_i(v) = \frac{100}{\left(1 + \frac{\rho}{100}\right)\sqrt{d(v,z_i) \times b(v,z_i)}}.$$

Indeed, if the epicentre z_i of hazard is located in a large open space L, then the number of obstructions $b(v, z_i)$ for all nodes v in L is equal to zero, which means $H_i(v) = 100$ for all nodes in L and for any value of ρ. Thus, in this particular case, we have a non-discrimination problem; that is, the above formula does not distinguish between nodes in close proximity to the hazard and nodes which are further away. Actually, in a large open space with an epicentre, the distance should be used as a criterion of hazard proximity within this space. This can be achieved if the term $b(v, z_i)$ is replaced by $[1 + b(v, z_i)]$ in the aforementioned formula. Now, if $b(v, z_i) = 0$, then the hazard proximity number $H_i(v)$ does depend on the distance, and $H_i(v) = 100$ only if the node v coincides with the epicentre z_i or $\rho = 0$. Thus, the updated formula for $H_i(v)$ in Step 3(a) extends the previous one by improving hazard propagation in large open spaces. The formula for the proximity ratios w.r.t. z_i in Step 6(b) was updated in a similar way.

The second binary search (Step 4) of the algorithm is carried out with respect to ρ, taking into account link complexity and hazard proximity. Basically, this step is similar to the previous one, however instead of distances, link complexities are used. The main difference is that some link complexities cannot be calculated in advance because the angle between two links is not a property of a given link but rather a pair of adjacent links. Hence, Dijkstra's algorithm cannot be directly used here. Instead, we apply the simplest path algorithm by Duckham and Kulik [21]. In their algorithm, one can use any weighting function

$$f \colon \mathcal{E} \to \mathbb{R}^+, \quad \text{where } \mathcal{E} = \{(ab, bc) \in (E\,(G^*) \cup \mathrm{nil}) \times E\,(G^*)\}$$

and $E(G^*)$ is the set of links in G^*. This function is defined below and it is based on specific weights for the complexity attributes. Let us label the five complexity attributes as follows:

A1: Staircase link going up;
A2: Staircase link going down;
A3: Door;
A4: Turn of 1 radian;
A5: Distance (10 metres).

Although the first four attributes are obviously elements of complexity, the inclusion of the distance attribute as a complexity criterion should be explained. Let us consider two straight routes inside a corridor without doors, say 5 metres long and 50 metres long. The first four attributes add no complexity to the routes, so without the distance attribute it is impossible to distinguish between the two routes from the viewpoint of their complexities. This would contradict the common sense: the former route is obviously 'simpler' than the latter, which is only reflected in their lengths. For assigning appropriate weights to the attributes, we apply one particular step from the AHP (see [55]): first, construct a comparison matrix, where the rows and columns correspond to the attributes A1–A5:

$$
\begin{array}{c c c c c c}
 & \text{A1} & \text{A2} & \text{A3} & \text{A4} & \text{A5} \\
\text{A1} & 1 & 1.25 & 2 & 8 & 8 \\
\text{A2} & 1/1.25 & 1 & 1.6 & 6.4 & 6.4 \\
\text{A3} & 1/2 & 1/1.6 & 1 & 4 & 4 \\
\text{A4} & 1/8 & 1/6.4 & 1/4 & 1 & 1 \\
\text{A5} & 1/8 & 1/6.4 & 1/4 & 1 & 1
\end{array}.
$$

The entries in this matrix reflect the relative importance of the attributes according to the rules of the AHP. For instance, the entry '8' means that A1 is nearly extremely more important than A4. Further, the largest eigenvalue is 5 and the corresponding normalized eigenvector is

$$(0.3922, 0.3137, 0.1961, 0.0490, 0.0490).$$

This eigenvector provides the required weights for the attributes, whereas the Consistency Ratio of 0 indicates that the comparison matrix is perfectly consistent. Thus, the complexity attributes are assigned the following weights:

A1: 0.3922 (staircase link going up);
A2: 0.3137 (staircase link going down);
A3: 0.1961 (door);
A4: 0.0490 (turn of 1 radian);
A5: 0.0049 (distance of 1 metre).

Now, the aforementioned function f is defined as follows:

$f(ab, bc) = [0.3922 + 0.0049D(bc)]H(bc)$ if bc is a staircase link going up (ab may be 'nil'),

$f(ab, bc) = [0.3137 + 0.0049D(bc)]H(bc)$ if bc is a staircase link going down (ab may be 'nil'),

$f(ab, bc) = [0.1961 + 0.0049D(bc)]H(bc)$ if bc is a link going through a door (ab may be 'nil'),

$f(ab, bc) = [0.0490\alpha + 0.0049D(bc)]H(bc)$ where α is the angle (in radians) between the non-staircase links ab and bc, $ab \neq$ nil, and bc is not a 'door' link.

Note that in the first three cases it is allowed to have $ab =$ nil. Hence, if the first link e in a route is a door or a staircase link, then the value of the function $f(\text{nil}, e)$ is calculated for such a link. It may be pointed out that the first run of the binary search for $\rho = 0$ will produce the simplest (p, q)-route because in this case $H(e) = 100$ for any link e. For other values of ρ, both link complexities and hazard proximities are taken into account when generating the corresponding (p, q)-routes, which is reflected in the definition of the function f. For instance, the value of the function $f(ab, bc)$ for a door bc increases if it is closer to the epicentre of hazard because $H(bc)$ will be larger. This binary search is terminated if at least one of the stopping criteria is satisfied: a specified number of generated routes included in R, a specified length of the widest interval and a running time limit.

The Second Part of PIR-Algorithm—the AHP

In Step 5, all the routes in the set R are denoted by $P_1, P_2, ..., P_n$ for simplicity of presentation. These routes form the input for the AHP. Next, in Step 6, the following parameters are calculated for each route in R: the route length, the proximity ratios, the proximity index and the route complexity. The calculation of complexity is consistent with that of Step 4(c). The proximity ratios and indices are similar to those used in [77] with the only difference in the term $[1 + b(v, z_i)]$, which is needed to avoid the aforementioned problem of non-discrimination in large open spaces. The necessary statistical characteristics are computed in Step 7.

PIR-Algorithm: Prioritization of (p, q)-routes in a building, where an extreme event is occurring.

MODULE 2

---

(5) Denote all the routes in R by $P_1, P_2, ..., P_n$.
(6) For each route $P_j \in R$, calculate the following:
  (a) The length of P_j: $D_j = \sum_{e \in P_j} D(e)$.
  (b) The proximity ratios w.r.t. z_i:

$$r_i(e) = \frac{\sqrt{d(u, z_i) \times [1 + b(u, z_i)]} + \sqrt{d(v, z_i) \times [1 + b(v, z_i)]}}{2D(e)}$$

for each $i = 1, 2, ..., k$ and each link $e = uv$ in P_j.

(c) The proximity ratios $r(e) = \min_{1 \leq i \leq k} r_i(e)$ for each link $e = uv$ in P_j.

(d) The proximity index

$$PI_j = \frac{l\,(P_j)}{\sum_{e \in P_j} \frac{1}{r(e)}},$$

where PI is the harmonic mean of $r(e)$, and $l\,(P_j)$ is the number of links in P_j.

(e) The complexity of P_j:
$$C_j = 0.3922\Omega_j + 0.3137\omega_j + 0.1961\delta_j + 0.0490\chi_j + 0.0049D_j,$$
where Ω_j is the number of staircase links going up in P_j; ω_j is the number of staircase links going down; δ_j is the number of doors; χ_j is the total turning angle (in radians) and D_j is the length of P_j (in metres).

(7) For the routes in R, compute the means and the sample standard deviations of distances, proximity indices and complexities: μ_D, σ_D, μ_{PI}, σ_{PI}, μ_C, σ_C.

(8) Create the 3×3 matrix M_1 of relative importance of distance (D), hazard proximity (HP) and route complexity (RC), based on the specified preferential ranking. The row and columns of M_1 correspond to D, HP and RC, respectively. By default, HP $>$ D $>$ RC; that is,

$$M_1 = \begin{pmatrix} 1 & 1/2 & 2 \\ 2 & 1 & 4 \\ 1/2 & 1/4 & 1 \end{pmatrix}.$$

(9) Create three $n \times n$ comparison matrices M_2, M_3, M_4 for the relative importance of the routes in the set R w.r.t. distance, proximity index and complexity as follows:

- For each entry $M_2[i, j]$, compute $\beta = (D_j - D_i)/\sigma_D$.
  If $\sigma_D = 0$, then we set $\beta = 0$. Calculate $M_2[i, j] = 9^{0.2\beta}$.
  Set $M_2[i, j] = 1/9$ if $\beta < -5$; set $M_2[i, j] = 9$ if $\beta > 5$.
- For each entry $M_3[i, j]$, compute $\beta = (PI_i - PI_j)/\sigma_{PI}$.
  If $\sigma_{PI} = 0$, then we set $\beta = 0$. Calculate $M_3[i, j] = 9^{0.2\beta}$.
  Set $M_3[i, j] = 1/9$ if $\beta < -5$; set $M_3[i, j] = 9$ if $\beta > 5$.
- For each entry $M_4[i, j]$, compute $\beta = (C_j - C_i)/\sigma_C$.
  If $\sigma_C = 0$, then we set $\beta = 0$. Calculate $M_4[i, j] = 9^{0.2\beta}$.
  Set $M_4[i, j] = 1/9$ if $\beta < -5$; set $M_4[i, j] = 9$ if $\beta > 5$.

(10) For each matrix M_i, run the Direct Iteration Algorithm to calculate its largest eigenvalue λ_i and the corresponding normalized eigenvector w_i with precision 0.0001.

(11) For each route $P_j \in R$, calculate the aggregate score:

$$AS(P_j) = w_1[1]w_2[j] + w_1[2]w_3[j] + w_1[3]w_4[j].$$

(12) Determine the route P_j with the highest aggregate score. Report P_j and information about its parameters. Algorithm stops.

The initial stage of the AHP (see [55]) is to set up a hierarchy tree, which is shown in Figure 3.19. The first level of hierarchy in the tree consists of three criteria: distance (D), hazard proximity (HP) and route complexity (RC), whereas the lowest three levels comprise the routes in the set R. In Step 8 of PIR-Algorithm, the comparison matrix M_1 is created, which is based on the specified preferential ranking of the criteria. The row and columns of M_1 correspond to D, HP and RC, respectively. By default, HP > D > RC, that is,

$$M_1 = \begin{pmatrix} 1 & 1/2 & 2 \\ 2 & 1 & 4 \\ 1/2 & 1/4 & 1 \end{pmatrix}.$$

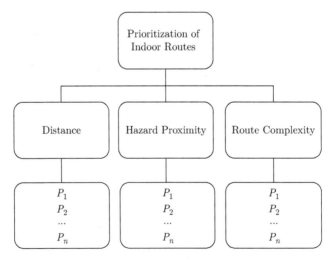

Fig. 3.19 Hierarchy tree for the AHP.

In general, there are seven possibilities for the preferential ranking:
$D > HP > RC$, $D > RC > HP$, $HP > D > RC$, $HP > RC > D$, $RC > D > HP$,
$RC > HP > D$ and $D = HP = RC$. For example, if $D > HP > RC$, then

$$M_1 = \begin{pmatrix} 1 & 2 & 4 \\ 1/2 & 1 & 2 \\ 1/4 & 1/2 & 1 \end{pmatrix} ;$$

if $D = HP = RC$, then M_1 consists of 1s. Notice that the matrix M_1 is always
perfectly consistent.

Three $n \times n$ comparison matrices M_2, M_3 and M_4 are constructed in Step 9.
They represent the relative importance of the routes in the set R w.r.t. distance,
proximity index and complexity, respectively, and the construction of the
matrices is based on the corresponding parameters and standard deviations.
For instance, the (i, j)-element of the matrix M_2 is calculated as follows: we
first compute $\beta = (D_j - D_i)/\sigma_D$, where D_j and D_i are the lengths of the routes
P_j and P_i in metres; if $\sigma_D = 0$, then we set $\beta = 0$. Next, $M_2[i, j] = 1/9$ if
$\beta < -5$, $M_2[i, j] = 9$ if $\beta > 5$, and $M_2[i, j] = 9^{0.2\beta}$ otherwise. Thus, the (i, j)-
element of the matrix M_2 represents the relative importance of lengths of the
routes P_i and P_j in terms of standard deviations. For example, if the difference
in lengths is 2.5 standard deviations, then one route is weakly more important
than the other, which will be reflected by the entry '3' in the comparison
matrix. It may be pointed out that different functions for constructing the
above matrices have been tested, and it turned out that the best one is the
exponential function $9^{0.2\beta}$. Also, because of the way the matrices M_2, M_3 and
M_4 are constructed, the corresponding Consistency Indices and Consistency
Ratios are always very small, or equal to zero. Hence, the matrices are very
consistent and there is no need to adjust them.

In Step 10, the standard Direct Iteration Algorithm is used to calculate the
largest eigenvalue λ_i for each matrix M_i and the corresponding normalized
eigenvector w_i with precision 0.0001. For each route $P_j \in R$, the aggregate
score is computed in Step 11 as follows:

$$AS(P_j) = w_1[1]w_2[j] + w_1[2]w_3[j] + w_1[3]w_4[j].$$

Finally, in Step 12, the route P_j with the highest aggregate score is determined
and reported, together with the information about the parameters of this route.

3.2.2 Testing PIR-Algorithm

We start testing PIR-Algorithm with a rather complex building shown in Figure 3.20. This nine-floor building has five stairwells and two exits (indicated by arrows in Figures 3.20 and 3.22), and it is based on the typical floor of the Doha World Trade Center depicted in Figure 3.21. All the stairwells are highlighted in red in Figure 3.22, which represents the navigable unified network G^* of the building.

The epicentre of an extreme event is on the ground floor and it is labelled by a red rectangle in Figures 3.20 and 3.23. The start node p is the artificial node outside the building (not shown in the unified network), and the destination node q represents a large room located on the last floor in the right part of the building. Thus, we are looking for the optimal route going from one of the exits to the room q. In what follows, we set $\rho_{max} = 100$ for one epicentre because our numerous tests showed that 'unreasonable' routes are often produced for $\rho > 100$, which do not belong to the efficient frontier. Also, the default preferential ranking of the criteria is used, that is, HP > D > RC.

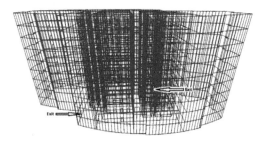

Fig. 3.20 The nine-floor building model based on the typical floor of the Doha World Trade Center.

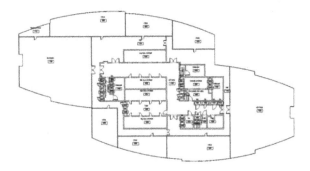

Fig. 3.21 First floor of the Doha World Trade Center.

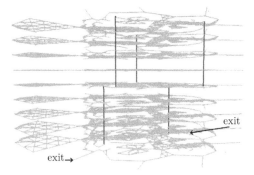

Fig. 3.22 The unified network G^* of the model.

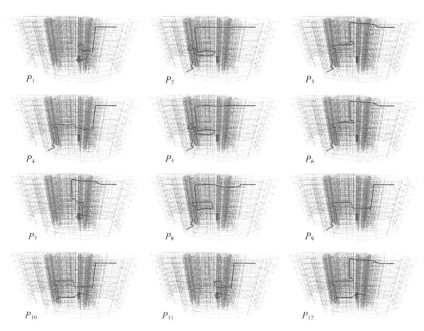

Fig. 3.23 The routes $P_1 - P_{12}$ generated by the binary searches.

The first binary search in Step 3 of PIR-Algorithm produces six routes $P_1 - P_6$, whereas the second binary search in Step 4 also generates six routes $P_7 - P_{12}$. These routes are depicted in Figure 3.23, and their parameters are summarized in Table 3.4. Note that in both binary searches one stopping criterion was used: the length of the widest interval is 0.01. It may be pointed out that different exits are used as an entrance point in the generated routes.

As can be seen in Table 3.4, the shortest route P_1 corresponds to the propagation coefficient $\rho = 0$ in the first binary search. The safest route P_2 has

Table 3.4 Parameters of routes $P_1 - P_{12}$.

Route	Propagation Coefficient ρ	Length D_j	Proximity Index PI_j	Complexity C_j	Aggregate Score $AS(P_j)$
			1st binary search		
P_1	0	154.61	1.89	6.67	0.085
P_2	42.481	222.79	4.29	9.28	0.090
P_3	15.478	194.88	3.76	7.47	0.089
P_4	7.190	183.30	3.79	7.98	0.092
P_5	29.576	222.58	4.24	9.00	0.088
P_6	26.485	196.45	3.50	7.83	0.080
			2nd binary search		
P_7	0	168.65	1.69	6.37	0.074
P_8	14.451	212.40	3.56	7.73	0.076
P_9	8.632	186.42	3.96	7.31	0.099
P_{10}	5.660	199.07	3.12	6.99	0.075
P_{11}	3.284	159.43	1.82	6.38	0.082
P_{12}	5.602	208.84	3.00	7.02	0.069

the largest proximity index 4.29 (2 dp) and the corresponding propagation coefficient is 42.481, which is given to 3 dp and rounded up to keep the correspondence to P_2. More precisely, the route P_2 would be generated for all values of ρ between 42.481 and 100, but the lowest value is only reported in the table. Thus, formally speaking, the route P_2 corresponds to the interval [42.481, 100] of propagation coefficients, whereas the route P_1 corresponds to the interval [0, 7.190), where the number 7.190 is excluded from the interval. Note that the safest route is generated for the largest propagation coefficient in the first binary search because maximum emphasis is put on the epicentre of hazard in this case. Also, it may be pointed out that the safest route P_2 is the longest and most complex one, which is a typical picture unless the safest route coincides with the shortest. The simplest route P_7 corresponds to the propagation coefficient $\rho = 0$ in the second binary search, which is always the case because there is no propagation of hazard for $\rho = 0$ and only complexity of the route is minimized.

Note that the route P_{10} is better than P_{12} for all attributes, and also the routes P_3, P_6, P_8 are 'dominated' by P_9. Therefore, the efficient frontier consists of eight routes: P_1, P_2, P_4, P_5, P_7, P_9, P_{10}, P_{11}. The efficient frontier is illustrated in Figure 3.24, where the above routes are labelled by numbers 1, 2, 4, 5, 7, 9, 10, 11. Its lower part comprises three routes: P_1, P_7 and P_{11}, which have very good lengths and complexities, but the proximity indices are

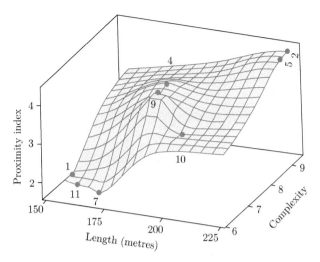

Fig. 3.24 The efficient frontier.

rather low. In the upper part there are two routes: P_2 and P_5, which are very long and complex, but safer than other routes. Notice that the routes P_2 and P_5 look similar in Figure 3.23, but there is some difference on the top floor. The middle part of the efficient frontier consists of three routes: P_4, P_9 and P_{10}. The first two routes form two local maxima and have high proximity indices and very reasonable lengths and complexities; hence, based on common sense, they are good candidates for being the optimal routes for the default preferential ranking.

The optimal route for the preferential ranking HP > D > RC is P_9 because it has the largest aggregate score of 0.099 (3 dp). This route has the third highest proximity index (3.96), which is close to the best value (4.29), and also reasonable length (186.42 m) and complexity (7.31), which are slightly better than the corresponding average values shown below. Here are the statistical parameters (means and sample standard deviations):

$$\mu_D = 192.45, \quad \sigma_D = 22.90,$$
$$\mu_{PI} = 3.22, \quad \sigma_{PI} = 0.94,$$
$$\mu_C = 7.50, \quad \sigma_C = 0.93.$$

Note that the normality test (Anderson–Darling) showed that the datasets in Table 3.4 displayed in the columns 'Length', 'Proximity Index' and 'Complexity' are normally distributed. For completeness, we also give the normalized eigenvectors:

$w_1 = (0.286, 0.571, 0.143),$

$w_2 = (0.157, 0.043, 0.073, 0.091, 0.043, 0.071, 0.120, 0.052, 0.086, 0.067,$
$\quad 0.143, 0.056),$

$w_3 = (0.041, 0.127, 0.099, 0.101, 0.124, 0.088, 0.038, 0.090, 0.109, 0.073,$
$\quad 0.040, 0.070),$

$w_4 = (0.114, 0.033, 0.078, 0.061, 0.038, 0.066, 0.131, 0.069, 0.084, 0.098,$
$\quad 0.131, 0.097).$

The eigenvectors for length (w_2), proximity index (w_3) and complexity (w_4) are visualized in the bar chart of Figure 3.25, together with the aggregate score, which is the weighted average of the elements of w_2, w_3 and w_4 using the weights of the eigenvector w_1. Also, the largest eigenvalues of the 12×12 comparison matrices M_2, M_3 and M_4 are 12, and hence the corresponding Consistency Indices and Consistency Ratios are equal to zero, so that the matrices are perfectly consistent.

Because the input of PIR-Algorithm includes the preferential ranking of the main criteria, we provide the optimal routes for different preferential rankings: the route P_1 for D > HP > RC and D > RC > HP; the route P_9 for HP > D > RC and HP > RC > D; the route P_{11} for RC > D > HP, RC > HP > D and D = HP = RC. Thus, if hazard proximity is not the most important attribute, then the optimal route is either P_1 or P_{11}. This can be explained by the fact that P_1 is the shortest route with the third-best complexity (6.67), which is very

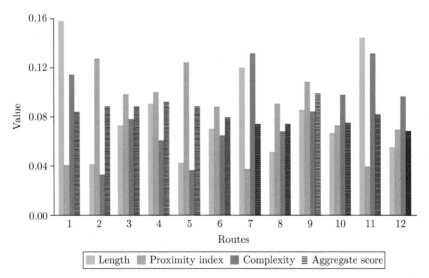

Fig. 3.25 The bar chart for eigenvectors and aggregate score.

close to the best complexity (6.37), so P_1 is the optimal route if distance is the most important criterion. The route P_{11} exhibits the second-best complexity (6.38), which is extremely close to the best value (6.37), and it is the second shortest (159.43 m), which is also very close to the best length (154.61 m). Hence, P_{11} is the optimal route if complexity is the most important attribute or all the criteria are of the same importance.

It may be pointed out in conclusion of this subsection that the proposed method is quite robust with respect to different parameters used in the algorithm; that is, the method is not sensitive to small changes in the parameters. For example, the sensitivity analysis was applied to the default matrix M_1, where two elements '2' are adjusted by at most ± 0.5 in such a way that the matrix remains perfectly consistent. More precisely, the elements $M_1[2, 1] = 2$ and $M_1[1, 3] = 2$, representing the relative importance of HP to D and D to RC, may be any numbers in the interval $[1.5, 2.5]$. Next, $M_1[1, 2] = 1/M_1[2, 1]$ by definition and $M_1[1, 2] \times M_1[2, 3] = M_1[1, 3]$ to keep the matrix perfectly consistent, that is,

$$M_1[2, 3] = M_1[1, 3]/M_1[1, 2] = M_1[1, 3] \times M_1[2, 1].$$

Also, $M_1[3, 1]$ and $M_1[3, 2]$ are the reciprocals of $M_1[1, 3]$ and $M_1[2, 3]$ by definition. Thus, the entire matrix is determined by its two elements $M_1[2, 1]$ and $M_1[1, 3]$. Now, if these elements are any numbers in the interval $[1.5, 2.5]$, then the corresponding optimal route is P_9; that is, it remains unchanged. Similar sensitivity analysis was carried out for the weights used in the complexity function in Step 4(c) of the algorithm. It showed that small changes in the weights do not change the set of generated routes in Step 4, and therefore the optimal route remains unchanged. The same comment can be made for the exponential function $9^{0.2\beta}$ used in Step 9 of PIR-Algorithm.

3.2.3 Incorporating 'Time' into PIR-Algorithm

To take into consideration the time required to travel from the start node to the destination node in a building, it is necessary to know the length and speed for all links in the route. Indeed, the travel time $T(e)$ along link e is calculated as follows:

$$T(e) = D(e)/S(e),$$

where $D(e)$ is the length of e in metres, and $S(e)$ is the speed along the link e (metres per second). The most reliable formula for people's indoor speed is Nelson–MacLennan–Pauls' relationship between speed and density (see [48, 52]):

$$S\left(e\right) = \begin{cases} 0.856K & \text{if } 0 \leq \Delta < 0.54, \\ \left(1 - 0.266\Delta\right)K & \text{if } 0.54 \leq \Delta \leq 3.75, \end{cases} \tag{3.1}$$

where Δ is the population density (persons per square metre) in the area corresponding to the link e, and the constant K is defined as follows:

$K = 1.40$ for horizontal movement,

$K = 1.08$ for moving downstairs,

$K = 0.81$ for moving upstairs.

Notice that the value of the constant K for moving upstairs was derived from the results of Fruin [25], whereas its value for moving downstairs depends on the characteristics of stairs such as the length of riser and tread (see [48]). The fundamental formula (3.1) provides the linear relationship between people's speed and density. More precisely, if the population density Δ is less than 0.54 persons/m^2, then people's movement would be dependent on their personal characteristics, and so the average constant speed of $0.856K$ m/s is used. If $\Delta > 3.75$ persons/m^2, then no movement is possible until the density is reduced. Between the density values of 0.54 and 3.75 persons/m^2, the speed is given by the linear function $\left(1 - 0.266\Delta\right)K$.

Thus, for calculation of the travel time along a given route it is necessary to know the distribution of people in the building, that is, the population densities in its areas. In some cases, such a distribution is known, and for emergency situations this can be achieved by the simulation of evacuation flows of people. There are software packages for modelling the distribution of people in a building. For example, Kisko and Francis [32] proposed the EVACNET+ software application for evacuation scenarios planning. Other alternative solutions are available as well.

Therefore, in the modified version of PIR-Algorithm presented below, we assume that the population densities are given for all cells in the building. Because each cell is represented by a node in the unified network G^*, the densities are associated with nodes in the set $V(G^*)$. In the new Step 0, we first calculate the density for the link uv, that is, the average density for two areas associated with the nodes u and v. Then, the speed $S(e)$ along the link e is computed using formula (3.1). Here we assume that the agent is moving together with the flow of people. In emergency situations, this typically happens when moving downstairs. If the agent is moving in a counter-flow, for example upstairs, then one should use formula (3.2), which will be discussed later. Finally, the travel time $T(e)$ along link e is calculated in Step 0(c).

PIR-Algorithm2 (with Time): Prioritization of (p, q)-routes in a building with an extreme event.

    Input: ...

                Preferential ranking of travel time (TT), hazard proximity (HP) and route complexity (RC) (optional, by default HP > TT > RC).
                The population density $\Delta(v)$ (persons/m^2) for the area corresponding to the node $v \in V(G^*)$.

(0) For each link $e = uv \in E(G^*)$, compute the following:
    (a) The population density for the area corresponding to e:

$$\Delta(e) \;=\; 0.5[\Delta(u) + \Delta(v)].$$

    (b) The speed $S(e)$ along the link e using formula (3.1), or (3.2) for counter-flow movement.
    (c) The travel time $T(e)$ along link e: $T(e) = D(e)/S(e)$.
    ...

(3) (b) For each link $e = uv \in E(G^*)$, compute the hazard proximity numbers for links:

$$HD(e) = 0.5[H(u) + H(v)] \times T(e),$$

    where $T(e)$ is the travel time along e in seconds.

    ...

(6) For each route $P_j \in R$, calculate the following:
    (a) The length $D_j = \sum_{e \in P_j} D(e)$ and the travel time $T_j = \sum_{e \in P_j} T(e)$.
    (b) The proximity ratios w.r.t. z_i:

$$r_i(e) = \frac{\sqrt{d(u, z_i) \times [1 + b(u, z_i)]} + \sqrt{d(v, z_i) \times [1 + b(v, z_i)]}}{2T(e)}$$

    for each $i = 1, 2, ..., k$ and each link $e = uv$ in P_j.

    ...

(7) For the routes in R, compute the means and the sample standard deviations of travel times, proximity indices and complexities: μ_T, σ_T, μ_{PI}, σ_{PI}, μ_C, σ_C.

...

(9) Create three $n \times n$ comparison matrices M_2, M_3, M_4 for the relative importance of the routes in the set R w.r.t. travel time, proximity index and complexity as follows:
- For each entry $M_2[i,j]$, compute $\beta = (T_j - T_i)/\sigma_T$.
  If $\sigma_T = 0$, then we set $\beta = 0$. Calculate $M_2[i,j] = 9^{0.2\beta}$.
  Set $M_2[i,j] = 1/9$ if $\beta < -5$; set $M_2[i,j] = 9$ if $\beta > 5$.

...

Further, in Step 3(b), the formula for the hazard proximity numbers is updated by using travel time instead of link length. Thus, if there is no hazard propagation, that is, $\rho = 0$, then the fastest route will be generated in the first binary search instead of the shortest one. In the modified Step 6, in addition to the route length, we calculate the travel time for each route in the set R. Also, the formula for proximity ratios is updated with travel time $T(e)$ because longer travel time for a link should decrease the proximity ratio for this link and the proximity index for the route, thus making it worse. Next, in Step 7, the statistical parameters are now calculated for travel times, and the comparison matrix M_2 in Step 9 represents the relative importance of routes in the set R with respect to travel time.

Testing PIR-Algorithm2

Before we start testing PIR-Algorithm2, let us apply the original PIR-Algorithm to the scenario described in Section 3.2.2 with the start and destination nodes swapped; that is, the start node is the large room in the right part of the last floor and the destination node is the artificial node outside the building. Note that there is no symmetry in moving downstairs and upstairs because the complexity of the former is lower. The result is summarized in Table 3.5. The first binary search produces the same routes as in Table 3.4 because it is independent of complexity, but the complexity figures are lower than the corresponding numbers in Table 3.4. The second binary search generates five routes, which are basically similar to the routes in the second half of Table 3.4. The optimal route is X_{11}, which is similar to the optimal route P_9 for moving upstairs depicted in Figure 3.23.

Now, let us consider the following four typical scenarios:

Table 3.5 Parameters of routes $X_1 - X_{11}$ for moving downstairs.

Route	Propagation Coefficient ρ	Length D_j	Proximity Index PI_j	Complexity C_j	Aggregate Score $AS(X_j)$
			1st binary search		
X_1	0	154.61	1.89	6.03	0.0908
X_2	42.481	222.79	4.29	8.56	0.0903
X_3	15.478	194.88	3.76	6.91	0.0900
X_4	7.190	183.30	3.79	7.35	0.0942
X_5	29.576	222.58	4.24	8.28	0.0893
X_6	26.485	196.45	3.50	7.32	0.0803
			2nd binary search		
X_7	0	157.66	1.79	5.68	0.0904
X_8	30.018	207.35	4.02	7.25	0.0909
X_9	10.625	185.73	3.98	6.71	0.1022
X_{10}	5.878	198.38	3.12	6.36	0.0776
X_{11}	29.236	184.44	4.01	6.69	0.1040

Scenario 1 There are no people in the building; that is, the population densities for all nodes are equal to 0.

Scenario 2 The initial stage of uncontrolled evacuation, when people have just left their offices. Thus, $\Delta = 0$ for any office, $\Delta = 1$ for all staircases and landings, and $\Delta = 0.5$ for all other spaces (halls, corridors etc.).

Scenario 3 The middle stage of uncontrolled evacuation: $\Delta = 3$ for all staircases and landings, $\Delta = 2$ in the areas on the ground floor between staircases and entrances to the building, $\Delta = 2$ in the areas on Floor 4 where people switch staircases, $\Delta = 0$ for all other spaces.

Scenario 4 The final stage of uncontrolled evacuation: $\Delta = 3$ for staircases and landings between the ground floor and Floor 4, $\Delta = 2$ in the areas on the ground floor between staircases and entrances to the building, $\Delta = 0$ for all other spaces.

In the first scenario, we assume that there are no people inside the building, for example during weekend or night. Notice that this scenario also covers the sparse distribution of people ($\Delta < 0.54$) when an agent is moving together with people's flow because in this case the speed given by formula (3.1) is a constant, and hence the travel time for a given link or route is fixed. Scenario 2 describes an initial stage of evacuation, when all offices are empty, the crowd condition in the staircases is moderate, and other spaces have the minimum crowd condition. Scenario 3 is for a middle stage of evacuation, when all staircases experience the crush crowd condition, and the medium densities are on the ground floor in the specified areas and on Floor 4 in the areas between staircases. This scenario also covers the situation where there are some queues near the entrances to the staircases on some floors because the generated routes will remain unchanged. The final stage of evacuation is given in Scenario 4, where the staircases between the ground floor and Floor 4 have the crush crowd condition and the medium densities are on the ground floor between staircases and entrances to the building.

The results of PIR-Algorithm2 applied to Scenarios 1–4 are summarized in Table 3.6. Compared to the set of routes in Table 3.5, there is a new route of length 206.52 m for Scenario 1, two new routes (206.52 m, 206.38 m) for Scenarios 2 and 3, and two new routes (206.52 m, 183.28 m) for Scenario 4. The optimal route for all the scenarios is the same: it is the route X_{11} of length 184.44 m (it is similar to the route P_9 depicted in Figure 3.23). This demonstrates some robustness of the algorithm and its insensitivity to different

Table 3.6 Results of PIR-Algorithm2 for Scenarios 1–4 (moving downstairs).

Scenario 1					Scenario 2				
Length	Time	PI	C	AS	Length	Time	PI	C	AS
154.61	153	1.98	6.03	0.0908	154.61	171	1.80	6.03	0.0915
222.79	210	4.55	8.56	0.0909	222.79	227	4.19	8.56	0.0919
206.52	196	3.95	7.69	0.0813	206.52	214	3.63	7.69	0.0820
183.30	177	3.96	7.35	0.0932	183.30	195	3.61	7.35	0.0933
194.88	186	3.93	6.91	0.0892	206.38	214	3.63	7.65	0.0822
222.58	210	4.49	8.28	0.0899	222.58	227	4.14	8.28	0.0909
157.66	155	1.87	5.68	0.0904	157.66	173	1.70	5.68	0.0911
207.35	197	4.24	7.25	0.0909	207.35	215	3.89	7.25	0.0916
185.73	179	4.16	6.71	0.1012	185.73	197	3.80	6.71	0.1013
198.38	189	3.38	6.36	0.0795	198.38	207	3.16	6.36	0.0814
184.44	178	4.19	6.69	0.1028	184.44	196	3.82	6.69	0.1028

Scenario 3					Scenario 4				
Length	Time	PI	C	AS	Length	Time	PI	C	AS
154.61	525	0.62	6.03	0.0927	154.61	338	0.74	6.03	0.0849
222.79	617	1.52	8.56	0.0932	222.79	400	2.09	8.56	0.0847
206.52	592	1.29	7.69	0.0825	206.52	386	1.77	7.69	0.0739
206.38	592	1.29	7.65	0.0827	183.30	367	1.73	7.35	0.0820
183.30	569	1.25	7.35	0.0886	183.28	367	1.78	7.43	0.0843
222.58	617	1.50	8.28	0.0921	194.88	377	1.74	6.91	0.0800
157.66	530	0.58	5.68	0.0919	222.58	400	2.06	8.28	0.0838
207.35	594	1.39	7.25	0.0921	157.66	341	0.70	5.68	0.0846
185.73	572	1.32	6.71	0.0963	207.35	387	1.91	7.25	0.0825
198.38	581	1.24	6.36	0.0903	185.73	369	1.83	6.71	0.0895
184.44	570	1.32	6.69	0.0976	198.38	378	1.65	6.36	0.0793
					184.44	368	1.83	6.69	0.0905

scenarios of uncontrolled evacuation, even though there is a dramatic change in travel times for Scenarios 1 and 3.

It may be pointed out that for all the scenarios, the travel time, the proximity index and the complexity of the optimal route X_{11} are better than the corresponding mean values. The only other route possessing this property is the route X_9 of length 185.73 m, which is very similar to X_{11}. As can be seen in Table 3.6, the route X_9 is the second-best route in all the scenarios. The third-best route is of length 183.30 m for Scenarios 1 and 2; it is the route P_4 shown in Figure 3.23. Also, the third-best routes for Scenarios 3 and 4 are of length 222.79 m and 154.61 m, respectively. The former is the safest available route P_2 and the latter is the shortest/quickest route P_1; both are depicted in Figure 3.23.

In the previous test, we assumed that an agent is moving together with the flow of people. However, an SAR team or a single rescuer can overtake people's flow and move with a higher speed than one given by formula (3.1). Let us now consider the situation when an agent is moving with speed increased by 25%; that is, the new speed is $1.25S(e)$, where $S(e)$ is calculated according to (3.1). The results of this test are very similar to the previous one. For all the scenarios, exactly the same routes are generated as in Table 3.6. The travel times for the optimal route (184.44 m) are, of course, different: 142 s, 156 s, 456 s, 294 s for Scenarios 1–4, respectively.

Moving Upstairs in a Counter-flow
Let us consider the situation when an SAR team or a single rescuer is moving upstairs in Scenarios 1–4. In this case, the movement is in a counter-flow, so formula (3.1) is not applicable. The rescuers' speed is dependent on their personal characteristics, whether they carry heavy equipment etc. Let us assume that for horizontal movement, the speed is 1.5 m/s for density $\Delta = 0$ p/m^2, which is consistent with the previous example. Also, suppose that the speed is 0.9, 0.5 and 0.3 m/s for densities 1, 2 and 3 p/m^2, respectively. This dataset can be described by the following function:

$$S(e) = \hat{K}0.6^{\Delta},\tag{3.2}$$

where $S(e)$ is the speed along link e, Δ is the population density in the area corresponding to e, and the constant $\hat{K}$ is defined as follows:

$\hat{K} = 1.50$ for horizontal movement,
$\hat{K} = 1.16$ for moving downstairs,
$\hat{K} = 0.87$ for moving upstairs.

Here we assume that the ratios (1.50:1.16:0.87) of the coefficient $\hat{K}$ for different types of movement are the same as the ratios of the similar constant K in (3.1). Formula (3.2) should be considered as a first approximation of a rescuer's speed in a counter-flow. Additional research is needed to develop this formula further. In contrast to formula (3.1) with a linear relationship, formula (3.2) represents an exponential relationship between density and speed in a counter-flow. The graphical illustration of formula (3.2) is given in Figure 3.26.

The results of this test for Scenarios 1–4 described above are summarized in Table 3.7. Compared to the set of routes in Table 3.4, there are two new routes (206.52 m, 206.38 m) for all the scenarios. In addition, Scenario 3 has a new route of length 185.85 m, and Scenario 4 has two new routes (173.30 m, 183.28 m). Note that the shortest route of length 154.61 m is also the quickest in Scenarios 1 and 2. However, in Scenarios 3 and 4, this route does not show up because different routes are fastest: 185.85 m and 173.30 m, respectively. This happens because the speed of horizontal movement is very different for the ground floor and Floors 4 and 8. Indeed, the speed is 1.5 m/s if density is 0, and it is 0.54 m/s if density is 2.

The optimal route for Scenarios 1, 2 and 4 is the route P_9 of length 186.42 m shown in Figure 3.23, whereas the optimal route for Scenario 3 is the safest available route P_2 of length 222.79 m. It may be pointed out that the optimal route P_9 for Scenarios 1, 2 and 4 possesses the property that its travel time, proximity index and complexity are better than the corresponding mean values. In contrast, the route P_9 does not have this property in Scenario 3 because its travel time (667 s) is worse than the corresponding mean value

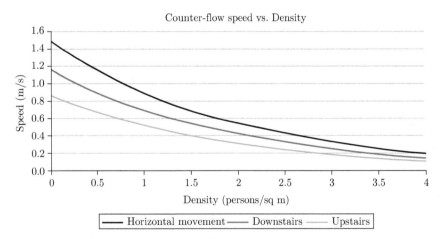

Fig. 3.26 Graphical representation of formula (3.2).

Table 3.7 Results of PIR-Algorithm2 for Scenarios 1–4 (moving upstairs).

Scenario 1					Scenario 2				
Length	Time	PI	C	AS	Length	Time	PI	C	AS
154.61	150	2.09	6.67	0.0854	154.61	235	1.36	6.67	0.0853
222.79	195	4.88	9.28	0.0901	222.79	291	3.26	9.28	0.0902
206.52	184	4.22	8.25	0.0806	206.52	276	2.81	8.25	0.0813
183.30	169	4.18	7.98	0.0901	183.30	258	2.75	7.98	0.0897
206.38	184	4.22	8.24	0.0807	206.38	276	2.81	8.24	0.0814
222.58	195	4.82	9.00	0.0891	222.58	291	3.22	9.00	0.0892
168.65	159	1.86	6.37	0.0752	168.65	247	1.21	6.37	0.0748
212.40	188	4.02	7.73	0.0767	212.40	282	2.68	7.73	0.0769
186.42	171	4.39	7.31	0.0967	186.42	261	2.89	7.31	0.0953
199.07	179	3.71	6.99	0.0792	199.07	273	2.51	6.99	0.0797
159.43	153	2.01	6.38	0.0830	159.43	239	1.31	6.38	0.0828
208.84	186	3.57	7.02	0.0732	208.84	282	2.42	7.02	0.0735

Scenario 3					Scenario 4				
Length	Time	PI	C	AS	Length	Time	PI	C	AS
185.85	604	0.71	7.09	0.0859	173.30	377	0.94	7.32	0.0818
222.79	712	1.31	9.28	0.0894	222.79	432	1.89	9.28	0.0836
206.52	682	1.12	8.25	0.0790	206.52	421	1.60	8.25	0.0724
206.38	682	1.12	8.24	0.0791	183.30	406	1.54	7.98	0.0777
194.88	656	1.09	7.47	0.0867	183.28	406	1.59	8.06	0.0799
222.58	712	1.29	9.00	0.0883	206.38	421	1.60	8.24	0.0725
168.65	607	0.48	6.37	0.0814	222.58	432	1.87	9.00	0.0826
212.40	688	1.07	7.73	0.0766	168.65	388	0.58	6.37	0.0706
186.42	667	1.13	7.31	0.0879	212.40	425	1.54	7.73	0.0703
199.07	674	1.07	6.99	0.0839	186.42	408	1.63	7.31	0.0848
159.43	617	0.51	6.38	0.0775	199.07	415	1.48	6.99	0.0760
208.84	664	1.05	7.02	0.0843	159.43	382	0.62	6.38	0.0767
					208.84	421	1.43	7.02	0.0711

(664 s). Also, in terms of standard deviations, the proximity index of P_9 is further away from the best value in Scenario 3 compared to other scenarios. Hence, the aggregate score of this route in Scenario 3 loses some points, so that the aggregate score of the safest route P_2 becomes marginally better by just 0.0015. In the situation of Scenario 3 when all routes have rather low proximity indices, the AHP decision to choose the safest available route as optimal does look reasonable. This test demonstrates that PIR-Algorithm2 is not very sensitive to different scenarios of uncontrolled evacuation: the optimal route (P_9) for three scenarios is the same as one generated in Section 3.2.2 by PIR-Algorithm. Only in one scenario a different optimal route is produced, but P_9 has a very high aggregate score, hence it can be considered as near-optimal.

Finally, let us apply sensitivity analysis to the speed of rescuers because it is rather variable, depending on visibility, their personal characteristics, whether heavy equipment is needed etc. More precisely, we consider two situations when the speed given by formula (3.2) is increased by 25% and decreased by

25% for the above Scenarios 1–4. Both tests generate the same sets of routes and the same optimal routes. Thus, even though formula (3.2) represents a first approximation of the real speed in a counter-flow, the method demonstrates some robustness with respect to variability in the speed.

Comparison with TDOR-Algorithm

In this subsection, PIR-Algorithm2 will be compared to the time-dependent optimal routing algorithm (TDOR-Algorithm) developed by Park et al. [51]. The main differences are summarized in Table 3.8. As explained in Section 3.2.1, PIR-Algorithm2 is based on the 3D model of a building, which is represented by the logical and unified networks. In addition, it is applicable to any extreme event with epicentres. In contrast, TDOR-Algorithm was only tested for a 2D model, and its applicability is limited to fire/smoke assuming that sensor information is available.

Another important feature of PIR-Algorithm2 is its ability to generate an optimal route depending on user's needs, thus producing a set of optimal

Table 3.8 Comparison of PIR-Algorithm2 with TDOR-Algorithm.

	PIR-Algorithm2	TDOR-Algorithm
Dimension of model	3D (represented by two networks)	2D (tested for one floor only)
Applicability	Any extreme event with epicentres	Fire/smoke only if sensor information is available
Output	A set of routes (the optimal route depends on user's needs)	A single route (not dependent on user's needs)
Route proximity to hazard	It is measured by proximity index	Not measured
Hazard propagation	Yes (based on distances and the number of obstructions)	No (sensor information is needed)
Link proximity to hazard	Yes	Yes
Travel time	Yes	Yes, but along a flow only
Route complexity	Yes (based on five attributes)	No
Speed along a flow	Yes	Yes
Speed in a counter-flow	Yes	No
Dynamic aspects	Partial	Partial
Reliability/robustness	It is implied by the AHP, which possesses these properties	It has to be proved yet
Sensitivity	Insensitive to different parameters and scenarios of evacuation	It has to be proved yet

routes. The hazard proximity of a route is measured by the proximity index. TDOR-Algorithm generates a single route, and no measure of its hazard proximity is given. The algorithms determine link proximity to the hazard differently: the former is based on hazard propagation, whereas the latter relies on sensor information. Further advantages of PIR-Algorithm2 are that the route complexity is taken into account as well as the formula for an agent's speed in a counter-flow. Both algorithms are applicable to various scenarios of evacuation, and hazard epicentres can be easily updated if sensor information is available. However, the dynamics of hazardous events should be developed further for situations without sensors. Finally, PIR-Algorithm2 is reliable and insensitive to different parameters and evacuation scenarios as shown in previous sections, whereas the same properties for TDOR-Algorithm have not been demonstrated yet.

In the following example, we illustrate the feature of PIR-Algorithm2 to generate an optimal route depending on user's needs. Let us consider Scenario 4 from the previous section when moving upstairs. We introduce five epicentres as shown in Figure 3.27, where the left red cylinder represents two epicentres on different floors. The destination node q is on the seventh floor in the middle part of the building. Because there are many epicentres, we set $\rho_{\max} = 200$. Depending on the user's preferential ranking, PIR-Algorithm2 produces the following routes shown in Figure 3.27: route Y_1 for TT > HP > RC, route Y_8 for TT > RC > HP, route Y_6 for HP > TT > RC, route Y_9 for HP > RC > TT, route Y_8 for RC > TT > HP, route Y_{11} for RC > HP > TT and route Y_8 for TT = HP = RC. Note that route Y_8 appears several times because it is the simplest route with travel time close to the fastest route Y_1.

To apply TDOR-Algorithm to this scenario, we have to make some assumptions. First, we assume that a correct formula for the movement in a counter-flow is used in this algorithm. Second, the algorithm was only tested in a 2D model, so we assume that it is also applicable to a 3D model with a proper 3D hazard propagation. Under these assumptions, our calculations show that TDOR-Algorithm would produce the route Y_2 (*Time* = 442 s; $PI = 1.42$; $C = 10.27$), which is similar to the route Y_9 (*Time* = 443 s; $PI = 1.43; C = 9.28$) generated by PIR-Algorithm2 with input HP > RC > TT. Note that travel times and proximity indices of these routes are very similar, however the former route is more complex. Thus, PIR-Algorithm2 produces a set of optimal routes depending on user's needs, whereas TDOR-Algorithm only generates a single route. In addition, PIR-Algorithm2 takes into account route complexity and calculates the proximity index, which is important in deciding how dangerous the route is.

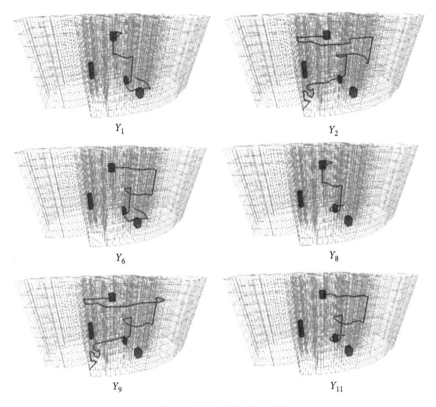

Y_1

Y_2

Y_6

Y_8

Y_9

Y_{11}

Fig. 3.27 Moving upstairs in Scenario 4 with five epicentres.

Let us now consider the test building used for TDOR-Algorithm, which is shown in Figure 3.28. This is a 2D model of the Student Hall at the University of Seoul with three exits A, B and C. In this scenario, the rescuers should find the optimal route from one of the exits to the place located in the corridor on the right side of exit B. However, the rescuers can only use exits A and C. The hazard is represented by smoke, detected by sensors, and its epicentres are marked by red circles. If all nodes in a room are epicentres, then the entire room is coloured in red. Under this scenario, TDOR-Algorithm generates the single route F_2 shown in Figure 3.28. If the travel time is the most important criterion, for example there is protective equipment, then the optimal route produced by PIR-Algorithm2 is the route F_1 depicted in Figure 3.28. This situation corresponds to the user's input TT > RC > HP or TT > HP > RC. However, if the hazard proximity is the most important attribute (HP > RC > TT or HP > TT > RC), then the optimal route is F_2, which agrees with TDOR-Algorithm. In the situation when the route complexity is the most important, or the three criteria are equally important, the optimal route is

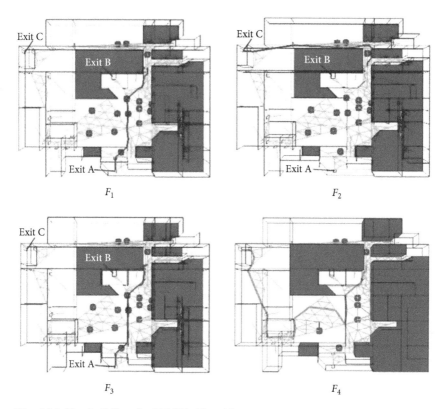

Fig. 3.28 Test building for TDOR-Algorithm.

F_3. It may be pointed out that with fewer epicentres and when the hazard proximity is the most important, another rather unexpected optimal route F_4 is generated; it is shown in Figure 3.28. This can be explained by the fact that all other routes go through the epicentres or are extremely close to the epicentres. The limitations of this scenario are that it is a simple 2D model, where there are relatively few possible routes from the exits to the destination. Nevertheless, it is still possible to see that even in this simple scenario PIR-Algorithm2 is able to produce the optimal route depending on user's needs.

3.3 Automated Construction of Variable Density Navigable Networks

Destructive disasters caused by events such as fire, storm or explosion result in damages and structural instability of buildings. Such disasters are rare but inevitable, and in the urban environment with high-rise and big public buildings they lead to casualties and occupants being trapped inside. In this type of crisis, the challenge is to assist the search and rescue personnel in quickly preparing a plan to locate and rescue the surviving victims. Disasters such as the fire at Villagio shopping mall in Doha in May 2012, where 19 people were killed including 13 children, are reminders that effective preparedness and response might eliminate or reduce human errors and panic which lead to the loss of lives [58].

Tashakkori et al. [63] emphasized the importance of situation awareness available for first responders in emergency response systems. The knowledge about indoor structure, occupancy or location of fire utilities is minimal prior their arrival to the emergency scene. An assessment of the scene and wrong decisions increase the response time. Therefore, actual and constantly updated information about the indoor situation, nearby outdoor area and optimal routes to selected points inside a building is crucial. This information includes, to name a few: floor plans, material types, location of hazardous materials, occupancy information, space ownership, location of water resources, fire and emergency utilities etc. Such up-to-date information is needed for coordination, communication and efficient decision-making in an emergency [16]. Availability of information about the indoor structure is critical for simulation of people movement and egress way finding for individual pedestrians and groups of people [48, 52]. Usually, the focus is put on evacuation paths, flow of people and time necessary to leave a building [34]. However, the opposite paths from outdoor to trapped people at certain locations in a building are taken by search and rescue teams. In such a scenario, rapid identification of where first responder and occupiers are located is of great importance in search and rescue operations [42]. In case of emergency, standard pathfinding using normal navigation routes may not be sufficient as they may be too risky or not available due to damage [45]. In these circumstances, the ability for emergency and rescue personnel to identify alternative and optimal paths becomes critical.

In Section 3.1, we discussed a new method for calculating safe paths which takes into account hazard proximity. Standard navigable networks are not sufficient to perform hazard analysis. They include only connections available for pedestrians, whereas topological relations between adjacent rooms without doors in between are not reflected. This applies not only for adjacency in horizontal direction but also vertical through slabs. Indeed, these connections are important in simulation of several hazard types such as fire spread or explosion impact. Thus, 3D properties included in a model (e.g. spatial relationship in all directions) are essential for emergency response applications which go beyond evacuation pathfinding.

This section discusses the development of a new automated method for deriving a navigable network in a 3D environment, including a full 3D topological model, which may be used for finding alternative egress routes and simulating phenomena associated with emergency situations. The motivation of this study is to provide better situation awareness to the first responders prior to their arrival to the scene.

Problem Statement

Figure 3.29a illustrates a close-up of a single level (the ground floor) in the building shown in Figure 3.4. The corridor (the grey area) is connected to adjacent rooms by doors and to the outdoors by exits A–C; there are three staircases S1–S3. The logical network representing connections among rooms at the ground floor is shown in Figure 3.29b. Other vertical connections between adjacent rooms at different levels are also included in the logical network (see Figure 3.4c). In a non-emergency scenario, only standard navigation routes via doors are used. These routes are represented by a sub-network of the logical network shown in Figure 3.29c.

The network can be effectively used for navigation routes in simple models; for example, if it consists of 'cubical' rooms with one door or two doors

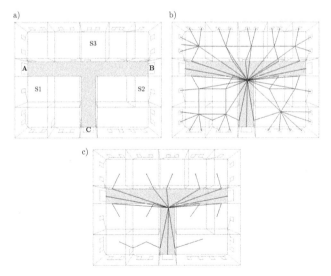

Fig. 3.29 Ground floor of the building represented with DHE: a) grey area represents a corridor with three exits A–C; S1–S3 are staircases; b) logical network; c) sub-network reflecting navigable connections through doors.

a)   b)

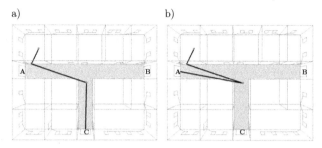

Fig. 3.30 Paths (black lines) from a selected room through the corridor (grey area) to one of the exits (A–C are exits from a building): a) C is wrongly calculated as the closest exit; b) A is the closest exit but the path is not a 'natural' way of navigation.

located on opposite walls. A problem arises when rooms have a complex shape and many doors (e.g. long corridors, concave rooms, big open spaces). If the shortest path from a room to the closest exit goes through a corridor node located in the geometric centre, an incorrect exit may be selected (see Figure 3.30a). In reality, the shortest path between the graph nodes is calculated properly, but navigation routes within a corridor are not represented correctly. In addition, if the pathfinding algorithm is forced to select the correct closest exit (A in Figure 3.30b), then the path is not a 'natural' way of navigation—going to the centre of a corridor first and then taking the correct exit. A solution to the problem is to enhance the model by providing a better representation of 'natural' ways of indoor navigation.

The closest model is the one proposed by Liu and Zlatanova [45], where the logical network representing a building structure is combined with navigable networks within single rooms: the network generation algorithm is based on the location of doors and concave corners. However, the navigable network is built from a spatial subdivision, the Voronoi tessellation, which is related to the method proposed by Lee [38].

In this section, we study a novel approach of navigable network generation, the Variable Density Network (VDN), based on the Voronoi Diagrams (VD). This approach allows a new automated method for deriving a navigable network in a 3D indoor environment, including a full 3D topological model, which may be used not only for standard navigation but also for finding alternative egress routes and simulating phenomena associated with disasters such as fire spread and heat transfer. The main application for the proposed network is calculation of egress paths in a dangerous environment to assist rescue teams.

3.3.1 Methodology and the Algorithm

In this research, two mock data models and a real-life case study were created to illustrate the proposed new method for generation of VDNs. The first model is a high-rise office building with cubical rooms connected to a centrally located corridor (see Figure 3.31a). It is used to illustrate how the method works in narrow corridors of a complex shape with several doors to adjacent rooms. The second model is a two-storey building with a central big open space (see Figure 3.31b). It is used to show applicability of the same method to open spaces. The third model represents a floor plan of the Doha World Trade Center in Qatar. This model is selected to carry out a comparative study between the proposed VDN and other prevailing models. The models were built in the Autodesk Revit design environment and exported to the gbXML data format. A gbXML file was imported and analysed using in-house software developed for testing purposes. Models are represented using the DHE data structure, which is an integral part of the in-house system. Some models may require validation in order to preserve consistency between geometry and topology which, in particular, concerns detecting and fixing two problems: missing faces and overlapping between adjacent rooms. The workflow is shown in Figure 3.32.

The model enhancement, discussed in this section, introduces a novel network for indoor navigation in spaces along navigation routes, such as corridors, big open spaces and rooms with several doors. A floor is tessellated using a VD construction algorithm, where doors and concave corners are initial Voronoi vertices (nucleation points), and additional points are added to create a denser tessellation. The nucleation points from adjacent cells are connected by links, thus forming a network, which is better suited for navigation, than the network without such innovation.

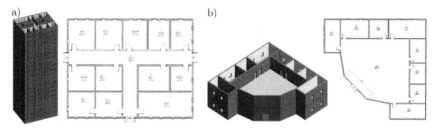

Fig. 3.31 Mock data models—3D views and ground floor plans: a) high-rise building; b) two-storey building.

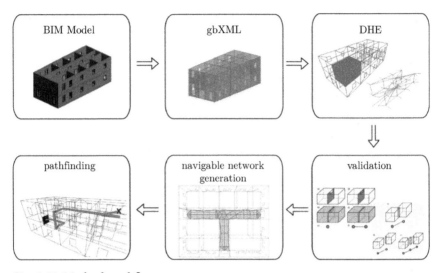

Fig. 3.32 Method workflow.

There are several advantages to this strategy:

1) Pathfinding algorithms for the additional network give more accurate results within a single room.
2) The additional network is generated using a 2D algorithm and does not affect the original network representing the 3D topology of a building.
3) The same method may be applied for corridors and big open spaces of an arbitrary shape.
4) 'Natural' paths for human movement are reflected in the network.

The proposed method consists of the following steps:

A) Spaces (called *rooms* in the following description) with more than one door are selected.
B) The floor of the selected room is detected based on semantic information from the original BIM model or, if this information is not available, then surfaces with the normal vector pointing upwards are considered as a floor.
C) Dual nodes representing doors are projected on the floor and together with concave vertices of the floor surface are selected as nucleation points for the tessellation algorithm—they are called *constraint nucleation points* in the following description.

D) The VD algorithm is performed iteratively and additional points are added; the nucleation points of neighbour cells are connected by links forming a network.

E) The locations of additional nucleation points occurring on constraint edges are moved towards the geometric centre of the associated cell, while the original location of the constraint points is not changed.

The process of the VD construction (see Step D above) requires a detailed description. An iterative algorithm for the constraint VD was implemented based on the Green and Sibson algorithm [28], see Figure 3.33.

The floor is represented by a polygon of an arbitrary shape. The original polygon edges are constraint edges, which are not modified in the tessellation process. In the first iteration, dual nodes, representing doors projected on the floor, and concave corners are selected as constraint nucleation points. The locations of the constraint points are not changed later. The nucleation point is always enclosed by a single Voronoi cell. Subsequently, each node is connected by links with nodes from adjacent cells; all these links and nodes form a network. In the next step, new nucleation points are added at the middle of each link if the link is bounded by constraint points, or if both bounding points lie on constraint edges, which are not co-linear. The process is repeated until no new points may be added because they would be too close, or a certain number of iterations have been reached.

The algorithm for VD construction is presented below as Algorithm 3.3. In the examples considered here, the following input parameters were used: the threshold $T_1 = 6$ (see Figure 3.34c and Figure 3.35b), the threshold $T_2 = 0.7$ (see Figure 3.36) and the maximum tessellation level $TL_{max} = 10$. The threshold T_2 is used if an edge e in the tessellation is close to the hazard

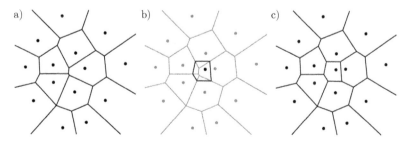

Fig. 3.33 Process of Voronoi Diagram construction: a) existing tessellation; b) new vertex and its Voronoi cell; c) tessellation after point insertion.

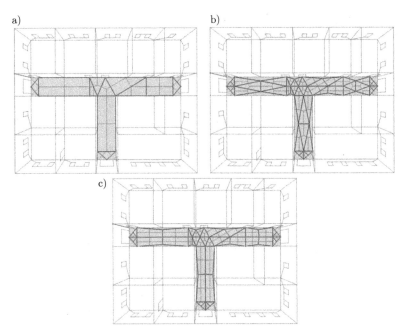

Fig. 3.34 Floor tessellation (grey cells) and associated network (black lines): a) first iteration; b) second iteration; c) final tessellation.

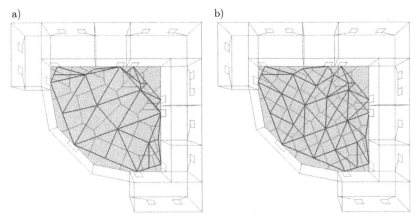

Fig. 3.35 Tessellation of big open space (grey area): a) tessellation without a maximum link length threshold; b) tessellation with a maximum link length threshold (in this case 6 m).

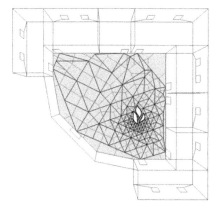

Fig. 3.36 Local densification of the navigable network around a hazardous area.

location. In this study, the edge e is considered to be close to the hazard location if

$$3 \times \text{length}(e) > \text{distance}(u, h) + \text{distance}(v, h),$$

where u and v are end-points of e and h is the hazard location.

For the sake of algorithm clarity, only one polygon representing a floor is considered. If a floor is represented as a set of adjacent polygons in an original model, they may be merged into one and the same algorithm can be used.

Algorithm 3.3: Navigable network generation (tessellation of a floor).

Input:	The cell representing room R.
	The floor polygon for R.
	Threshold values for edge length: T_1 and T_2.
	Maximum tessellation level $TL_{\max}$.
Output:	Tessellation f.

(1) Make a copy of the floor polygon, denoted by f.

(2) Calculate the number of doors in room R, denoted by d. If $d \leq 1$, then stop the algorithm (no network generation is necessary).

(3) For each door dual node, make its projection on the boundary of f.

(4) Create a list N consisting of constraint nucleation points for VD in order of their appearance on the boundary of f. The constraint nucleation points are door nodes and concave corner nodes.

(5) Create a list V of VD cells and set $V_1 = f$. (The floor polygon f is the initial Voronoi cell associated with N_1.)

(6) Set $TL = 1$, where TL is the current tessellation level.

(7) Take the first unprocessed element from N, denoted by N_j.

(8) Create a list of cells for testing, denoted by CT, and add to it a cell from V, which includes node N_j.

(9) Take the first unprocessed element from CT and denote it by V_k (it is associated with N_k).

(10) Calculate a bisector line bi of the segment (N_j, N_k).

(11) Calculate intersection points of bi with V_k. The set of such points is denoted by IP.

(12) If the number of points in IP is greater than 2, then do the following:
   (a) Find in IP two intersection points whose line segment divides the cell V_k into two parts such that each part includes one of the nodes N_j and N_k.
   (b) Remove the rest of the intersection points from IP.

(13) If the number of points in IP is less than 2, then go to Step 9.

(14) Create the edge (j, k) bounded by intersection points from IP, and insert it into the existing tessellation of the polygon f. The edge (j, k) divides V_k into two cells, C_j and C_k, such that N_j is located in C_j and N_k in C_k.

(15) Add C_j to V (i.e. $V_j = C_j$) and set $V_k = C_k$. (The edge (j, k) is part of the boundary of the newly created cell for N_j.)

(16) Add those cells to the list CT, which are adjacent to the edges intersected by the edge (j, k), taking into account the following:
   (a) If an intersected edge is a constraint edge (the initial boundary of f), then add all cells adjacent to V_k.
   (b) If a cell is already in CT, then do not add it.

(17) If there are unprocessed elements in CT, then go to Step 9. (Otherwise, the boundary of the new cell for N_j was created.)

(18) Remove all edges from inside the boundary of the newly created cell for N_j. (This is illustrated in Figure 3.33, the transition from b) to c).)

(19) If there are unprocessed elements in N, then go to Step 7. (Otherwise, the current level of tessellation is completed.)

(20) If $TL < TL_{\max}$, then for each edge e connecting nucleation points of adjacent cells in VD, do the following. Calculate the bisector point B for e only if:
   (a) e is longer than T_1; or
   (b) e is close to the hazard location and it is longer than T_2; or

(c) e is bounded by two constraint nucleation points (doors or con-cave corners); or

(d) both end-points of e lie on constraint edges, which are not co-linear. Add the point B into N.

(21) If $TL < TL_{\mathrm{max}}$ and there are unprocessed elements in N, then $TL = TL + 1$ and go to Step 7.

(22) If there are nucleation points in N, other than constraint nucleation points, which are located on constraint edges (the initial boundary of f), then move locations of those points towards the centre of the associated cells.

(23) Report f. Algorithm stops.

The first iteration of the tessellation (grey cells) and the corresponding network (black edges) are shown in Figure 3.34a. Because the tessellation is not dense, the network consists of few links. Many of them lie on constraint floor edges. Therefore, several navigation routes go from door to door along walls. The process stopped after the second iteration generates shorter links (see Figure 3.34b). Shorter links are the result of new point insertion in the midpoint of the original link, which leads into generation of a Voronoi cell with associated dual edges (i.e. shorter links) connecting the new cell with neighbour cells. The locations of new nucleation points occurring on constraint edges are modified by moving them towards the centre of the associated cell (see Step E), which overcomes the problem of navigation along walls. However, there are some links in the network, which are perpendicular to the walls, and their bounding points occur on constraint edges. Such links may produce sharp turns on navigation routes. Additional nodes and links are added in the next iterations, thus solving this problem. The final result is shown in Figure 3.34c. The tessellation and corresponding network are attached to the dual node representing the room. The network is used for precise path calculation within a room.

An advantage of the proposed solution is that it can be applied not only to corridors but also to any room shape (e.g. irregular spaces). This is an important property, which helps to avoid the need for classifying a room as a corridor or an open space, and then performing different algorithms for network generation. An example of an indefinable area is a narrow corridor which, at some point, turns into a bigger area (a hub) with many doors or a junction with other corridors.

However, in this scenario, long links in the network may be introduced because nucleation points in the central area of a room are enclosed by relatively bigger cells and the distance to neighbours is bigger (see Figure 3.35). This may produce 'wobbling' paths. Consequently, for links that are longer than a specified threshold, new nucleation points are introduced in the midpoints of these links.

The higher density of the network is helpful for finding alternative or safest paths when a hazardous event takes place within an open space: some nodes may be prohibited for navigation, while the rest of the space with a lower risk is still considered as safe. This applies to buildings such as airports or shopping malls.

The network might be densified around a dangerous area using the same method. For links located closer to the hazard source, a smaller threshold for links is used. Therefore, new nucleation points are introduced in the midpoints of these links and new shorter links are generated. This operation provides local network densification (see Figure 3.36). Densification is commonly applied in the finite element mesh generation and refinement [20, 31, 53].

3.3.2 Results and Discussion

The main reason for introducing additional nucleation points is to improve the network for a pathfinding application. This increases the number of cells in the tessellation, thus the number of network links is increased and they are shorter. If the tessellation process is finished after the first iteration, a path between two locations goes through constraint points, which is not a 'natural' route for navigation—it is like walking from door to door along walls (see Figure 3.37a). The second iteration produces a network better suited for navigation, but the path is slightly 'wobbling' (see Figure 3.37b), mainly because there are links that cross the corridor, perpendicular to the constraint edges. The path based on the final network is still affected by 'wobbling' but less than in the previous iterations (see Figure 3.37c), and it is acceptable in the presented study, since a very precise path calculation is not required.

The proposed method solves the problem of incorrect paths shown in Figure 3.30. When the additional network within the corridor is used, the result of the shortest pathfinding algorithm is correct (see Figure 3.38). The egress route from the selected room goes through the door straight to the nearby exit.

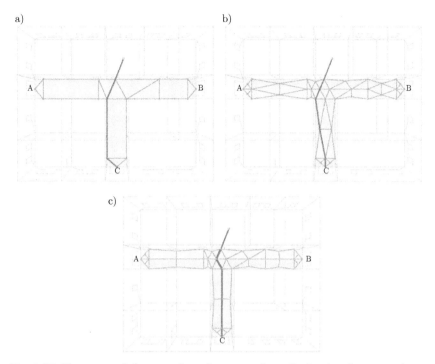

Fig. 3.37 Shortest path from a selected room to door C: a) using the network produced by the first iteration; b) using the network produced by the second iteration; c) using the final network.

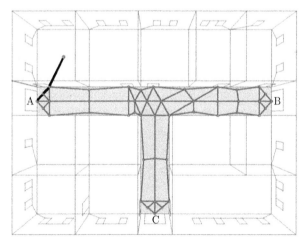

Fig. 3.38 The shortest path from a selected room to exit A (A–C are exits from a building) based on the original network and the network produced by the corridor floor tessellation.

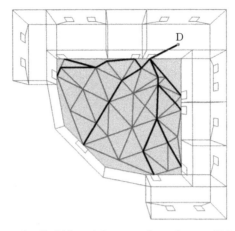

Fig. 3.39 Shortest paths (bold lines) from a selected room (D) to all doors of an open space.

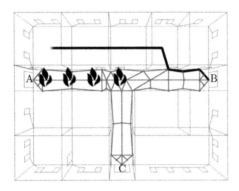

Fig. 3.40 Alternative pathfinding.

A satisfactory improvement is also achieved in the case of open spaces. All shortest paths from a selected room to all doors of the open space are shown in Figure 3.39. The network within the open space is a better approximation of 'natural' navigation routes than the original logical network.

The logical network may be used for alternative pathfinding in case of emergency, when the conventional navigation routes are blocked. In the hypothetical scenario shown in Figure 3.40, a section of the corridor may be on fire. Therefore, it is not safe to use a conventional route to leave the selected room and reach exit A as a result of fire behind the door. Thus, an alternative path is calculated, which goes through walls where a hole can be possibly made. Once the room with a door located outside the dangerous area is reached, a conventional egress route to exit B is used.

Alternative paths can be found automatically if there is no safe connection from a room through doors. A connection from the logical network with the smallest weight is selected. The weight for each wall in the room may differ depending on a construction material. Therefore, walls that are easier to demolish are selected as candidates for navigation; for example, partitions made of plaster panels are favoured over walls made of concrete.

It is important to note that windows may be considered as alternative exits if they are located close above the external ground, for example windows on the ground floor. In addition, rescue teams are able to get inside the building through windows located high above the ground using ladders. In order to show navigation through walls using the logical network, windows were not allowed to be used in the illustrated simulation.

3.3.3 Comparison of VDN with Prevailing Indoor Navigable Networks

Two state-of-the-art methods of the navigable network reconstruction used in emergency response research [38, 44] were selected for a comparison. The reason for this selection is the close relationship to the VDN method discussed here. Three configurations of rooms shown in Figure 3.34, Figure 3.35 and Figure 3.41 are used as examples for this comparison: T-shaped corridor (also see Figure 3.42a, d, g and j); open space (also see Figure 3.42b, e, h and k); and Z-shaped corridor (also see Figure 3.42c, f, i and l), respectively. The first two examples are mock data, while the latter is a real building—the World Trade Center in Doha, Qatar.

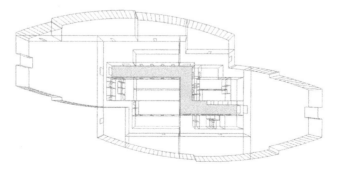

Fig. 3.41 Floor plan—the World Trade Center in Doha, Qatar. Grey-shaded corridor is used for comparison with other methods.

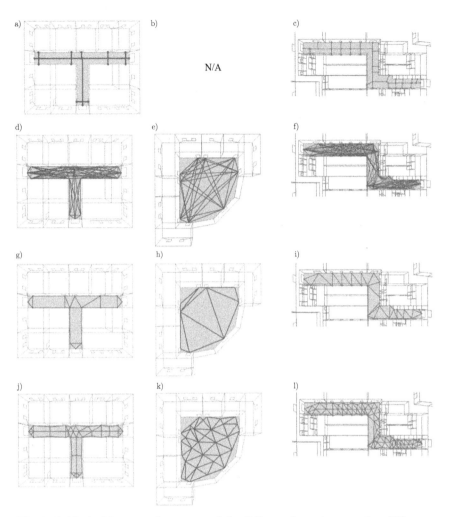

Fig. 3.42 Navigable networks generated for different floor shapes using different methods: a)–c) NRS; d)–f) DtD; g)–i) VDN (1st level); j)–l) VDN (maximum level with the threshold).

Lee [38] proposed the Node-Relation Structure (NRS) as the topological model. The Straight Medial Axis Transformation is applied for generating a navigable network within a corridor. A skeleton of the corridor floor is calculated; this is based on VD construction. The result is a linear representation of a polygon—a 'backbone' representing a navigation route through the corridor. Nodes representing rooms adjacent to the corridor with a door in between are connected to the skeleton by additional links. The advantage of this method is that it is a simple network with a relatively small number of links (see Table 3.9).

Table 3.9 Number of nodes and links in navigable networks generated using different methods.

	NRS [38]	DtD [44]	VDN (1st level)	VDN (max level)
T-shaped corridor	20 nodes 20 links	16 nodes 77 links	16 nodes 26 links	45 nodes 84 links
Open space	N/A	9 nodes 36 links	9 nodes 15 links	32 nodes 74 links
Z-shaped corridor	47 nodes 46 links	30 nodes 187 links	30 nodes 56 links	85 nodes 190 links
Comments	*NRS*: Simple network with a small number of links; limited applicability to open spaces. *DtD*: Non-planar graph; minimal number of links on door-to-door routes. *VDN*: Variable number of links depending on densification level; local network densification possible.			

However, the applicability of NRS to big open spaces was not presented, possibly because of unsatisfactory representation of navigation routes. Consequently, it is suggested that it is not practical to use the aforementioned method for spaces other than simple-shaped corridors. Additionally, because the skeleton is located in the middle of the corridor and the adjacent rooms are connected to the skeleton by perpendicular links, the network does not reflect 'natural' ways of navigation—if two doors are coplanar, that is, located on the same wall, one must go from the first door to the middle of a corridor, turn $90°$, walk in the middle of the corridor towards the second door, turn $90°$ again and then get to the second door. This may be not a significant drawback, especially for narrow corridors when the shortest paths are calculated. However, as complexity increases, it may become significant in computation of the simplest paths, where the sharpness of turn angle is considered.

The door-to-door (DtD) method [44] is based on an idea of navigation from a selected door directly to another door if two doors are in direct visibility. If there is no direct visibility, then intermediate concave corridor corners are introduced to the path. Taking all possible connections among all the doors within a corridor, including concave corners, the navigable network consists of more links than NRS (see Table 3.9). The network is a non-planar graph in the considered examples. The method may be easily applied for big irregular spaces.

These two methods produce a fixed network depending on the geometry of a room and location of doors. This may be a drawback when alternative paths within a big room are necessary to calculate, for instance to avoid hazard located in the room. This is critical when big open spaces in a shopping mall or airport are considered. In contrast, the proposed VDN method is more flexible, mainly because the density of tessellation may be easily increased resulting in more links in the network and thus more flexible routes (see Table 3.9).

It should be noted that the number of links calculated for the DtD method is based on the following assumption: links connecting the same nodes are not duplicated in the network. For example, considering three doors $d_1 - d_3$ and one concave corner c_1, if doors d_1 and d_2 are in a direct line of sight and d_3 is connected with d_1 and d_2 through c_1, then there are four links: (d_1, d_2), (d_1, c_1), (d_2, c_1) and (c_1, d_3).

It should be also noted that, in the NRS construction, the skeleton of the corridor is calculated based on Straight Medial Axis Transformation. Door nodes are connected to the skeleton by line segments, which are perpendicular to the skeleton, and new nodes at an intersection point are added to the skeleton. However, if the distance from a door to an existing node in the skeleton is shorter than the perpendicular connection, the door node is connected to that node. Conversely, in order to avoid very short links in the network, a door node is connected to an existing node if a new intersection point is located within a range of 10 cm from the existing node.

In order to compare the precision of different methods for tessellation, we consider the main corridor of the actual floor plan of the Doha World Trade Center (see Figure 3.41). There are 27 doors in the corridor as depicted in Figure 3.43. Ten pairs of doors were randomly generated, which represent ten random possibilities for moving from one door to another in the corridor (see Table 3.10). For each pair of doors, the length of the actual walking path between them was calculated as well as the lengths of paths generated by

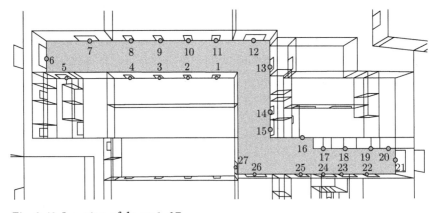

Fig. 3.43 Location of doors 1–27.

Table 3.10 Ten random routes and their lengths (metres) for different methods.

Path	From Room	To Room	Actual Route	NRS	DtD	VDN (1st level)	VDN (max level)
P_1	11	23	18.8	24.1	18.0	19.3	19.5
P_2	7	9	7.4	10.1	6.6	6.6	6.8
P_3	5	16	28.1	34.3	25.4	26.0	27.3
P_4	17	12	15.7	18.8	14.1	14.8	16.1
P_5	15	20	14.1	18.1	12.0	12.2	13.2
P_6	25	8	22.7	28.3	21.4	23.1	23.1
P_7	10	26	15.7	18.7	14.8	17.1	15.6
P_8	20	18	5.1	6.2	4.1	4.1	4.2
P_9	9	1	6.8	8.9	6.4	7.1	6.6
P_{10}	22	2	22.9	29.0	20.6	22.0	22.6

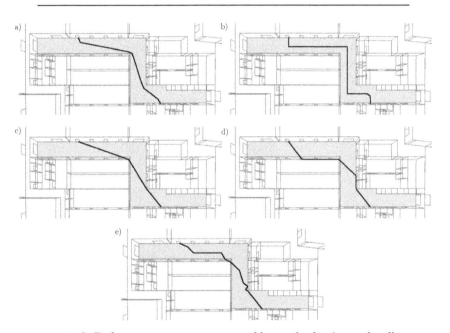

Fig. 3.44 Path P_6 from room 25 to 8 generated by methods: a) actual walking path; b) NRS; c) DtD; d) VDN (1st level); e) VDN (maximum level with the threshold).

the NRS method, the DtD approach and the VDN method for the first and maximal levels of tessellation. The corresponding results are summarized in Table 3.10. Examples of path P_6 generated by the aforementioned methods are shown in Figure 3.44.

The actual path, which is a point of reference in this comparison, was generated manually considering the following rules: the path should not run closer than 0.8 m to the concave corners (see Figure 3.45a); the next node after a door node is located 0.8 m from the door frame into direction of the next move (see Figure 3.45b); the direction of door opening is not taken into consideration.

Furthermore, three standard errors were calculated: the mean error (ME), the mean absolute error (MAE) and the mean squared error (MSE). Note that the former represents the bias, whereas the last two refer to the accuracy of methods. The results are given in Table 3.11. The NRS method very often overestimates the length of the actual path, which is reflected in the ME equal to -3.9 m, whereas the DtD approach tends to underestimate the length with $ME = 1.4$ m. On average, there is some underestimation in the VDN method with the $ME = 0.5$ m and 0.2 m for the first and maximal levels. The MAE figures for the above methods are similar to MEs: 3.9 m, 1.4 m, 1.0 m and 0.5 m, respectively. The same numbers expressed as percentages of the MAE to the average length (15.7 m) of actual routes are as follows: 24.8%, 8.8%, 6.4% and 3.3%, thus the 5% standard threshold for errors is achieved by the maximal level of tessellation in the VDN method. Finally, the MSE measure, which penalizes large errors, is rather high (18.1) for the NRS method, whereas the DtD and VDN approaches exhibit small MSEs as can be seen in Table 3.11. Thus, the VDN method with the maximal level of tessellation demonstrates the best characteristics for the bias and accuracy.

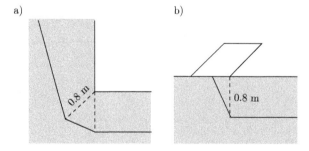

Fig. 3.45 Actual path generation: a) distance from a wall; b) distance from a door.

Table 3.11 Errors of the four methods.

	NRS	DtD	VDN (1st level)	VDN (max level)
ME	−3.9	1.4	0.5	0.2
MAE	3.9	1.4	1.0	0.5
MSE	18.1	2.4	1.3	0.4

3.4 Related Work

The typical research in emergency response is devoted to evacuation and rescue, with a focus on indoor navigation and route finding.

Kwan and Lee [35] investigated possible improvements of navigable networks for the particular purpose of facilitating quick emergency response to terrorist attacks in the integrated system of the ground transportation system and multistory office buildings. They concluded that extending the standard 2D GIS (Two-Dimensional Geographic Information System) to a real-time 3D GIS has 'considerable potential for improving the speed of emergency response after terrorist attacks on multilevel structures in urban areas'. Meijers et al. [47] developed a semantic model representing interior spaces in a building. Their model can be used for an intelligent computation of evacuation routes.

Lee [39] reviewed 3D models and building evacuation models and developed a pedestrian-based indoor navigation model using 3D GIS. The human behaviour was studied by Choi and Lee [15] using a social force model. Recently, an advanced configurable crowd model for different behaviours and scenarios was developed by Sun and Wu [62], in particular the simulation of evacuation in a building was implemented. Liu and Zlatanova [44] proposed a new door-to-door approach for finding routes between rooms and also a detailed route in a single room. Their algorithm was tested on a 2D floor plan of a building with complex indoor structure. Vanclooster et al. [65] developed a capacity-constrained flow algorithm on a 3D geometric network model. Also, Vanclooster et al. [66] applied Grum's least risk path algorithm to an indoor space for minimizing risks of getting lost, and proposed several improvements to Grum's algorithm to make it more compatible with indoor networks. The role of elevators and stairs in efficient evacuation was investigated by Lay [37].

A spatial model always underlies any algorithms for indoor navigation. A good example of such a model was given by Kwan and Lee ([35], Figure 5). The structure of a building is represented as a logical network, where the nodes represent spatial objects such as rooms, corridors and other navigable areas. The edges represent navigable connections between adjacent objects. The network can be further extended to a geometric network to model precise geometric properties (e.g. distance between nodes and their locations) and provide real navigation routes. This representation can be used for graph algorithms such as Dijkstra's or A^* algorithms for finding shortest routes in evacuation planning (Dijkstra [18]; Hart et al. [30]). An advanced and efficient strategy for the shortest path problem with uncertain travel cost can be found in Shahabi et al. [59]. Their approach might be particularly relevant for evacuation algorithms in the built environment because the distribution of people in a building during an extreme event is typically unknown, so that the link travel time function is uncertain.

As a rule, 2D floor plans are used for reconstruction of horizontal navigable networks, and a 3D building is obtained by linking contiguous floors at some connection points, for example staircases. Spatial relationships between the rooms in the vertical direction are not reflected in the model. This solution is sufficient for a simple analysis of indoor human movements, but other important phenomena related to emergency response, for example fire spread or heat propagation, cannot be simulated in such models. Also, many 3D navigation models represent buildings with a simple, often regular, structure, which is not sufficient for complex interiors. Liu and Zlatanova [45] proposed an interesting concept of automatic navigable network generation based on the geometry and semantics of a building. Their method requires a valid spatial model with preserved consistency between the geometry and topology, which is not always readily available.

In their pioneering work, Kisko and Francis [32] proposed the EVACNET+ software application for evacuation scenarios planning. By formulating the task at hand as an optimization problem, they developed a user-friendly interface allowing a user interaction for the computation of egress times and the determination of problematic building locations. For the sake of efficient emergency evacuation, workload relaxation techniques from queueing theory were used by Deng et al. [17] to deal with the curse of complexity in large buildings. For this purpose, the authors developed tools for modelling and optimizing the occupancy evolution during evacuation scenarios. They derived complexity lower bounds on the evacuation time and consolidated their findings by providing realistic building simulations. Pursals and Garzon [54] presented a new formulation for the problem of building evacuation through the development of a model for occupants' movement, allowing a good choice of evacuation paths during emergency scenarios. Some important surveys in the area of evacuation are worth mentioning. A review of 16 evacuation models was given by Gwynne et al. [29], and Kuligowski [34] reviewed 28 egress models. Further reviews of fire and evacuation models can be found in Friedman [24], Olenick and Carpenter [50], and Watts [70].

Park et al. [51] developed a time-dependent optimal routing algorithm based on a 2D network representing the building configuration, which has been enriched by relevant information about the facility. The focus of their research is on computing optimal routes leading search and rescue personnel to disaster locations taking into account the location of evacuees and smoke density, in contrast to the approach for selection of evacuation routes leading trapped occupants to main building exit points. The method requires detecting positions of people in a building per time period. For this purpose, an evacuation simulation system for identifying the movement patterns of people in emergency situations was used. Even though the algorithm had considerable potential, the authors concluded that the method needs 'further improvements to fully apply to real-time evacuation systems'.

The simplest path algorithm for a road network was proposed by Duckham and Kulik [21]. The purpose of their method is to 'minimize the complexity of a route description, based on the amount of information required to negotiate each decision point'. It is interesting to note that, unlike shortest paths, simplest paths are neither symmetric nor do they satisfy the triangle inequality. Although the simplest path algorithm was developed for road networks, one of the advantages of the algorithm is that any weighting function can be used in its input as a complexity function. This property was used in this chapter for developing a new complexity function for indoor navigation.

Navigable Models

Available navigable models in emergency response research focus on standard pathfinding using doors and corridors [26, 35, 45, 66, 73]. Most of these approaches retrieve a navigable model from 2D floor plans, however the 3D properties of the original environment are not always reflected in the model, even when the original 3D model is available. Very often, adjacent floors are connected only by vertical connections representing staircases, and spatial relationships between adjacent rooms sharing a ceiling/floor or a wall without a door are not reflected in those models. The resulting networks are often simplified in order to reduce the storage cost and improve pathfinding computation time, which makes them unsuitable for alternative pathfinding or simulation of phenomena related to emergency.

Various researchers proposed more sophisticated models [36, 38, 45, 60, 72], where a complex network is generated for special spaces intensively used for pedestrian navigation such as corridors or complex open spaces. This improves the navigable network within a building because the original logical network, which reflects connections among adjacent rooms, is not an appropriate representation of navigation routes. The proposed automated methods are based on door and concave corner locations and visibility [45], Straight Medial Axis Transformation [38], Delaunay Triangulation spatial subdivision [33, 36] and convex sub-regions partitioning [60]. These methods often require input models with preserved consistency between the geometry and topology.

In the case of incomplete or inaccurate floor plans, a network model may be interactively generated [46], but this may not be efficient in the case of emergency situations when models for decision support should be available in a real or near-real time.

Indoor navigation including emergency navigation models have to meet several requirements [13, 69]. Indoor spaces and their geometry included in an original building model should be classified based on predefined types and they must be represented in the navigable model. Subdivision of spaces is one of the required functionalities, which makes the model suitable for accurate routing computation. Semantic information defining function, use or occupancy should also be reflected and attached to the relevant spaces. Static and moving objects that are obstacles for navigation are other elements which should be incorporated in the model. Finally, connectivity among spaces based on their type has to be derived. These requirements are the most fundamental and are usually supplemented by other, more specific rules.

A slightly different approach to navigable network generation is utilized in robotic motion planning [67]. It is based on the Voronoi Diagrams (VD) [2], where vertices and edges of the VD are used respectively to represent nodes and links in the network. This method may be used in an unknown environment, where the network is dynamically and automatically updated while the indoor scene is explored. Hierarchical structures are used to store networks in different granularity: from the coarsest to more detailed networks reflecting geometric details of the indoor environment. Different scales are used depending on required precision of path planning.

Higher level of granularity may be achieved through a network densification similar to mesh densification utilized in the finite element mesh generation and refinement [20, 31, 53]. Usually, mesh elements (e.g. triangles) are divided into smaller elements based on their centroids and original edges until the required level of granularity is achieved. Different criteria may be considered: the area of a mesh element or edge length. The size of an element is determined in the process (often iterative) of minimizing an error between the exact solution and its finite element approximation [20, 22]. A denser mesh may be generated locally in the vicinity of selected points [14]. Another field where densification of mesh is applied is geo-information. Triangulated Irregular Networks (TINs) can be used to represent a digital model of terrain generated from point clouds, which can be obtained by laser scanning. A TIN is often densified using the Progressive TIN Densification (PTD) method [3, 75]. New points are interpolated into a mesh representing the terrain, thus triggering division of a selected mesh triangle into smaller parts. The selection of a new point and the right triangle is based on specialized filters. Point candidates that do not belong to the terrain (buildings, trees etc.) are calculated from their distance to TIN, and an angle between a mesh facet and a line connecting the point with the closest vertex of the facet.

Other models for indoor navigation systems were presented by Afyouni et al. [1]. The biggest drawback of these methods, in the context of applications other than standard indoor navigation, is that not all spatial relations among adjacent rooms are available in the models. Usually, 2D plans are used to reconstruct a navigable network within an individual building level. A link between two rooms is created only if they share a door. Otherwise, where adjacent rooms share a slab or a wall without doors, adjacency relationship is not reflected in a model. Navigable networks for consecutive floors are connected at certain points by additional links representing staircases [40].

Some studies tried to utilize fully the 3D properties available from model reconstructions [5, 9]. Their methods are based on the Poincaré duality: in practice, the geometry of a volume is represented in the primal space using the boundary representation (B-Rep) [61], while a volume object (e.g. a room) is represented as a dual node. A dual representation of a facet (e.g. a wall) is a dual edge bounded by dual nodes representing adjacent rooms; it is called a *link*. Therefore, the dual structure is a graph of connections among 3D objects. Attributes assigned to links may determine the navigability of the connections. These links with attributes form a subgraph of the initial graph of connections and may be called a network.

The dual half-edge (DHE) proposed by Boguslawski [6] is one of few data structures that are able to store simultaneously both the primal and dual graphs. The construction of a model is realized using CAD-like operators— Euler operators—where only the primal structure is explicitly constructed while the dual is automatically updated [8]. Figure 3.4 provides an example of a building represented with DHE. In the model, doors and windows are represented as zero-volume objects with associated dual nodes, which are part of the graph; windows, which are embedded in a wall, are connected to the wall boundary using 'bridge' edges in order to fulfil B-Rep requirements and avoid disconnections in the graph. A characteristic property of the DHE model is the external volume enclosing the indoor model, which represents outside space. The indoor model can be combined with a terrain model or a transportation network by the external cell [7]. The external volume is also considered in automatic exit detection: a door adjacent to internal and external volumes is marked as an exit. One of the advantages of the DHE representation is that not only links between adjacent rooms in the horizontal direction are present in the model but all possible adjacency relations are reflected including those in the vertical direction. This is an important property, which is primarily used in hazard propagation simulation. The possibility of vertical navigation through staircases is automatically included in the model. Additionally, different weights for pathfinding algorithms may be attached to links going upstairs and downstairs.

Regardless the method used for navigable network generation, in order to share the resulting network and exchange between different applications, it should be represented in a common schema. IndoorGML [49] is a new standard designed for indoor navigation applications, which complements other standards such as CityGML, KML and IFC. It consists of two main data models: topology of indoor environment and indoor navigation, which describe a network topology in built environment.

3.5 Exercises

3.5.1

(a) Let us recall that the graph of connections G is the dual model of a building. Explain why the graph G cannot be used for proper navigation inside the building.

(b) Suppose that for each floor of a building we have a navigable network, and then all such networks are correctly connected together by links representing staircases. Explain why the resulting network cannot be used for proper propagation of hazard (e.g. fire) inside the building.

(c) Assume that we have a navigable network representing a building and we want to use the following approach (which is alternative to hazard proximity numbers): if a link is deemed to be too dangerous (e.g. it is in close proximity to a hazard epicentre), then such a link is forbidden for navigation. Why is this approach not appropriate?

3.5.2

This exercise is devoted to Algorithm 3.1 from Section 3.1.3.

(a) What does the combined network G^+ represent?

(b) Explain the meaning of the parameters $d(v, z_i)$ and $b(v, z_i)$ and how they are calculated in DB_Procedure. What is the Breadth-First Search Algorithm and how is it used in DB_Procedure?

(c) Why are the hazard proximity numbers in HP_Procedure based on the geometric mean of $d(v, z_i)$ and $b(v, z_i)$?

(d) Explain how the hazard proximity numbers $H_i(v)$ and $H(v)$ are calculated. In particular, what is the propagation coefficient?

(e) How are hazard proximity numbers for links calculated? Why is the length of a link used in the formula?

(f) In general, what can be determined using Dijkstra's algorithm (in terms of its input and output)? How is it used in Algorithm 3.1?

(g) What is the main limitation of Algorithm 3.1 in terms of the generated safest route, and how can it be overcome?

3.5.3

In a building, where an extreme event with one epicentre z_1 is occurring, there are two routes between locations p and q:

$$P_1 = (p, a, q) \quad \text{and} \quad P_2 = (p, u_1, u_2, q),$$

where

$$D(p, a) = D(a, q) = 20\,\text{m} \quad \text{and} \quad D(p, u_1) = D(u_1, u_2) = D(u_2, q) = 30\,\text{m}.$$

Apply Algorithm 3.1 from Section 3.1.3 to find the safest (p, q)-route assuming that the following dataset is given:

$$d(p, z_1) = d(a, z_1) = d(q, z_1) = 32\,\text{m},$$
$$d(u_1, z_1) = d(u_2, z_1) = 50\,\text{m}.$$

Also, $\rho_{\max} = 100$ and $b(x, z_1) = 2$ for any $x \in P_1 \cup P_2$.

3.5.4

This exercise is devoted to Algorithm 3.2 from Section 3.1.3.

(a) What is the basic idea underlying Algorithm 3.2? What does the hazard tolerance (t) mean?

(b) How does the binary search work in Step 3 of the algorithm?

(c) What is the proximity index of a route? Why is the harmonic mean used for calculation of proximity indices?

(d) How is the aggregate score calculated for a route? What does the aggregate score mean?

3.5.5

Apply Algorithm 3.2 to the following scenario to find the balanced (p, q)-route in a building, where an extreme event with one epicentre z_1 is occurring. It is assumed that the binary search finds the routes

$$P_i = (p, v_i, q), \quad i = 1, 2, 3, 4$$

with the following parameters:

$$b(p, z_1) = b(v_1, z_1) = 1, \ b(v_2, z_1) = b(v_3, z_1) = 2,$$
$$b(v_4, z_1) = 3, \ b(q, z_1) = 4;$$
$$d(p, z_1) = 10, \ d(v_1, z_1) = 14, \ d(v_2, z_1) = 15,$$
$$d(v_3, z_1) = 24, \ d(v_4, z_1) = 22, \ d(q, z_1) = 20;$$
$$D(p, v_1) = 14, \ D(p, v_2) = 15, \ D(p, v_3) = 17, \ D(p, v_4) = 15,$$
$$D(v_1, q) = 10, \ D(v_2, q) = 15, \ D(v_3, q) = 16, \ D(v_4, q) = 22.$$

Notice that the parameters $D(x, y)$ and $d(w, z_1)$ are in metres.

3.5.6

Briefly describe the differences between Algorithm 3.2 from Section 3.1.3 and PIR-Algorithm from Section 3.2.1. In particular, formulate five complexity attributes and summarize four steps of the AHP.

3.5.7

(a) During evacuation of people from a building, Mr White has to walk 40 m in a corridor until he enters a staircase, where he should go 30 m downstairs. Finally, he has to walk 20 m on the ground floor before reaching the exit. Calculate the total evacuation time for Mr White assuming that the population density in the corridor before entering the staircase is 0.5 people/m^2, the density in the staircase is 2 people/m^2 and on the ground floor 2.8 people/m^2.

Hint: formula (3.1) should be used.

(b) Mr White was given instructions to stay in his office until a rescuer will reach him with necessary equipment. How long approximately will it take for the rescuer to reach Mr White's office assuming that the rescuer uses the aforementioned route, but in the opposite direction?

3.5.8

The people's flow F (on a link) depends on the speed along the link, the population density in the link area (which might affect the speed) and the effective width:

$$F = S\Delta W_e,$$

where the effective width W_e is the actual width of the area (e.g. a corridor) reduced by some distance from walls, door frames and railings: 15 cm for each door frame and staircase bounding rail, and 20 cm from walls.

Suppose that during evacuation of people from a building, you can regulate the population density by controlling the number of people entering the corridor on the ground floor from staircases. Calculate the optimal density such that the corresponding flow of people is maximized. What is the maximal flow in the corridor if its width is 3.4 m and the corridor is bounded by two walls? Calculate the corresponding speed of people in the corridor.

3.5.9

Briefly describe how PIR-Algorithm can be modified to take into account travel time; that is, explain the differences between PIR-Algorithm and PIR-Algorithm2.

3.5.10

The following floor polygon and dual door nodes are defined with the well-known WKT format, which uses a Cartesian coordinate system:

Floor polygon: POLYGON ((0 0, 240 0, 240 150, 200 150, 200 50, 120 50, 120 150, 80 150, 80 50, 40 50, 40 150, 0 150, 0 0))

Dual door nodes:

D_1: POINT (20 150), D_2: POINT (100 150), D_3: POINT (220 150)

This polygon represents a corridor on some floor in a building, and dual nodes show the locations of doors in the corridor.

(a) For the given configuration of the corridor and doors, apply Algorithm 3.3 to construct the polygon tessellation and the navigable network. Assume that there is no hazard in the building and the following parameters are used in the algorithm: $T_1 = 3$ and $T_2 = 1$. For the parameter TL_{max}, there are three cases to consider:

   (1) $TL_{max} = 1$,   (2) $TL_{max} = 2$ and   (3) $TL_{max} = 3$.

   Thus, three tessellation diagrams should be constructed.

(b) How many cells were generated for different values of the parameter TL_{max} in cases (1)–(3)?

(c) Calculate the length of the shortest path between doors at far ends of the corridor ((D_1, D_3)-path) for different values of the parameter TL_{max} in cases (1)–(3). Explain why the length of the shortest paths is changing.

(d) What is the meaning of the parameter T_2? How would the tessellation change if there is a new hazard epicentre located in the corridor?

(e) How is closeness to a hazard location defined in this study? Can you provide other measurements, which can be applied instead?

3.6 Solutions

3.5.1

(a) For navigation purposes, all corridors and large open spaces in a building should be partitioned automatically to generate a navigable network reflecting real navigation routes. Without a proper partition, an incorrect distance between nodes (cells) might be calculated, and hence wrong connections (links) might be selected for evacuation. This is illustrated in Figure 3.5, where a person has to go from Room 11 to the nearest exit. Using the graph of connections G, where the corridor is modelled by just one node, the person would follow a dashed line, while the actual walking pattern should be different; one possible path is shown in Figure 3.6 (the bold line).   ∎

(b) As a rule, 2D floor plans are used for reconstruction of horizontal navigable networks, and a 3D model of a building is obtained by linking contiguous floors at some connection points, for example staircases. Spatial relationships between the rooms in the vertical direction are not reflected in the model. This solution is sufficient for a simple analysis of indoor human movements, but other important phenomena related to emergency response, for example fire spread or heat propagation, cannot be simulated in such models. ∎

(c) First, if such a link e is removed from a navigable networks, then the network may become disconnected, so that there is no route to the destination. Second, when comparing the aggregate hazard proximity of two routes, it might happen that a route with the link e is less dangerous than another route without 'forbidden' links because the latter has many links with average hazard proximity (or worse), which are not forbidden for navigation. ∎

3.5.2

(a) Let G denote the original 3D dual model of the building without tessellation; that is, the graph of connections between cells. The combined network G^+ represents the 3D model of the building with all the necessary tessellations. Thus, G^+ has some additional nodes such as corner points and new nodes added during the tessellation process of corridors and large open spaces. Examples of the graphs G and G^+ are given in Figure 3.7, where only navigable links are shown. Note that non-navigable links, which are not shown in Figure 3.7, represent physical obstructions, for example the link between the nodes in the left and right top corners. ∎

(b) The parameter $d(v, z_i)$ denotes the direct distance from node v to the epicentre z_i in metres, and $b(v, z_i)$ is the minimum number of obstructions (walls, floors, ceilings) between v and z_i. These parameters are calculated in DB_Procedure. Using 3D coordinates of the nodes, the direct distance $d(v, z_i)$ from node v to the epicentre z_i in metres is calculated for all nodes $v \in V(G^+)$ and all epicentres. Then, the Breadth-First Search Algorithm (see Section 1.3.1) is run in the graph G from the node z_i. It finds the number of links in shortest paths from z_i to all other nodes in G. Because each link in those paths represents a physical obstruction, the number of links in such a (z_i, v)-path is the minimum number of obstructions $b(v, z_i)$ between z_i and v. Note that $b(v, z_i)$ should be further adjusted for doors and exit doors.   ∎

(c) The hazard proximity numbers are based on the parameters $d(v, z_i)$ and $b(v, z_i)$, which should first be replaced by one variable representing their average. For two variables with different numerical ranges it is appropriate to use a (weighted) geometric mean. Since $d(v, z_i)$ and $b(v, z_i)$ have different ranges, the weighted geometric mean is applied:

$$WGM = \left(d(v, z_i)^{w_1} b(v, z_i)^{w_2} \right)^{\frac{1}{w_1 + w_2}}$$

where w_1 and w_2 are the relative weights for the direct distance and the number of obstructions. Because the direct distance is as important as the number of obstructions, we can assume that the corresponding weights for the two variables are in proportion 50:50, that is, $w_1 = 0.5$ and $w_2 = 0.5$. However, these weights can be adjusted if necessary, for example for buildings with many large open spaces and few obstructions. Thus, the above formula is simplified to the standard geometric mean:

$$GM = \sqrt{d(v, z_i) \times b(v, z_i)}.$$ ∎

(d) The values of aforementioned geometric means should be transformed to the scale going from 100 to 0 taking into account the propagation coefficient $\rho \geq 0$. This is achieved by using the formula for the hazard proximity numbers with respect to the epicentre z_i:

$$\frac{100}{\left(1 + \frac{\rho}{100}\right)^{\sqrt{d(v,z_i) \times b(v,z_i)}}}.$$

For instance, if $\rho = 100$, then hazard proximity numbers for nodes propagate quickly from 100 (in the epicentres) to small positive numbers (far from the epicentres). This puts a strong emphasis on the epicentre and the rooms in its close proximity. In contrast, if ρ is a small positive number, then the propagation is slow, thus putting less emphasis on the epicentre and the nearby rooms. In the extreme case $\rho = 0$ there is no propagation; that is, all hazard proximity numbers for nodes are equal to 100.

Having calculated $H_i(v)$ for all the nodes in G^+ and all values of $i = 1, 2, ..., k$, the following formula is used to compute the final hazard proximity numbers for nodes:

$$H(v) = \max_{1 \leq i \leq k} H_i(v).$$

Here we assume that the hazard at a particular node is equal to the maximal hazard proximity number at this node for all the epicentres; this approach is justified in many cases.   ■

(e) For each link $e = uv$ in the graph G^+, the hazard proximity number $HD(e)$ for e is determined by calculating the arithmetic average of the hazard proximity numbers of its end-nodes, and then by multiplying the resulting number by the length of e in metres:

$$HD(e) = 0.5[H(u) + H(v)] \times D(e).$$

The use of $D(e)$ in this formula is important because G^+ is not a homogeneous network. For example, let us suppose that one link is 2 metres long and another is 10 metres long, and they both have the same hazard proximity number, say 10. If they both are used in a navigation route, then it is natural to assume that a travel time for a longer link would be approximately five times longer, so the hazard proximity number for the longer link should be 50. In other words, if we subdivided the longer link into five 2-metre-long links to make the network more homogeneous, then those five links would approximately contribute 50 to the total hazard proximity of the route.   ■

(f) Dijkstra's algorithm (see Section 1.3.4) finds a shortest route between nodes in a weighted graph. Its input is a graph/network where each link has some 'cost' (e.g. distance, time), and also the starting node and a target node (or nodes) are specified. The output is a route from the starting node to the target node with the smallest possible cost.

The final stage of Algorithm 3.1 is to run Dijkstra's algorithm in the graph G^+ from node p with link weights $HD(e)$. It produces the safest available (p, q)-route P in the building with respect to two criteria: distance and the number of obstructions.   ■

(g) The safest route can be too long and/or go upstairs, as illustrated in Figure 3.12. To overcome this problem, we can introduce a penalty for going upstairs or use the balanced route generated by Algorithm 3.2. ■

3.5.3

According to HP_Procedure, $\tau = 2$ and

$$H_1(p) = H_1(a) = H_1(q) = \frac{100}{2\sqrt{64}} = 0.3906,$$

$$H_1(u_1) = H_1(u_2) = \frac{100}{2\sqrt{100}} = 0.0977.$$

Further,

$$HD(p, a) = HD(a, q) = 0.5 \times 2 \times 0.3906 \times 20 = 7.81$$

and

$$HD(P_1) = 15.62.$$

Also, $HD(u_1, u_2) = 2.93$ and

$$HD(p, u_1) = HD(u_2, q) = 0.5 \times (0.3906 + 0.0977) \times 30 = 7.32.$$

We obtain

$$HD(P_2) = 17.57.$$

Because there exist two (p, q)-routes only, there is no need to run Dijkstra's algorithm—the safest route is P_1 (for $\rho_{max} = 100$). The route P_2 is a bit further away from the epicentre of hazard, but this does not 'justify' its length (90 m), which makes it more dangerous than P_1 (40 m) for the given criteria.   ∎

3.5.4

(a) The shortest route might be too dangerous, whereas the safest route might be too long. The idea of Algorithm 3.2 is to find a balanced route, where a trade-off between route length and hazard proximity is made. One of the input parameters of the algorithm is the hazard tolerance coefficient t. The hazard tolerance is a trade-off between distance and safety and it can be equal to 0, 0.5 or 1. For example, if $t = 1$, then the shortest route will be generated. If $t = 0$, then the algorithm finds the safest available route. If t is not specified, then a route with the 50/50 balance of distance/safety will be reported as the balanced route (if there are enough generated routes).

A member of the rescue team, who is fully protected from the hazard, may wish to use the shortest route even if it is the most dangerous one; that is, their hazard tolerance t is equal to 1. In contrast, an unprotected person with a respiratory disease may want to use the safest evacuation route, whatever its length, in which case the hazard tolerance t is 0. The hazard tolerance is an optional parameter; by default $t = 0.5$. The default value of 0.5 simply means that the hazard tolerance has not been specified, and this number will be used as a relative weight for the distance attribute, so the relative weight for the hazard proximity will be 0.5 too. Thus, if t is not specified, then a route with the 50/50 balance of distance/hazard proximity will be chosen. ■

(b) In Step 3, the binary search is carried out with respect to the propagation coefficient ρ. The first run is for $\rho = 0$, producing the shortest (p, q)-route P_0 because all hazard proximity numbers for links are $100D(e)$. The route P_0 is included in the set R. The next run is for $\rho = \rho_{max}$. If the resulting route coincides with P_0, then there is no interval for the binary search and it is terminated. Otherwise, the route is different from P_0 and it is included in R. The next run is for $\rho = 0.5\rho_{max}$. There are three possibilities here. If the resulting route is a new one, then it is included in R and the binary search continues for two intervals $(0; 0.5\rho_{max})$ and $(0.5\rho_{max}; \rho_{max})$. If the resulting route coincides with one of the routes in the set R, then one of the intervals is removed from the search and the other interval is used in the binary search. For example, if the route coincides with P_0, then the binary search continues for the interval $(0.5\rho_{max}; \rho_{max})$, whereas the interval $(0; 0.5\rho_{max})$ is removed. This procedure is terminated if at least one of the stopping criteria is satisfied: a specified size of R, a specified length of the widest interval and a running time limit. ■

(c) In Step 4 of the algorithm, the total length of each route in R is calculated. Then, the proximity ratios are computed for each link in a route. They are based on the geometric averages of parameters $d(v, z_i)$ and $b(v, z_i)$ for end-nodes of the link and its length. In contrast to hazard proximity numbers, the proximity ratios do not depend on the propagation coefficient ρ, and a small proximity ratio means a close proximity to one of the epicentres. For a given link, the final proximity ratio $r(e)$ is the smallest proximity ratio for that link:

$$r(e) = \min_{1 \leq i \leq k} r_i(e).$$

The proximity index for a route P is the harmonic mean of $r(e)$ for all links in the route:

$$PI\left(P\right) = \frac{l\left(P\right)}{\sum_{e \in P} \frac{1}{r(e)}}$$

where $l(P)$ is the number of links in P. Note that the proximity index is an average of *rates*. Also, the proximity index should not be dominated by sections of a route with large proximity ratios, and actually the impact of small proximity ratios is important. Therefore, the harmonic mean is an appropriate measure for the proximity index. Since the proximity index is independent on the propagation coefficient, it can be used for comparison of the routes from the set R. ■

(d) In the final part of the algorithm, a multi-attribute decision-making technique is used to rank the routes in the set R and choose a balanced (p, q)-route. First of all, the maximal and minimal values of the lengths and proximity indices are calculated for all routes in R:

$$d^- = \min_{P \in R} D\left(P\right); \qquad d^+ = \max_{P \in R} D\left(P\right);$$
$$p^- = \min_{P \in R} PI\left(P\right); \qquad p^+ = \max_{P \in R} PI\left(P\right).$$

Then, quadratic value functions are applied for rating the routes with respect to two attributes, the distance and the proximity index. For each route in R, the scores for these attributes are given by the following formulae, respectively:

$$S_D\left(P\right) = 100 \left(1 - \left(\frac{D\left(P\right) - d^-}{d^+ - d^-}\right)^2\right)$$

and

$$S_{PI}\left(P\right) = 100 \left(1 - \left(\frac{p^+ - PI\left(P\right)}{p^+ - p^-}\right)^2\right).$$

Finally, the aggregate score is calculated as a weighted average of the routes' scores, where the weights depend on the tolerance coefficient t. More precisely, the weights are t and $1 - t$ for the scores S_D and S_{PI}, respectively. The balanced (p, q)-route is one with the highest aggregate score. For example, if $t = 1$, then the shortest route is chosen. If $t = 0$, then the algorithm returns the route with the highest proximity index; that is, the safest available route. If t is not specified and there are enough routes in the set R, then a route with the 50/50 balance of distance/proximity will be reported as the balanced route. ■

3.5.5

We obtain

$$r_1(p, v_1) = \frac{\sqrt{10 \times 1} + \sqrt{14 \times 1}}{2 \times 14} = 0.247.$$

In a similar way,

$$r_1(p, v_2) = 0.288, \ r_1(p, v_3) = 0.297, \ r_1(p, v_4) = 0.376,$$
$$r_1(v_1, q) = 0.634, \ r_1(v_2, q) = 0.481,$$
$$r_1(v_3, q) = 0.496, \ r_1(v_4, q) = 0.388.$$

Hence,

$$PI(P_1) = \frac{2}{1/0.247 + 1/0.634} = 0.355.$$

Also,

$$PI(P_2) = 0.360, \ PI(P_3) = 0.371, \ PI(P_4) = 0.382$$

and

$$D(P_1) = 24, \ D(P_2) = 30, \ D(P_3) = 33, \ D(P_4) = 37.$$

We have

$$d^- = 24, \quad d^+ = 37, \quad p^- = 0.355, \quad p^+ = 0.382.$$

Therefore, the aggregate scores are as follows (rounded to the nearest integer):

$$AS(P_1) = 50, \quad AS(P_2) = 56, \quad AS(P_3) = 68, \quad AS(P_4) = 50.$$

Thus, the balanced route is P_3. ∎

3.5.6

Here are the new elements introduced in PIR-Algorithm:

(a) A new criterion—link/route complexity, which has the following attributes: staircase link going up; staircase link going down; door; turning angle; distance.

(b) The second binary search, which is based on two criteria: hazard proximity and link complexity.

(c) An updated formula for $H_i(v)$ (i.e. hazard proximity numbers for nodes w.r.t. z_i).

(d) The AHP, a more advanced multi-attribute rating technique. This method can be summarized into four main steps as follows:

    (1) Structuring a problem into a decision hierarchy consisting of criteria and alternatives.

    (2) Establishing pairwise comparisons between decision elements at each hierarchy level.

    (3) Transforming comparison matrices into sets of weights.

    (4) Aggregating the weights to rank the alternatives. ∎

3.5.7

(a) Using formula (3.1), we obtain:

    Speed = 1.198 m/s for the corridor outside the office (40 m)

    Speed = 0.505 m/s for the staircase (going down 30 m)

    Speed = 0.357 m/s for the corridor on the ground floor (20 m)

Now, the corresponding evacuation times are: 33.4 s, 59.4 s and 56.0 s.

Thus, the total evacuation time is 149 s. ∎

(b) Using formula (3.2), we have:

    Speed = 1.162 m/s for the corridor outside the office (40 m)

    Speed = 0.313 m/s for the staircase (going down 30 m)

    Speed = 0.359 m/s for the corridor on the ground floor (20 m)

Now, the corresponding times for the rescuer's movements are: 34.4 s, 95.8 s and 55.7 s.

Thus, the total time to reach the office is 186 s. ∎

3.5.8

Using formula (3.1), we obtain:

$$F = S\Delta W_e = (\Delta - 0.266\Delta^2)KW_e.$$

Hence,

$$\frac{dF}{d\Delta} = (1 - 0.532\Delta)KW_e = 0$$

and

$$\Delta = \frac{1}{0.532} = 1.88.$$

Thus, the flow of people is maximal possible if the density is

$$1.88 \text{ people/m}^2.$$

If the width of the corridor is 3.4 m, then

$$W_e = 3.4 - 0.2 - 0.2 = 3\,\text{m},$$

and the maximal flow is calculated as follows:

$$F_{\max} = S\Delta W_e = (1 - 0.266 \times 1.88) \times 1.4 \times 1.88 \times 3 = 3.95.$$

Thus, the capacity (maximal flow) is ca. 4 people/second. The corresponding speed is

$$(1 - 0.266 \times 1.88) \times 1.4 = 0.70\,\text{m/s}$$

(it is an optimal speed, not maximal). ∎

3.5.9

It is assumed in PIR-Algorithm2 that the population densities are given for all cells in the building. Because each cell is represented by a node in the unified network G^*, the densities are associated with nodes in the set $V(G^*)$.

In the new Step 0, we first calculate the density for all links uv; that is, the average density for two areas associated with the nodes u and v. Then, the speed $S(e)$ along the link e is computed using formula (3.1). Here we assume that the agent is moving together with the flow of people. In emergency situations, this typically happens when moving downstairs. If the agent is moving in a counter-flow, for example upstairs, then one should use formula (3.2). Finally, the travel time $T(e)$ along link e is calculated in Step 0(c).

Further, in Step 3(b), the formula for the hazard proximity numbers is updated by using travel time instead of link length. Thus, if there is no hazard propagation (i.e. $\rho = 0$), then the fastest route will be generated in the first binary search instead of the shortest one. In the modified Step 6, in addition to the route length, we calculate the travel time for each route in the set R. Also, the formula for proximity ratios is updated with travel time $T(e)$ because longer travel time for a link should decrease the proximity ratio for this link and the proximity index for the route, thus making it worse. Next, in Step 7, the statistical parameters are now calculated for travel times, and the comparison matrix M_2 in Step 9 represents the relative importance of routes in the set R with respect to travel time. ∎

3.5.10

(a) Polygon tessellations:

$$TL_{\max} = 1 \qquad TL_{\max} = 2 \qquad TL_{\max} = 3$$

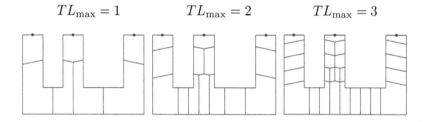

The dual door nodes are shown as blue dots.
Navigable networks (green colour):

$$TL_{\max} = 1 \qquad TL_{\max} = 2 \qquad TL_{\max} = 3$$

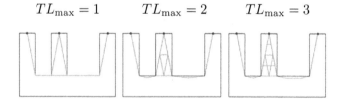

∎

(b) The number of cells is 7, 14 and 26 for cases (1)–(3), respectively. ∎
(c) The lengths of the shortest paths (dark green lines) are:
(1) 36.4, (2) 36.58, (3) 36.64.

$$TL_{\max} = 1 \qquad TL_{\max} = 2 \qquad TL_{\max} = 3$$

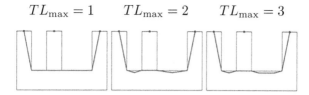

When $TL_{\max}$ increases, the shortest path is getting longer because of network densification, which reflects more natural ways of movement. Indeed, some links overlap with corridor corners and walls, especially in case (1). The reason is that concave corners are used as nucleation points in the initial tessellation and hence they become network nodes. In order to avoid collisions with the corridor corners, the network may be improved by shifting the corner nodes by a certain distance from the original corner towards the corridor centreline. ∎

(d) The parameter T_2 is responsible for densification of a tessellation around the hazard epicentre. In case there is a new epicentre located in the corridor, the tessellation will be densified in a close proximity to the epicentre: cells will be divided into smaller ones. This will result in a denser structure of the navigable network around the epicentre and better flexibility when choosing an escape path. ∎

(e) The close proximity to the hazard location can be defined depending on an application. In this particular study, the close proximity area is defined using a distance between two Voronoi nucleation points represented as an edge in the navigable network and a distance between the hazard epicentre and the end-points of the edge. If the length of the edge multiplied by three is greater than a sum of distances from the epicentre to each end of the edge, then the edge is considered to be close to hazard. Close proximity can be measured using a circle of a certain radius with a centre at the hazard location. Edges included in or intersecting with the circle would be considered as located close to the hazard. ∎

Acknowledgements

This research was made possible by a National Priority Research Program NPRP award [NPRP-06-1208-2-492] from the Qatar National Research Fund, a member of The Qatar Foundation. The statements made herein are solely the responsibility of the authors. Section 3.1 is based on the article by Zverovich et al. [77] (© 2016 *Computer-Aided Civil and Infrastructure*

Engineering), with permission from John Wiley and Sons. Section 3.2 is based on the article by Zverovich et al. [78], released under the Creative Commons Attribution License. Section 3.3 is based on the article in *Automation in Construction*, **72**, P. Boguslawski, L. Mahdjoubi, V. Zverovich and F. Fadli, Automated construction of variable density navigable networks in a 3D indoor environment for emergency response, 115–128, © 2016, with permission from Elsevier.

References

[1] I. Afyouni, R. Cyril and C. Christophe, Spatial models for context-aware indoor navigation systems: a survey, *Journal of Spatial Information Science*, **1** (4) (2012), 85–123.

[2] F. Aurenhammer, Voronoi diagrams—a survey of a fundamental geometric data structure, *ACM Computing Surveys*, **23** (3)(1991), 345–405.

[3] P. Axelsson, DEM generation from laser scanner data using adaptive TIN models, in D. Fritsch and M. Molenaar (eds), *XIXth ISPRS Congress*, ISPRS, Amsterdam, 2000, 110–117.

[4] H. Barki, F. Fadli, A. Shaat, P. Boguslawski and L. Mahdjoubi, BIM models generation from 2D CAD drawings and 3D scans: an analysis of challenges and opportunities for AEC practitioners, in L. Mahdjoubi, C. Brebbia and R. Laing (eds), *Building Information Modelling (BIM) in Design, Construction and Operations*, Southampton: WIT Press, 2015, **149**, 369–380.

[5] T. Becker, C. Nagel and T. H. Kolbe, A multilayered space-event model for navigation in indoor spaces, *Geo-Information Sciences 3D*, 2009, 61–77.

[6] P. Boguslawski, *Modelling and Analysing 3D Building Interiors with the Dual Half-Edge Data Structure*, PhD thesis, University of Glamorgan, Pontypridd, Wales, 2011.

[7] P. Boguslawski and C. Gold, Buildings and terrain unified—multidimensional dual data structure for GIS, *Geo-spatial Information Science*, **18** (4)(2015), 151–158.

[8] P. Boguslawski and C. Gold, Euler operators and navigation of multi-shell building models, in T. Neutens and P. Maeyer (eds), *Developments in 3D Geo-Information Sciences*, Lecture Notes in Geoinformation and Cartography, Basel: Springer, 2010, 1–16.

[9] P. Boguslawski, C. Gold and H. Ledoux, Modelling and analysing 3D buildings with a primal/dual data structure, *ISPRS Journal of Photogrammetry and Remote Sensing*, **66** (2)(2011), 188–197.

[10] P. Boguslawski, L. Mahdjoubi, V. Zverovich and F. Fadli, Automated construction of variable density navigable networks in a 3D indoor environment for emergency response, *Automation in Construction*, **72** (2)(2016), 115–128.

[11] P. Boguslawski, L. Mahdjoubi, V. Zverovich and F. Fadli, A dynamic approach for evacuees' distribution and optimal routing in hazardous environments, *Automation in Construction*, **94** (2018), 11–22.

[12] P. Boguslawski, L. Mahdjoubi, V. Zverovich, H. Barki and F. Fadli, BIM-GIS modelling in support of emergency response applications, in L. Mahdjoubi, C. Brebbia and R. Laing (eds), *Building Information Modelling (BIM) in Design, Construction and Operations*, Southampton: WIT Press, **149** (2015), 381–392.

[13] G. Brown, C. Nagel, S. Zlatanova and T. H. Kolbe, Modelling 3D topographic space against indoor navigation requirements, in J. Pouliot, S. Daniel, F. Hubert and

A. Zamyadi (eds), *Progress and New Trends in 3D Geoinformation Sciences*, Berlin: Springer, 2013, 1–22.

[14] R. Chedid and N. Najjar, Automatic finite-element mesh generation using artificial neural networks—Part I: prediction of mesh density, *IEEE Transactions on Magnetics*, **32** (5)(1996), 5173–5178.

[15] J. Choi and J. Lee, 3D geo-network for agent-based building evacuation simulation, in J. Lee and S. Zlatanova (eds), *3D Geo-Information Sciences*, Berlin: Springer, 2009, 283–299.

[16] C. DeCapua and B. Bhaduri, *Applications of Geospatial Technology in International Disasters and During Hurricane Katrina*, 2007. (Available at the Project Site of Capturing Hurricane Katrina Data for Analysis and Lessons-Learned Research) www.gri. msstate.edu/research/katrinalessons/Documents/ GeoSp_Tech_Applications.pdf.

[17] K. Deng, W. Chen, P. Mehta and S. Meyn, Resource pooling for optimal evacuation of a large building, in *The 47th IEEE Conference on Decision and Control*, Cancun, 2008, 5565–5570.

[18] E. W. Dijkstra, A note on two problems in connexion with graphs, *Numerische Mathematik*, **1** (1959), 269–271.

[19] H. A. Donegan and F. J. Dodd, A note on Saaty's random indexes, *Mathematical and Computer Modelling*, **15** (10)(1991), 135–137.

[20] Q. Du and M. Gunzburger, Grid generation and optimization based on centroidal Voronoi tessellations, *Applied Mathematics and Computation*, **133** (2–3)(2002), 591–607.

[21] M. Duckham and L. Kulik, 'Simplest' paths: automated route selection for navigation, in W. Kuhn, M. F. Worboys and S. Timpf (eds), *Spatial Information Theory: Foundations of Geographic Information Science*, Berlin: Springer, 2003, 169–185.

[22] D. N. Dyck, D. A. Lowther and S. McFee, Determining an approximate finite element mesh density using neural network techniques, *IEEE Transactions on Magnetics*, **28** (2)(1992), 1767–1770.

[23] E. H. Foreman, Facts and fictions about the analytic hierarchy process, *Mathematical and Computer Modelling*, **17** (4–5)(1993), 19–26.

[24] R. Friedman, An international survey of computer models for fire and smoke, *Journal of Fire Protection Engineering*, **4** (1992), 81–92.

[25] J. J. Fruin, Pedestrian planning and design, *Elevator World Inc.* (Educational Services Division), Alabama, 1987, LIB.CON 70-159312.

[26] M. Goetz and A. Zipf, Formal definition of a user-adaptive and length-optimal routing graph for complex indoor environments, *Geo-spatial Information Science*, **14** (2)(2011), 119–128.

[27] E. G. Gomes and M. P. E. Lins, Integrating geographical information systems and multi-criteria methods: a case study, *Annals of Operations Research*, **116** (2002), 243–269.

[28] P. J. Green and R. Sibson, Computing Dirichlet tessellations in the plane, *The Computer Journal*, **21** (2)(1978), 168–173.

[29] S. Gwynne, E. R. Galea, P. J. Lawrence, M. Owen and L. Filippidis, A review of the methodologies used in the computer simulation of evacuation from the built environment, *Building and Environment*, **34** (1999), 741–749.

[30] P. E. Hart, N. J. Nilsson and B. Raphael, A formal basis for the heuristic determination of minimum cost paths, *IEEE Transactions on Systems Science and Cybernetics*, SSC4, **4** (2)(1968), 100–107.

[31] K. Ho-Le, Finite element mesh generation methods: a review and classification, *Computer-Aided Design*, **20** (1)(1988), 27–38.

[32] T. Kisko and R. Francis, EVACNET+: A computer program to determine optimal building evacuation plans, *Fire Safety Journal*, **9** (2)(1985), 211–220.

[33] M. Krūminaitė and S. Zlatanova, Indoor space subdivision for indoor navigation, *Proceedings of the Sixth ACM SIGSPATIAL International Workshop on Indoor Spatial Awareness*, ACM, Dallas/Fort Worth, TX, 2014, 25–31.

[34] E. D. Kuligowski, Review of 28 egress models, in R. D. Peacock and E. D. Kuligowski (eds), *The Workshop on Building Occupant Movement During Fire Emergencies*, 10–11 June 2004, National Institute of Standards and Technology, Gaithersburg, MD, 2004, 68–90.

[35] M.-P. Kwan and J. Lee, Emergency response after 9/11: the potential of real-time 3D GIS for quick emergency response in micro-spatial environments, *Computers, Environment and Urban Systems*, **29** (2005), 93–113.

[36] F. Lamarche and S. Donikian, Crowd of virtual humans: a new approach for real time navigation in complex and structured environments, *Computer Graphics Forum*, **23** (3)(2004), 509–518.

[37] S. Lay, Alternative evacuation design solutions for high-rise buildings, *The Structural Design of Tall and Special Buildings*, **16** (2007), 487–500.

[38] J. Lee, A spatial access-oriented implementation of a 3-D GIS topological data model for urban entities, *GeoInformatica*, **8** (3)(2004), 237–264.

[39] J. Lee, A three-dimensional navigable data model to support emergency response in microspatial built-environments, *Annals of the Association of American Geographers*, **97** (3)(2007), 512–529.

[40] J. Lee and M. P. Kwan, A combinatorial data model for representing topological relations among 3D geographical features in micro-spatial environments, *International Journal of Geographical Information Science*, **19** (10)(2005), 1039–1056.

[41] J. Lee and S. Zlatanova, A 3D data model and topological analyses for emergency response in urban areas, in S. Zlatanova and J. Li (eds), *Geo-information Technology for Emergency Response*, London: Taylor and Francis, 2008, 143–168.

[42] N. Li, B. Becerik-Gerber, B. Krishnamachari and L. Soibelman, A BIM centered indoor localization algorithm to support building fire emergency response operations, *Automation in Construction*, **42** (2014), 78–89.

[43] P. Lienhardt, Topological models for boundary representation: a comparison with *n*-dimensional generalized maps, *Computer Aided Design*, **23** (1)(1991), 59–82.

[44] L. Liu and S. Zlatanova, A 'door-to-door' path-finding approach for indoor navigation, in *Proceedings of the Gi4DM 2011: GeoInformation for Disaster Management*, International Society for Photogrammetry and Remote Sensing (ISPRS), Antalya, 2011.

[45] L. Liu and S. Zlatanova, Towards a 3D network model for indoor navigation, in S. Zlatanova, H. Ledoux, E. Fendel and M. Rumor (eds), *Urban and Regional Data Management*, London: CRC Press/Taylor and Francis Group, 2012, 79–92.

[46] F. Luo, G. Cao and X. Li, An interactive approach for deriving geometric network models in 3D indoor environments, *Proceedings of the Sixth ACM SIGSPATIAL International Workshop on Indoor Spatial Awareness*, ACM, Dallas/Fort Worth, TX, 2014, 9–16.

[47] M. Meijers, S. Zlatanova and N. Pfeifer, 3D geo-information indoors: structuring for evacuation, in G. Groeger and T. H. Kolbe (eds), *The First International Workshop on Next Generation 3D City Models*, Bonn, 2005, 11–16.

[48]   H. E. Nelson and H. A. MacLennan, Emergency movement, in P. J. DiNenno, C. L. Beyler, R. L. P. Custer, W. D. Walton, J. M. Watts, D. Drysdale and J. R. Hall (eds), *SFPE Handbook of Fire Protection Engineering*, Society of Fire Protection Engineers, Boston, MA, 1995, 3/286–3/295.

[49]   OGC, OGC IndoorGML, Open Geospatial Consortium Inc., 2014.

[50]   S. M. Olenick and D. J. Carpenter, Updated international survey of computer models for fire and smoke, *Journal of Fire Protection Engineering*, **13** (2003), 87–110.

[51]   I. Park, G. Jang, S. Park and J. Lee, Time-dependent optimal routing in micro-scale emergency situation, in *Proceedings of the Tenth International Conference on Mobile Data Management: Systems, Services and Middleware*, Taipei, 2009, 714–719.

[52]   J. Pauls, Movement of people, in P. J. DiNenno, C. L. Beyler, R. L. P. Custer, W. D. Walton, J. M. Watts, D. Drysdale and J. R. Hall (eds), *SFPE Handbook of Fire Protection Engineering*, Society of Fire Protection Engineers, Boston, MA, 1995, 3/263–3/285.

[53]   S. B. Petersen and P. A. F. Martins, Finite element remeshing: a metal forming approach for quadrilateral mesh generation and refinement, *International Journal for Numerical Methods in Engineering*, **40** (8)(1997), 1449–1464.

[54]   S. Pursals and F. Garzón, Optimal building evacuation time considering evacuation routes, *European Journal of Operational Research*, **192** (2)(2009), 692–699.

[55]   T. L. Saaty, *The Analytic Hierarchy Process: Planning, Priority Setting, Resource Allocation*, New York, NY: McGraw-Hill, 1980.

[56]   T. L. Saaty, *Multicriteria Decision Making: The Analytic Hierarchy Process*, Pittsburgh, PA: RWS Publications, 1990.

[57]   T. L. Saaty, *Decision Making for Leaders: The Analytic Hierarchy Process for Decision in a Complex World*, Pittsburgh, PA: RWS Publications, 1999.

[58]   N. Scharfenort, Urban development and social change in Qatar: the Qatar national vision 2030 and the 2022 FIFA world cup, *Journal of Arabian Studies*, **2** (2)(2012), 209–230.

[59]   M. Shahabi, A. Unnikrishnan and S. D. Boyles, Robust optimization strategy for the shortest path problem under uncertain link travel cost distribution, *Computer-Aided Civil and Infrastructure Engineering*, **30** (6)(2015), 433–448.

[60]   E. P. Stoffel, B. Lorenz and H. J. Ohlbach, Towards a semantic spatial model for pedestrian indoor navigation advances in conceptual modeling—foundations and applications, *Lecture Notes in Computer Science*, Chicago, IL, 2007, 328–337.

[61]   I. Stroud, *Boundary Representation Modelling Techniques*, New York, NY: Springer-Verlag, 2006.

[62]   Q. Sun and S. Wu, A configurable agent-based crowd model with generic behaviour effect representation mechanism, *Computer-Aided Civil and Infrastructure Engineering*, **29** (7)(2014), 531–545.

[63]   H. Tashakkori, A. Rajabifard and M. Kalantari, A new 3D indoor/outdoor spatial model for indoor emergency response facilitation, *Building and Environment*, **89** (2015), 170–182.

[64]   Z. Turskis, A. Daniūnas, E. K. Zavadskas and J. Medzvieckas, Multicriteria evaluation of building foundation alternatives, *Computer-Aided Civil and Infrastructure Engineering*, **31** (2016), 717–729.

[65]   A. Vanclooster, P. Maeyer and V. Fack, On the way of integrating evacuation approaches, in *Proceedings of the 5th International Conference on 3D Geoinformation*, Berlin, **38** (4/W15) (2010), 169–172.

[66]   A. Vanclooster, P. Maeyer, V. Fack and N. V. Weghe, Calculating least risk paths in 3D indoor space, in U. Isikdag (ed), *Innovations in 3D Geo-Information Sciences*, Cham: Springer, 2014, 13–31.

[67] J. Wallgrün, Autonomous construction of hierarchical Voronoi-based route graph representations, in C. Freksa, M. Knauff, B. Krieg-Brückner, B. Nebel and T. Barkowsky (eds), *Spatial Cognition IV. Reasoning, Action, Interaction*, Lecture Notes in Computer Science, Berlin: Springer, 2005, 413–433.

[68] Y. Wang, C. Zhang, L. Zhu and Q. Sun, Application of AHP in the evaluation of railway emergency plans, in *2010 International Conference on Networking, Sensing and Control (ICNSC)*, Chicago, IL, 10–12 April, 2010, 564–569.

[69] Z. Wang and S. Zlatanova, Taxonomy of navigation for first responders, in M. J. Krisp (ed), *Progress in Location-based Services*, Berlin: Springer, 2013, 297–315.

[70] J. M. Watts, Computer models for evacuation analysis, *Fire Safety Journal*, **12** (1987), 237–245.

[71] K. Weiler, The radial edge structure: a topological representation for non-manifold geometric modeling, in M. J. Wozny, H. McLaughlin and J. Encarnacao (eds), *Geometric Modeling for CAD Applications*, Amsterdam: Elsevier Science, 1988, 3–36.

[72] Y. Wenjie and M. Schneider, Supporting continuous range queries in indoor space, in *Eleventh International Conference on Mobile Data Management*, (MDM), Kansas, MO, 2010, 209–214.

[73] L. Yang and M. Worboys, Generation of navigation graphs for indoor space, *International Journal of Geographical Information Science*, **29** (10)(2015), 1737–1756.

[74] F. Zahedi, The analytic hierarchy process—a survey of the method and its applications, *Interfaces*, **16** (1986), 96–108.

[75] J. Zhang and X. Lin, Filtering airborne LiDAR data by embedding smoothness-constrained segmentation in progressive TIN densification, *ISPRS Journal of Photogrammetry and Remote Sensing*, **81** (2013), 44–59.

[76] W. Zhang, K. Sun, C. Lei, Y. Zhang, H. Li and B. F. Spencer, Jr., Fuzzy analytic hierarchy process synthetic evaluation models for the health monitoring of shield tunnels, *Computer-Aided Civil and Infrastructure Engineering*, **29** (9)(2014), 676–688.

[77] V. Zverovich, L. Mahdjoubi, P. Boguslawski, F. Fadli and H. Barki, Emergency response in complex buildings: automated selection of safest and balanced routes, *Computer-Aided Civil and Infrastructure Engineering*, **31** (8)(2016), 617–632.

[78] V. Zverovich, L. Mahdjoubi, P. Boguslawski and F. Fadli, Analytic prioritization of indoor routes for search and rescue operations in hazardous environments, *Computer-Aided Civil and Infrastructure Engineering*, **32** (9)(2017), 727–747.

4

Graph Models for Backbone Sets and Limited Packings in Networks

V. Zverovich, A. Gagarin and A. Poghosyan

In this chapter, a graph-theoretic approach will be applied to some problems in networks, for example in wireless sensor networks (WSNs). To maximize functional lifetime of a WSN, which depends on energy consumption, it is important to select appropriately some sensor nodes to behave as a backbone set to support routing communications in an efficient and fault-tolerant way. Backbone sets can be considered as dominating sets in the corresponding underlying graph of the network. Dominating sets of several different kinds have proved to be useful and effective for modelling backbone sets. In the first section, we consider four different types of multiple domination (k-, k-tuple, α- and α-rate domination) and discuss recent upper bounds for cardinality of these types of dominating sets. Also, we explicitly present randomized algorithms for finding multiple dominating sets whose expected size satisfies the aforementioned upper bounds. In the second section, we study limited packings in networks, in particular the k-limited packing number. One possible application of limited packings is a secure facility location problem in which there is a need to place as many resources as possible in a given network subject to some security constraints. The third section is devoted to two general frameworks for multiple domination, which are called $\langle \mathbf{r}, \mathbf{s} \rangle$-domination and parametric domination. They generalize and unify $\{k\}$-domination, k-domination, total k-domination and k-tuple domination. In addition, we consider different threshold functions for multiple domination.

4.1 Multiple Domination Models for Backbone Sets

Domination is one of the fundamental concepts in graph theory with various applications to wireless and ad hoc networks, in distributed computing, social networks and web graphs [1, 14, 15, 29]. Dominating sets are also used as models for facility location problems in operational research. An important

Modern Applications of Graph Theory. Vadim Zverovich, Oxford University Press (2021). © Vadim Zverovich.
DOI: 10.1093/oso/9780198856740.003.0004

role is played by multiple domination. For example, k- and k-tuple dominating sets can be used for balancing efficiency and fault tolerance in WSNs [15, 42, 45] and for optimizing the placement of charging stations for electric vehicles in road networks (see next chapter).

WSNs and ad hoc mobile networks can be considered as natural examples of applications of multiple domination. A WSN comprises a set of small autonomous devices, called sensors, for monitoring and measuring some physical parameters (e.g. pressure). Each sensor contains a microcontroller, a radio transmitter and a receiver for communications with its neighbours. A sensor is constrained in energy availability because of limitations in power of its battery. A *base station* is a sensor node that collects information gathered by other sensors and it connects the WSN with the outside world. WSNs have some military applications, for example in battlefield surveillance.

A routing algorithm allows the sensor nodes to self-organize into a WSN. As stated in [38], an important goal in WSN design is to maximize the functional lifetime of a sensor network by using energy efficient distributed algorithms, networking and routing techniques. To maximize the functional lifetime, it is necessary to select some sensor nodes to behave as a backbone set to support routing communications in an efficient and fault-tolerant way. The backbone set can be considered as a dominating set in the corresponding underlying graph of the network.

Dominating sets of several different kinds have proved to be useful and effective for modelling backbone sets. In the recent literature (e.g. see [15, 42, 45]), particular attention has been drawn to construction of m-connected k- and k-tuple dominating sets in WSNs. Several probabilistic, approximating and deterministic approaches have been proposed and analysed. The backbone set of sensor nodes needs to be selected as small as possible. On the other hand, it should guarantee high efficiency and reliability of networking and communications.

In this section, we discuss recently discovered upper bounds for the k- and k-tuple domination numbers and explicitly describe effective and efficient randomized algorithms to construct multiple dominating sets, whose expected orders satisfy the upper bounds. The algorithms arise from probabilistic constructions used to prove the corresponding bounds. All the presented algorithms can be implemented in parallel or as local distributed algorithms, which is particularly important in the context of WSNs [38]. Also, the aforementioned upper bounds generalize two classical bounds for the domination number and improve a number of known upper bounds for the multiple domination parameters from [8, 18, 22, 27, 40, 48].

4.1.1 Basic Definitions and Key Known Results

We consider networks represented by simple unweighted graphs. If G is a graph of order n, then $V(G) = \{v_1, v_2, ..., v_n\}$ is the set of vertices of G, and d_i denotes the degree of v_i, $1 \leq i \leq n$. The degree of a vertex v without an index is denoted by d_v. Denote by $\delta = \delta(G)$ and $\Delta = \Delta(G)$ the minimum and maximum vertex degrees of G, respectively. Let $N(v)$ denote the neighbourhood of a vertex v in G, and denote by $N[v] = N(v) \cup \{v\}$ the closed neighbourhood of v. A set $X \subseteq V(G)$ is called a *dominating set* if every vertex not in X is adjacent to at least one vertex in X. The minimum cardinality of a dominating set of G is the *domination number* $\gamma(G)$.

A set X is called a *k-dominating set* if every vertex not in X has at least k neighbours in X. The minimum cardinality of a k-dominating set of G is the *k-domination number* $\gamma_k(G)$. A set X is called a *k-tuple dominating set* of G if for every vertex $v \in V(G)$, $|N[v] \cap X| \geq k$. It is obvious that a k-tuple dominating set is also a k-dominating set. The minimum cardinality of a k-tuple dominating set of G is the *k-tuple domination number* $\gamma_{\times k}(G)$. The k-tuple domination number is only defined for graphs with $\delta \geq k - 1$. Clearly, $\gamma_{\times k}(G) \geq \gamma_k(G)$. The 2-tuple domination number $\gamma_{\times 2}(G)$ is called the *double domination number*, and the 3-tuple domination number $\gamma_{\times 3}(G)$ is called the *triple domination number*.

Let α be a real number satisfying $0 < \alpha \leq 1$. A set $X \subseteq V(G)$ is called an *α-dominating set* of G if for every vertex $v \in V(G) - X$, $|N(v) \cap X| \geq \alpha d_v$; that is, v is adjacent to at least $\lceil \alpha d_v \rceil$ vertices of X. The minimum cardinality of an α-dominating set of G is called the *α-domination number* $\gamma_\alpha(G)$. The α-domination was introduced by Dunbar et al. [17]. It is easy to see that for graphs without isolated vertices $\gamma(G) \leq \gamma_\alpha(G)$, and $\gamma_{\alpha_1}(G) \leq \gamma_{\alpha_2}(G)$ for $\alpha_1 < \alpha_2$. Also, $\gamma(G) = \gamma_\alpha(G)$ if α is sufficiently close to 0. In [21], we defined a set $X \subseteq V(G)$ to be an *α-rate dominating set* of G if $|N[v] \cap X| \geq \alpha d_v$ for any vertex $v \in V(G)$. The concept of α-rate domination is similar in a certain sense to the concept of k-tuple domination. The minimum cardinality of an α-rate dominating set of G is called the *α-rate domination number* $\gamma_{\times \alpha}(G)$. It is easy to see that $\gamma_\alpha(G) \leq \gamma_{\times \alpha}(G)$.

The following fundamental result was independently proved by Alon and Spencer [2], Arnautov [4], Lovász [33] and Payan [39]. Notice that a deterministic algorithm to construct a dominating set satisfying bound (4.1) can be found in [2].

Theorem 4.1 [2, 4, 33, 39] *For any graph G,*

$$\gamma(G) \leq \frac{\ln(\delta + 1) + 1}{\delta + 1} n. \tag{4.1}$$

Similar upper bounds for the double and triple domination numbers are known (see [27, 40]). For $t \leq \delta$, the *closed t-degree of a graph G* is defined as follows:

$$\tilde{d}_t = \tilde{d}_t(G) = \frac{1}{n} \sum_{i=1}^{n} \binom{d_i + 1}{t}.$$

Note that $\tilde{d}_1$ is the average degree $\bar{d} = \bar{d}(G)$ of G plus 1. Zverovich [48] and Chang [8] proved the following upper bound for the k-tuple domination number, which was originally stated as a conjecture by Rautenbach and Volkmann in [40]. Both proofs independently exploit the same model of random generation of a k-tuple dominating set [22].

Theorem 4.2 [8, 48] *For any graph G with $\delta \geq k - 1$,*

$$\gamma_{\times k}(G) \leq \frac{\ln(\delta - k + 2) + \ln \tilde{d}_{k-1} + 1}{\delta - k + 2} n. \tag{4.2}$$

Theorems 4.4–4.7 below generalize bound (4.1) and also the following Caro–Roditty bound (4.3), which is one of the strongest known upper bounds for the domination number in the class of all graphs:

Theorem 4.3 ([11], [29] p. 48) *For any graph G with $\delta \geq 1$,*

$$\gamma(G) \leq \left(1 - \frac{\delta}{(1 + \delta)^{1 + 1/\delta}}\right) n. \tag{4.3}$$

Note that the bounds of Theorems 4.1 and 4.3 are asymptotically the same, even though the latter is stronger for small values of δ.

4.1.2 k-Tuple Domination

We first consider k-tuple domination. The following theorem improves the upper bound of Theorem 4.2. Also, the probabilistic construction used in the proof of Theorem 4.4 implies randomized Algorithm 4.1 for finding a k-tuple dominating set, whose order satisfies the bound of Theorem 4.4 with positive probability. In other words, the expectation of the order of the set D returned by Algorithm 4.1 satisfies the upper bound of Theorem 4.4. Notice that Algorithm 4.1 is written on the same lines with the algorithm for finding an α-rate dominating set. For $t \leq \delta$, we define

$$\delta' = \delta - k + 1, \quad b_t = b_t(G) = \binom{\delta}{t} \quad \text{and} \quad \tilde{b}_t = \tilde{b}_t(G) = \binom{\delta+1}{t}.$$

Theorem 4.4 *For any graph G with $\delta \geq k$,*

$$\gamma_{\times k}(G) \leq \left(1 - \frac{\delta'}{\tilde{b}_{k-1}^{1/\delta'}(1+\delta')^{1+1/\delta'}}\right) n. \tag{4.4}$$

Proof: For each vertex $v \in V(G)$, we arbitrarily select δ vertices from its neighbourhood $N(v)$ and denote the resulting set by $N'(v)$. Let

$$p = 1 - 1/\left(\tilde{b}_{k-1}(1+\delta')\right)^{1/\delta'}$$

and let A be a set formed by an independent choice of vertices of G, where each vertex is selected with probability p. For $m = 0, 1, ..., k-1$, we denote

$$B_m = \{v_i \in V(G) - A : |N'(v_i) \cap A| = m\}.$$

Also, for $m = 0, 1, ..., k-2$, we denote

$$A_m = \{v_i \in A : |N'(v_i) \cap A| = m\}.$$

For each set A_m, we form a set A'_m in the following way. For every vertex $v \in A_m$, we take $k - m - 1$ neighbours from $N'(v) - A$ and add them to A'_m. Such neighbours always exist because $\delta \geq k$. It is obvious that $|A'_m| \leq (k - m - 1)|A_m|$. For each set B_m, we form a set B'_m by taking $k - m - 1$ neighbours from $N'(v) - A$ for every vertex $v \in B_m$. We have $|B'_m| \leq (k - m - 1)|B_m|$.

We construct the set D as follows:

$$D = A \cup \left(\bigcup_{m=0}^{k-2} A'_m\right) \cup \left(\bigcup_{m=0}^{k-1} B_m \cup B'_m\right).$$

It is easy to see that D is a k-tuple dominating set. The expectation of $|D|$ is

$$\mathbb{E}[|D|] \leq \mathbb{E}\left[|A| + \sum_{m=0}^{k-2}|A'_m| + \sum_{m=0}^{k-1}|B_m| + \sum_{m=0}^{k-1}|B'_m|\right]$$

$$\leq \mathbb{E}[|A|] + \sum_{m=0}^{k-2}(k-m-1)\mathbb{E}[|A_m|] + \sum_{m=0}^{k-1}(k-m)\mathbb{E}[|B_m|].$$

We have

$$\mathbb{E}[|A_m|] = \sum_{i=1}^{n}\mathbb{P}[v_i \in A_m]$$

$$= \sum_{i=1}^{n}p\binom{\delta}{m}p^m(1-p)^{\delta-m}$$

$$= p^{m+1}(1-p)^{\delta-m}b_m n$$

and

$$\mathbb{E}[|B_m|] = \sum_{i=1}^{n}\mathbb{P}[v_i \in B_m]$$

$$= \sum_{i=1}^{n}(1-p)\binom{\delta}{m}p^m(1-p)^{\delta-m}$$

$$= p^m(1-p)^{\delta-m+1}b_m n.$$

Taking into account that $b_{-1} = 0$, we obtain

$$\mathbb{E}[|D|] \leq pn + \sum_{m=0}^{k-2}(k-m-1)p^{m+1}(1-p)^{\delta-m}b_m n$$

$$+ \sum_{m=0}^{k-1}(k-m)p^m(1-p)^{\delta-m+1}b_m n$$

$$= pn + \sum_{m=1}^{k-1}(k-m)p^m(1-p)^{\delta-m+1}b_{m-1} n$$

$$+ \sum_{m=0}^{k-1}(k-m)p^m(1-p)^{\delta-m+1}b_m n$$

$$= pn + (1-p)^{\delta-k+2}n\sum_{m=0}^{k-1}(k-m)p^m(1-p)^{k-m-1}(b_{m-1}+b_m).$$

Furthermore, for $0 \le m \le k - 1$,

$$(k - m)(b_{m-1} + b_m) = (k - m)\binom{\delta + 1}{m} \le \prod_{j=1}^{\delta-k+2} \frac{(k - m + j - 1)}{j}\binom{\delta + 1}{m}$$

$$= \binom{\delta - m + 1}{\delta - k + 2}\binom{\delta + 1}{m} = \binom{k - 1}{m}\binom{\delta + 1}{k - 1} = \binom{k - 1}{m}\tilde{b}_{k-1}.$$

We obtain

$$\mathbb{E}[|D|] \le pn + (1 - p)^{\delta'+1} n\tilde{b}_{k-1} \sum_{m=0}^{k-1} \binom{k - 1}{m} p^m (1 - p)^{k-m-1}$$

$$= pn + (1 - p)^{\delta'+1} n\tilde{b}_{k-1} = \left(1 - \frac{\delta'}{\tilde{b}_{k-1}^{1/\delta'}(1 + \delta')^{1+1/\delta'}}\right) n,$$

as required. The proof of the theorem is complete.   ∎

The proof of Theorem 4.4 implies the following result, which improves the bound of Theorem 4.2 and generalizes the classical bound (4.1).

Corollary 4.1 *For any graph G with $\delta \ge k - 1$,*

$$\gamma_{\times k}(G) \le \frac{\ln(\delta - k + 2) + \ln \tilde{b}_{k-1} + 1}{\delta - k + 2} n.$$

Proof: Using the inequality $1 - p \le e^{-p}$, the proof of Theorem 4.4 implies a weaker upper bound for $\mathbb{E}[|D|]$:

$$\mathbb{E}[|D|] \le pn + e^{-p(\delta'+1)} n\tilde{b}_{k-1}.$$

The result easily follows if we set

$$p = \min\{1, \frac{\ln(\delta' + 1) + \ln \tilde{b}_{k-1}}{\delta' + 1}\}.$$

Note that if $p = 1$, then $\frac{\ln(\delta'+1)+\ln \tilde{b}_{k-1}}{\delta'+1} \ge 1$ and the upper bound is obviously true.   ∎

In some cases, Theorem 4.4 provides a much better upper bound than the bounds of Corollary 4.1 and Theorem 4.2. For example, let G be a 20-regular graph. Then, according to Corollary 4.1 (and Theorem 4.2), $\gamma_{\times 5}(G) < 0.738n$, while Theorem 4.4 yields $\gamma_{\times 5}(G) < 0.543n$. Thus, a k-tuple dominating set returned by Algorithm 4.1 in this case is expected to be much smaller than the upper bound of Theorem 4.2.

Algorithm 4.1: Randomized k-tuple dominating set {or α-rate dominating set}.

Input: A graph G and an integer k, $k \le \delta$ {or a real number α, $0 < \alpha \le 1$}.

Output: A k-tuple {or α-rate} dominating set D of G.

begin

Compute $p = 1 - 1/\left((1 + \delta')\tilde{b}_{k-1} \right)^{1/\delta'}$

{or $p' = 1 - 1/\left((1 + \widehat{\delta})\widetilde{d}_\alpha \right)^{1/\widehat{\delta}}$};

Initialize $A = \emptyset$;   /* Form a set $A \subseteq V(G)$ */

foreach vertex $v \in V(G)$ **do**

   with probability p {or p'}, decide if $v \in A$ or $v \notin A$;

end

Initialize $B = \emptyset$;   /* Form a set $B \subseteq V(G) - A$ */

foreach vertex $v \in V(G)$ **do**

   Compute $r = |N[v] \cap A|$;

   if $r < k$ {or $r < \alpha d_v$}   **then**

      if $v \in A$ **then**

         add any $k - r$ {or $\lceil \alpha d_v \rceil - r$} vertices from $N(v) - A$ into B;

      else   /* $v \notin A$ */

         add v and any $k - r - 1$ {or $\lceil \alpha d_v \rceil - r - 1$} vertices from $N(v) - A$ into B;

      end

   end

end

Set $D = A \cup B$;

   /* D is a k-tuple {or α-rate} dominating set */

   return D;

end

4.1.3 k-Domination

Let us now consider k-domination. Algorithm 4.2 presented below is a randomized algorithm for finding a k-dominating set, whose order satisfies the upper bound of Theorem 4.5 with positive probability. The algorithm is based on the probabilistic construction used in the proof of Theorem 4.5, and the expectation of the order of the set D returned by Algorithm 4.2 satisfies the upper bound of Theorem 4.5. Note that Algorithm 4.2 is written on the same lines with the algorithm for finding an α-dominating set.

Theorem 4.5 *For any graph G with $\delta \geq k$,*

$$\gamma_k(G) \leq \left(1 - \frac{\delta'}{b_{k-1}^{1/\delta'}(1+\delta')^{1+1/\delta'}} \right) n.$$

Proof: For each vertex $v \in V(G)$, we arbitrarily select δ vertices from its neighbourhood $N(v)$ and denote the resulting set by $N'(v)$. Let

$$p = 1 - 1/(b_{k-1}(1+\delta'))^{1/\delta'},$$

and let A be a set formed by an independent choice of vertices of G, where each vertex is selected with probability p. For $m = 0, 1, ..., k-1$, we denote

$$B_m = \{v_i \in V(G) - A : |N'(v_i) \cap A| = m\}.$$

Let us construct the set D as follows:

$$D = A \cup \left(\bigcup_{m=0}^{k-1} B_m \right).$$

It is easy to see that D is a k-dominating set. The expectation of $|D|$ is

$$\mathbb{E}[|D|] \leq \mathbb{E}\left[|A| + \sum_{m=0}^{k-1} |B_m| \right]$$

$$= \mathbb{E}[|A|] + \sum_{m=0}^{k-1} \mathbb{E}[|B_m|].$$

Algorithm 4.2: Randomized k-dominating set {or α-dominating set}.

Input: A graph G and an integer k, $k \le \delta$ {or a real number α,
             $0 < \alpha \le 1$}.

Output: A k-dominating {or α-dominating} set D of G.

begin

Compute $p = 1 - 1/\Big((1 + \delta') \, b_{k-1} \Big)^{1/\delta'}$

{or $p' = 1 - 1/\Big((1 + \widehat{\delta}) \widehat{d}_\alpha \Big)^{1/\widehat{\delta}}$};

Initialize $A = \emptyset$;   /* Form a set $A \subseteq V(G)$ */

foreach vertex $v \in V(G)$ **do**

  with probability p {or p'}, decide if $v \in A$ or $v \notin A$;

end

Initialize $B = \emptyset$;   /* Form a set $B \subseteq V(G) - A$ */

foreach vertex $v \in V(G) - A$ **do**

  if $|N(v) \cap A| < k$ {or $|N(v) \cap A| < \alpha d_v$}   **then**

    /* v is dominated by less than k {or αd_v}

    vertices of A */

    add v into B;

  end

end

Set $D = A \cup B$;

    /* D is a k-dominating {or α-dominating} set */

return D;

end

We have

$$\mathbb{E}[|B_m|] = \sum_{i=1}^{n} \mathbb{P}[v_i \in B_m] = \sum_{i=1}^{n} (1 - p) \binom{\delta}{m} p^m (1 - p)^{\delta - m}$$

$$= p^m (1 - p)^{\delta - m + 1} b_m n.$$

Therefore,

$$\mathbb{E}[|D|] \le pn + \sum_{m=0}^{k-1} p^m (1 - p)^{\delta - m + 1} b_m n$$

$$= pn + (1 - p)^{\delta - k + 2} n \sum_{m=0}^{k-1} p^m (1 - p)^{k - m - 1} b_m.$$

Furthermore, for $0 \leq m \leq k - 1$,

$$b_m = \binom{\delta}{m} \leq \left(\frac{\delta - m}{\delta - k + 1}\right)\binom{\delta}{m} = \binom{k-1}{m}\binom{\delta}{k-1} = \binom{k-1}{m}b_{k-1}.$$

We obtain

$$\mathbb{E}[|D|] \leq pn + (1-p)^{\delta'+1}nb_{k-1}\sum_{m=0}^{k-1}\binom{k-1}{m}p^m(1-p)^{k-m-1}$$

$$= pn + (1-p)^{\delta'+1}nb_{k-1}$$

$$= \left(1 - \frac{\delta'}{b_{k-1}^{1/\delta'}(1+\delta')^{1+1/\delta'}}\right)n,$$

as required. The proof of Theorem 4.5 is complete. ∎

An analogue of Theorem 4.2 and Corollary 4.1 for the k-domination number easily follows from Theorem 4.5:

Corollary 4.2 *For any graph G with $\delta \geq k$,*

$$\gamma_k(G) \leq \frac{\ln(\delta - k + 2) + \ln b_{k-1} + 1}{\delta - k + 2}n.$$

The proof of this corollary is similar to that of Corollary 4.1. It may be pointed out that Corollary 4.2 generalizes the classical bound (4.1).

4.1.4 α-Domination and α-Rate Domination

The concept of α-domination is different from k-domination in that a vertex must be dominated by a percentage of the vertices in its neighbourhood instead of a fixed number of its neighbours. However, the above randomized algorithms (Algorithm 4.2) for α-domination and k-domination are very similar. Intuitively, in a homogeneous WSN, since sensor nodes may fail or consume all of their energy resources in an unbalanced and poorly predictable way, it may be more effective and reasonable to dominate a sensor node by a certain percentage of its neighbourhood nodes instead of a fixed number of neighbours.

The problem of deciding whether $\gamma_\alpha(G) \leq q$ for a positive integer q is known to be NP-complete [17]. Therefore, it is important to have good upper bounds for the α-domination number and efficient algorithms for finding reasonably small α-dominating sets. The following bounds for the α-domination number were proved in [17]:

$$\frac{\alpha \delta n}{\Delta + \alpha \delta} \leq \gamma_\alpha(G) \leq \frac{\Delta n}{\Delta + (1 - \alpha)\delta} \tag{4.5}$$

and

$$\frac{2\alpha m}{(1 + \alpha)\Delta} \leq \gamma_\alpha(G) \leq \frac{(2 - \alpha)\Delta n - 2(1 - \alpha)m}{(2 - \alpha)\Delta}, \tag{4.6}$$

where m is the number of edges in G. Notice that the bounds in (4.6) can be rewritten without m taking into account that

$$2m = \sum_{i=1}^{n} d_i = (\tilde{d}_1 - 1)n = \bar{d}n.$$

For $0 < \alpha \leq 1$, the α-degree of a graph G is defined as follows:

$$\widehat{d}_\alpha = \widehat{d}_\alpha(G) = \frac{1}{n} \sum_{i=1}^{n} \left(\frac{d_i}{\lceil \alpha d_i \rceil - 1} \right).$$

Also, we denote

$$\widehat{\delta} = \lfloor \delta(1 - \alpha) \rfloor + 1.$$

The following theorem generalizes the upper bound (4.3) for the α-domination number. Indeed, if $d_i \geq 1$ are fixed for all $i = 1, \ldots, n$, and α is sufficiently close to 0, then $\widehat{\delta} = \delta$ and $\widehat{d}_\alpha = 1$. Notice that in some cases Theorem 4.6 provides a much better bound than the upper bounds in (4.5) and (4.6). For example, if G is a 1000-regular graph, then Theorem 4.6 gives $\gamma_{0.1}(G) < 0.305n$, while (4.5) and (4.6) yield only $\gamma_{0.1}(G) < 0.527n$.

Theorem 4.6 *For any graph G,*

$$\gamma_\alpha(G) \leq \left(1 - \frac{\widehat{\delta}}{(1 + \widehat{\delta})^{1 + 1/\widehat{\delta}} \, \widehat{d}_\alpha^{1/\widehat{\delta}}} \right) n. \tag{4.7}$$

Proof: Let A be a set formed by an independent choice of vertices of G, where each vertex is selected with probability

$$p = 1 - \left(\frac{1}{(1+\widehat{\delta})\widehat{d}_\alpha}\right)^{1/\widehat{\delta}}. \tag{4.8}$$

Let us denote

$$B = \{v_i \in V(G) - A \,:\, |N(v_i) \cap A| \leq \lceil \alpha d_i \rceil - 1\}.$$

It is obvious that the set $D = A \cup B$ is an α-dominating set. The expectation of $|D|$ is

$$\begin{aligned}
\mathbb{E}[|D|] &= \mathbb{E}[|A|] + \mathbb{E}[|B|] \\
&= \sum_{i=1}^{n} \mathbb{P}[v_i \in A] + \sum_{i=1}^{n} \mathbb{P}[v_i \in B] \\
&= pn + \sum_{i=1}^{n} \sum_{r=0}^{\lceil \alpha d_i \rceil - 1} \binom{d_i}{r} p^r (1-p)^{d_i - r + 1}.
\end{aligned}$$

It is easy to see that, for $0 \leq r \leq \lceil \alpha d_i \rceil - 1$,

$$\binom{d_i}{r} \leq \binom{d_i}{\lceil \alpha d_i \rceil - 1} \binom{\lceil \alpha d_i \rceil - 1}{r}.$$

Also,

$$d_i - \lceil \alpha d_i \rceil \geq \lfloor \delta(1 - \alpha) \rfloor.$$

Therefore,

$$\begin{aligned}
\mathbb{E}[|D|] &\leq pn + \sum_{i=1}^{n} \binom{d_i}{\lceil \alpha d_i \rceil - 1} (1-p)^{d_i - \lceil \alpha d_i \rceil + 2} \\
&\qquad \times \sum_{r=0}^{\lceil \alpha d_i \rceil - 1} \binom{\lceil \alpha d_i \rceil - 1}{r} p^r (1-p)^{\lceil \alpha d_i \rceil - 1 - r} \\
&= pn + \sum_{i=1}^{n} \binom{d_i}{\lceil \alpha d_i \rceil - 1} (1-p)^{d_i - \lceil \alpha d_i \rceil + 2} \\
&\leq pn + (1-p)^{\lfloor \delta(1-\alpha) \rfloor + 2} \widehat{d}_\alpha n \\
&= pn + (1-p)^{\widehat{\delta}+1} \widehat{d}_\alpha n \tag{4.9} \\
&= \left(1 - \frac{\widehat{\delta}}{(1+\widehat{\delta})^{1+1/\widehat{\delta}} \widehat{d}_\alpha^{1/\widehat{\delta}}}\right) n.
\end{aligned}$$

Note that the value of p in (4.8) is chosen to minimize the expression in line (4.9). Since the expectation is an average value, there exists a particular α-dominating set of order at most

$$\left(1 - \frac{\widehat{\delta}}{(1+\widehat{\delta})^{1+1/\widehat{\delta}}\,\widehat{d}_{\alpha}^{1/\widehat{\delta}}}\right) n,$$

as required. The proof of the theorem is complete. ∎

Algorithm 4.2, written on the same lines with the algorithm for finding a k-dominating set, is a randomized algorithm for finding an α-dominating set D, whose order satisfies the upper bound of Theorem 4.6 with positive probability. In other words, the expectation of the order of set D returned by Algorithm 4.2 satisfies the upper bound of Theorem 4.6. Also, Theorem 4.6 easily implies the following generalization of the well-known bound of Theorem 4.1:

$$\gamma_{\alpha}(G) \leq \frac{\ln(\widehat{\delta}+1) + \ln \widehat{d}_{\alpha} + 1}{\widehat{\delta}+1}\,n.$$

Now, let us consider α-rate domination, which combines the concepts of α-domination and k-tuple domination. For $0 < \alpha \leq 1$, the *closed α-degree of a graph G* is defined as follows:

$$\tilde{d}_{\alpha} = \tilde{d}_{\alpha}(G) = \frac{1}{n}\sum_{i=1}^{n}\left(\frac{d_i+1}{\lceil \alpha d_i \rceil - 1}\right).$$

In fact, the only difference between the α-degree and the closed α-degree is that to compute the latter, we choose from $d_i + 1$ vertices instead of d_i, that is, from the *closed* neighbourhood $N[v_i]$ of v_i instead of $N(v_i)$, $i = 1, 2, ..., n$.

Algorithm 4.1 above, written on the same lines with the algorithm for finding a k-tuple dominating set, is a randomized algorithm for finding an α-rate dominating set D. The expectation of the order of the α-rate dominating set D returned by Algorithm 4.1 satisfies the upper bound of Theorem 4.7. Also, Theorem 4.7 provides an analogue of the Caro–Roditty bound (Theorem 4.3) for the α-rate domination number.

Theorem 4.7 *For any graph G and $0 < \alpha \leq 1$,*

$$\gamma_{\times\alpha}(G) \leq \left(1 - \frac{\widehat{\delta}}{(1+\widehat{\delta})^{1+1/\widehat{\delta}}\,\tilde{d}_{\alpha}^{1/\widehat{\delta}}}\right) n. \qquad (4.10)$$

Proof: Let A be a set formed by an independent choice of vertices of G, where each vertex is selected with probability p, $0 \le p \le 1$. For $m \ge 0$, denote by B_m the set of vertices $v \in V(G)$ dominated by exactly m vertices of A and such that $|N[v] \cap A| < \alpha d_v$; that is,

$$|N[v] \cap A| = m \le \lceil \alpha d_v \rceil - 1.$$

Note that each vertex $v \in V(G)$ is in at most one of the sets B_m and $0 \le m \le \lceil \alpha d_v \rceil - 1$. We form a set B in the following way: for each vertex $v \in B_m$, select $\lceil \alpha d_v \rceil - m$ vertices from $N(v)$ that are not in A and add them to B. Consider the set $D = A \cup B$. It is easy to see that D is an α-rate dominating set. Taking into account that

$$(\lceil \alpha d_i \rceil - m) \binom{d_i + 1}{m} \le \binom{d_i + 1}{\lceil \alpha d_i \rceil - 1} \binom{\lceil \alpha d_i \rceil - 1}{m},$$

the expectation of $|D|$ is as follows:

$$\mathbb{E}[|D|] \le \mathbb{E}[|A|] + \mathbb{E}[|B|]$$

$$\le \sum_{i=1}^{n} \mathbb{P}[v_i \in A] + \sum_{i=1}^{n} \sum_{m=0}^{\lceil \alpha d_i \rceil - 1} (\lceil \alpha d_i \rceil - m) \mathbb{P}[v_i \in B_m]$$

$$= pn + \sum_{i=1}^{n} \sum_{m=0}^{\lceil \alpha d_i \rceil - 1} (\lceil \alpha d_i \rceil - m) \binom{d_i + 1}{m} p^m (1-p)^{d_i + 1 - m}$$

$$\le pn + \sum_{i=1}^{n} \sum_{m=0}^{\lceil \alpha d_i \rceil - 1} \binom{d_i + 1}{\lceil \alpha d_i \rceil - 1} \binom{\lceil \alpha d_i \rceil - 1}{m} p^m (1-p)^{d_i + 1 - m}$$

$$= pn + \sum_{i=1}^{n} \binom{sd_i + 1}{\lceil \alpha d_i \rceil - 1} (1-p)^{d_i - \lceil \alpha d_i \rceil + 2}$$

$$\times \sum_{m=0}^{\lceil \alpha d_i \rceil - 1} \binom{\lceil \alpha d_i \rceil - 1}{m} p^m (1-p)^{\lceil \alpha d_i \rceil - 1 - m}$$

$$= pn + \sum_{i=1}^{n} \binom{d_i + 1}{\lceil \alpha d_i \rceil - 1} (1-p)^{d_i - \lceil \alpha d_i \rceil + 2}$$

$$\le pn + (1-p)^{\lfloor \delta(1-\alpha) \rfloor + 2} \sum_{i=1}^{n} \binom{d_i + 1}{\lceil \alpha d_i \rceil - 1}$$

$$= pn + (1-p)^{\widehat{\delta} + 1} \widetilde{d}_\alpha n.$$

Thus,

$$\mathbb{E}[|D|] \leq pn + (1-p)^{\widehat{\delta}+1}\widetilde{d}_\alpha n. \tag{4.11}$$

Minimizing the expression (4.11) with respect to p, we obtain

$$\mathbb{E}[|D|] \leq \left(1 - \frac{\widehat{\delta}}{(1+\widehat{\delta})^{1+1/\widehat{\delta}}\,\widetilde{d}_\alpha^{1/\widehat{\delta}}}\right)n,$$

as required. This completes the proof.   ∎

Notice that, similar to Theorem 4.6, Theorem 4.7 implies the following generalization of the classical upper bound (4.1):

$$\gamma_{\times\alpha}(G) \leq \frac{\ln(\widehat{\delta}+1) + \ln\widetilde{d}_\alpha + 1}{\widehat{\delta}+1}n.$$

4.1.5 Complexity and Implementation of the Algorithms

For complexity analysis of the presented algorithms, we consider sequential implementation of the algorithms for k- and k-tuple dominating sets. The complexity analysis of the algorithms for α- and α-rate dominating sets can be done in a similar way.

An essential part of the algorithms is to compute the binomial coefficients $\binom{a}{b}$. By definition,

$$\binom{a}{b} = \frac{a!}{b!(a-b)!} = \frac{a(a-1)...(a-b+1)}{b!} = \frac{a(a-1)...(b+1)}{(a-b)!},$$

and this expression can be computed in $O(a)$ time in terms of elementary operations of multiplication, division, addition and subtraction. However, since in the worst-case scenario the required memory usage to store the products is $O(a\log a)$, writing to (reading from) the memory would require $O(a\log a)$ time. In practice, to overcome the memory and reading (writing) operations requirements (e.g. see pp. 93–96 in [37]), the commonly used approach is to compute the binomial coefficient by using dynamic programming and Pascal's triangle. In this case, the time complexity to compute the binomial coefficient would be $O(ab) = O(a^2)$, and the memory usage is $O(b) = O(a)$.

We assume that computing the binomial coefficient is done in $O(a^2)$ time and that an input graph G has no isolated vertices. It is easy to see that the minimum vertex degree δ of G can be computed in $O(m)$ time, where m is the number of edges in G. We will show that Algorithm 4.1 can take up to $O(m) = O(m + n)$ time, where $n = |V(G)|$. More precisely, in reference to Algorithm 4.1, a worst-case scenario when k is close to $\delta/2$ may require $O(\delta^2)$ steps to compute $\tilde{b}_{k-1}$, and δ' can be computed in $O(1)$. Therefore, in total, it takes $O(\delta^2)$ steps to compute the probability p. Note that $O(\delta^2)$ does not exceed $O(m)$. Clearly, it takes $O(n)$ time to find the set A. The numbers $r = |N[v] \cap A|$ for each $v \in V(G)$ can be computed separately or when finding the set A. In any case, we need to keep track of them only up to $r = k$. We may need to browse through all the neighbours of vertices in A, hence in total it can take $O(m)$ steps to calculate all the necessary r for each vertex $v \in V(G)$. Then, the set B can also be found in $O(m)$ steps. Thus, in total, Algorithm 4.1 runs in $O(m)$ time. For Algorithm 4.2, a complexity analysis similar to that of Algorithm 4.1 shows that it can take up to $O(m)$ steps to find a k-dominating set.

Algorithms 4.1 and 4.2 are presented here in a form consistent with the proofs of the corresponding theorems. However, when implementing these algorithms, the output sets D can be constructed more efficiently and effectively by a recursive extension of the corresponding initial set A. In other words, instead of adding missing vertices into the sets B, we can add them directly into A. This can result in a smaller k-tuple, k-, α- or α-rate dominating set D, respectively.

It is easy to see that, as soon as the probability p (resp., p') is known to all the vertices (sensor nodes in a WSN), Algorithms 4.1 and 4.2 can be easily and efficiently implemented in parallel or as local distributed algorithms. This is particularly important in case of WSNs (see [38] for details). To compute the probability p (resp., p') and to distribute its value to all the network nodes (graph vertices) in a WSN, one needs to use a data gathering round and a data distribution round coordinated from a base station or a selected supernode (vertex). To construct the corresponding multiple dominating set for the whole network (graph) after p is distributed to all the network nodes, each node (graph vertex) only needs to gather and communicate information locally in its own neighbourhood. It would be interesting to design reasonable online versions of these algorithms for a realistic case scenario, where the network changes dynamically and information about the whole network and local neighbourhoods is disclosed gradually in time.

4.2 Limited Packings in Graphs

In this section, we consider the classical packings and packing numbers of graphs as introduced in [34], and their generalization, called limited packings and limited packing numbers, respectively, as presented in [24]. Limited packing problems can be considered as secure facility location problems in networks. In the literature, the classical packings are often referred to under different names: (distance) 2-packings [34, 46], closed neighbourhood packings [41] or strong stable sets [31]. They can also be considered as generalizations of independent (stable) sets which, following the terminology of [34], would be (distance) 1-packings.

A vertex set X in a graph G is a *k-limited packing* if for every vertex $v \in V(G)$,

$$|N[v] \cap X| \le k,$$

where $N[v]$ is the closed neighbourhood of v. The *k-limited packing number* $L_k(G)$ of a graph G is the maximum size of a k-limited packing in G. In these terms, the classical (distance) 2-packings are 1-limited packings, and hence $\rho(G) = L_1(G)$, where $\rho(G)$ is the 2-packing number. The problem of finding a 2-packing (1-limited packing) of maximum size was proved to be NP-hard by Hochbaum and Schmoys [31]. As shown in [16], the problem of finding a maximum-size k-limited packing is NP-hard even for the classes of split and bipartite graphs.

A number of interesting application scenarios for limited packings were described in [24], including network security, market saturation and codes. These and other potential applications can be summarized as secure placement or distribution of facilities in a network. In a more general sense, these problems can be viewed as (maximization) facility location problems to place/distribute in a given network as many resources as possible subject to some (security) constraints. In advanced networks, node activity can be controlled to make 'damages' (e.g. radio interferences) to facilities or sensors of the 'enemy' side. In this scenario, to limit damaging influence on own sensors, it is necessary to choose a set of sensors that would have a small number of neighbouring own sensors, but have this set remain reasonably large. This property is achieved by a 1-limited packing set in the corresponding underlying graph of the network. If such a set is too small in a given network then one might wish to sacrifice more own sensors and parts of the network by considering larger k-limited packings for $k > 1$.

2-Packings (1-limited packings) are well studied in the literature from the structural and algorithmic points of view (e.g. see [31, 34, 41, 43]) and in connection with other graph parameters (e.g. see [7, 30, 34, 41, 46]). In particular, several papers discussed connections between packings and dominating sets in graphs (e.g. see [7, 16, 24, 30, 41]). Although the formal definitions for packings and dominating sets may appear to be similar, the problems have a very different nature: one of the problems is a maximization problem not to break some (security) constraints, and the other is a minimization problem to satisfy some reliability requirements. For example, given a graph G, the concept definitions imply a simple inequality $\rho(G) \leq \gamma(G)$ (e.g. see [41]). However, the difference between $\rho(G)$ and $\gamma(G)$ can be arbitrarily large as illustrated in [7]: $\rho(K_n \times K_n) = 1$ for the Cartesian product of complete graphs, but $\gamma(K_n \times K_n) = n$.

In this section, we discuss an application of the probabilistic method to k-limited packings in general and to 2-packings (1-limited packings) in particular. The probabilistic construction is used to derive two lower bounds for the k-limited packing number $L_k(G)$. Using a 'greedy' approach, an improved lower bound for the 2-packing (1-limited packing) number $\rho(G) = L_1(G)$ is obtained. The probabilistic construction implies a randomized algorithm for finding k-limited packings satisfying the lower bounds. Also, we show that the main lower bound is asymptotically sharp. The presented probabilistic construction and approach are different from the well-known probabilistic constructions used for independent sets (e.g. see [2], pp. 27–28). In terms of packings, an independent set in a graph G is a distance 1-packing: for any two vertices in an independent set, the distance between them in G is greater than 1.

4.2.1 The Probabilistic Construction and Lower Bounds

As usual, $\Delta = \Delta(G)$ denotes the maximum vertex degree in a graph G. Notice that $L_k(G) = n$ when $k \geq \Delta + 1$. We define

$$c_t = c_t(G) = \binom{\Delta}{t} \quad \text{and} \quad \tilde{c}_t = \tilde{c}_t(G) = \binom{\Delta + 1}{t}.$$

In what follows, we set $\binom{a}{b} = 0$ if $b > a$.

The following theorem gives a lower bound for the k-limited packing number. It may be pointed out that the probabilistic construction used in the proof of Theorem 4.8 implies a randomized algorithm for finding a k-limited packing set, whose size satisfies the bound of Theorem 4.8 with positive probability (see Algorithm 4.3 in the next section).

Theorem 4.8 *For any graph G of order n with $\Delta \geq k \geq 1$,*

$$L_k(G) \geq \frac{kn}{\tilde{c}_{k+1}^{1/k} \, (1+k)^{1+1/k}}. \tag{4.12}$$

Proof: Let A be a set formed by an independent choice of vertices of G, where each vertex is selected with probability

$$p = \left(\frac{1}{\tilde{c}_{k+1} \, (1+k)} \right)^{1/k}. \tag{4.13}$$

For $m = k, ..., \Delta$, let us denote

$$A_m = \{v \in A : |N(v) \cap A| = m\}.$$

For each set A_m, we form a set A'_m in the following way. For every vertex $v \in A_m$, take $m - (k-1)$ neighbours from $N(v) \cap A$ and add them to A'_m. Such neighbours always exist because $m \geq k$. It is obvious that

$$|A'_m| \leq (m - k + 1)|A_m|.$$

For $m = k+1, ..., \Delta$, let us consider

$$B_m = \{v \in V(G) - A : |N(v) \cap A| = m\}.$$

For each set B_m, we form a set B'_m by taking $m - k$ neighbours from $N(v) \cap A$ for every vertex $v \in B_m$. This implies

$$|B'_m| \leq (m - k)|B_m|.$$

Let us construct the set X as follows:

$$X = A - \left(\bigcup_{m=k}^{\Delta} A'_m \right) - \left(\bigcup_{m=k+1}^{\Delta} B'_m \right).$$

It is easy to see that X is a k-limited packing in G. The expectation of $|X|$ is

$$
\mathbb{E}[|X|] \geq \mathbb{E}\left[|A| - \sum_{m=k}^{\Delta} |A'_m| - \sum_{m=k+1}^{\Delta} |B'_m|\right]
$$

$$
\geq \mathbb{E}\left[|A| - \sum_{m=k}^{\Delta} (m-k+1)|A_m| - \sum_{m=k+1}^{\Delta} (m-k)|B_m|\right]
$$

$$
= pn - \sum_{m=k}^{\Delta} (m-k+1)\mathbb{E}[|A_m|] - \sum_{m=k+1}^{\Delta} (m-k)\mathbb{E}[|B_m|].
$$

Let us denote the vertices of G by $v_1, v_2, ..., v_n$ and the corresponding vertex degrees by $d_1, d_2, ..., d_n$. We will need the following lemma:

Lemma 4.1 *If* $p = \left(\frac{1}{\tilde{c}_{k+1}\,(1+k)}\right)^{1/k}$, *then for any vertex* $v_i \in V(G)$,

$$
\binom{d_i}{m}(1-p)^{d_i-m} \leq \binom{\Delta}{m}(1-p)^{\Delta-m}, \quad m \geq k. \tag{4.14}
$$

Proof: The inequality (4.14) holds if $d_i = \Delta$. It is also true if $d_i < m$ because in this case $\binom{d_i}{m} = 0$. Thus, we can assume that

$$
m \leq d_i < \Delta.
$$

It is easy to see that inequality (4.14) is equivalent to the following:

$$
(1-p)^{\Delta-d_i} \geq \frac{\binom{d_i}{m}}{\binom{\Delta}{m}} = \frac{(\Delta-m)!/(d_i-m)!}{\Delta!/d_i!} = \prod_{i=0}^{\Delta-d_i-1} \frac{\Delta-m-i}{\Delta-i}.
$$
$$\tag{4.15}$$

Further, $\Delta \geq k$ implies $\frac{\Delta}{k} \leq \frac{\Delta-i}{k-i}$, where $0 \leq i \leq k-1$. Taking into account that $\Delta > 0$, we obtain

$$
\left(\frac{\Delta}{k}\right)^k \leq \prod_{i=0}^{k-1} \frac{\Delta-i}{k-i} = c_k < \tilde{c}_{k+1}(1+k)
$$

or

$$\frac{1}{\tilde{c}_{k+1}(1+k)} < \left(\frac{k}{\Delta}\right)^k.$$

Thus,

$$p^k < \left(\frac{k}{\Delta}\right)^k \quad \text{or} \quad p < \frac{k}{\Delta} \le \frac{m}{\Delta}.$$

We have $p < \frac{m}{\Delta}$, which is equivalent to $1 - p > \frac{\Delta - m}{\Delta}$. Therefore,

$$(1-p)^{\Delta - d_i} > \left(\frac{\Delta - m}{\Delta}\right)^{\Delta - d_i} \ge \prod_{i=0}^{\Delta - d_i - 1} \frac{\Delta - m - i}{\Delta - i},$$

as required in (4.15). ∎

Let us continue the proof of Theorem 4.8. By Lemma 4.1,

$$\mathbb{E}[|A_m|] = \sum_{i=1}^{n} \mathbb{P}[v_i \in A_m]$$
$$= \sum_{i=1}^{n} p \binom{d_i}{m} p^m (1-p)^{d_i - m}$$
$$\le p^{m+1} \sum_{i=1}^{n} \binom{\Delta}{m} (1-p)^{\Delta - m}$$
$$= p^{m+1} (1-p)^{\Delta - m} c_m n,$$

where $p \binom{d_i}{m} p^m (1-p)^{d_i - m}$ is the probability of having vertex v_i, $i = 1, ..., n$, in the set A_m, $m = k, ..., \Delta$. Again, by Lemma 4.1,

$$\mathbb{E}[|B_m|] = \sum_{i=1}^{n} \mathbb{P}[v_i \in B_m]$$
$$= \sum_{i=1}^{n} (1-p) \binom{d_i}{m} p^m (1-p)^{d_i - m}$$
$$\le p^m \sum_{i=1}^{n} \binom{\Delta}{m} (1-p)^{\Delta - m + 1}$$
$$= p^m (1-p)^{\Delta - m + 1} c_m n,$$

where $(1-p)\binom{d_i}{m}p^m(1-p)^{d_i-m}$ is the probability of having vertex v_i, $i=1,...,n$, in the set B_m, $m=k+1,...,\Delta$.

Taking into account that $c_{\Delta+1}=\binom{\Delta}{\Delta+1}=0$, we obtain

$$\mathbb{E}[|X|]\geq pn-\sum_{m=k}^{\Delta}(m-k+1)p^{m+1}(1-p)^{\Delta-m}c_m n$$

$$-\sum_{m=k+1}^{\Delta+1}(m-k)p^m(1-p)^{\Delta-m+1}c_m n$$

$$=pn-\sum_{m=0}^{\Delta-k}(m+1)p^{m+k+1}(1-p)^{\Delta-m-k}c_{m+k}n$$

$$-\sum_{m=0}^{\Delta-k}(m+1)p^{m+k+1}(1-p)^{\Delta-m-k}c_{m+k+1}n$$

$$=pn-\sum_{m=0}^{\Delta-k}(m+1)p^{m+k+1}(1-p)^{\Delta-m-k}n\left(c_{m+k}+c_{m+k+1}\right)$$

$$=pn-p^{k+1}n\sum_{m=0}^{\Delta-k}(m+1)\tilde{c}_{m+k+1}p^m(1-p)^{\Delta-k-m}.$$

Furthermore,

$$(m+1)\tilde{c}_{m+k+1}=\binom{\Delta-k}{m}\frac{(m+1)!(\Delta+1)!}{(m+k+1)!(\Delta-k)!}$$

$$\leq\binom{\Delta-k}{m}\frac{(\Delta+1)!}{(k+1)!(\Delta-k)!}=\binom{\Delta-k}{m}\tilde{c}_{k+1}.$$

The binomial theorem implies

$$\mathbb{E}[|X|]\geq pn-p^{k+1}n\sum_{m=0}^{\Delta-k}\binom{\Delta-k}{m}\tilde{c}_{k+1}p^m(1-p)^{\Delta-k-m}$$

$$=pn-p^{k+1}n\tilde{c}_{k+1}$$

$$=pn(1-p^k\tilde{c}_{k+1})$$

$$=\frac{kn}{\tilde{c}_{k+1}^{1/k}(1+k)^{1+1/k}}.$$

Since the expectation is an average value, there exists a particular k-limited packing of size at least $\frac{kn}{\tilde{c}_{k+1}^{1/k}\,(1+k)^{1+1/k}}$, as required. The proof of the theorem is complete. ∎

The lower bound of Theorem 4.8 can be written in a simpler but weaker form as follows:

Corollary 4.3 *For any graph G of order n,*

$$L_k(G) > \frac{kn}{e(1+\Delta)^{1+1/k}}.$$

Proof: It is not difficult to see that

$$\tilde{c}_{k+1} \le \frac{(\Delta+1)^{k+1}}{(k+1)!}$$

and, using Stirling's formula,

$$(k!)^{1/k} > \left(\sqrt{2\pi k}\left(\frac{k}{e}\right)^k\right)^{1/k} = \sqrt[2k]{2\pi k}\,\frac{k}{e}.$$

By Theorem 4.8,

$$L_k(G) \ge \frac{kn\,((k+1)!)^{1/k}}{(\Delta+1)^{1+1/k}\,(1+k)^{1+1/k}}$$

$$> \frac{kn}{e(1+\Delta)^{1+1/k}} \times \frac{\sqrt[2k]{2\pi k}\,k}{1+k}$$

$$> \frac{kn}{e(1+\Delta)^{1+1/k}}.$$

Notice that $\frac{\sqrt[2k]{2\pi k}\,k}{1+k} = \frac{\sqrt[2k]{2\pi k}}{1+1/k} > 1$. The last inequality is obviously true for $k=1$, whereas for $k \ge 2$ it can be rewritten in the equivalent form: $2\pi k > (1+1/k)^{2k} = e^2 - o(1)$. ∎

Using an example based on projective spaces, Balister, Bollobás and Gunderson [5] showed that the lower bounds for $L_k(G)$ in Corollary 4.3 is asymptotically best possible up to the constant factor $1/e$, when k is fixed and $\Delta(G)$ tends to infinity.

In the case $k=1$, Theorem 4.8 gives the following lower bound for the 2-packing (1-limited packing) number:

Corollary 4.4 *For any graph G of order n with $\Delta \geq 1$,*

$$\rho(G) = L_1(G) \geq \frac{n}{2\Delta(\Delta + 1)}. \tag{4.16}$$

The lower bound of Corollary 4.4 can be improved as follows:

Theorem 4.9 *For any graph G of order n,*

$$\rho(G) = L_1(G) \geq \frac{n + \Delta(\Delta - \delta)}{\Delta^2 + 1} \geq \frac{n}{\Delta^2 + 1}. \tag{4.17}$$

Proof: Choose any vertex $v \in V(G)$ of the minimum degree δ in G. Then, add v to a set X and remove vertices of $N[N[v]]$ from the graph to obtain $G' = G - N[N[v]]$, where $N[N[v]] = \{w \ : \ w \in N[u]$ for some $u \in N[v]\}$ is the so-called *second closed neighbourhood* of v in G. Recursively apply the same procedure to the remaining graph G' until it is empty. It is not difficult to see that X is a 1-limited packing (distance 2-packing) of size at least $\left\lceil \frac{n + \Delta(\Delta - \delta)}{\Delta^2 + 1} \right\rceil$: we remove at most $1 + \Delta + \Delta(\Delta - 1) = 1 + \Delta^2$ vertices at each iteration, but at most $1 + \delta + \delta(\Delta - 1) = 1 + \delta\Delta$ vertices at the first iteration, and $(1 + \Delta^2) - (1 + \delta\Delta) = \Delta(\Delta - \delta)$. ∎

The proof of Theorem 4.9 provides a greedy algorithm for finding a distance 2-packing (1-limited packing) satisfying bound (4.17). We will explain later why the lower bound of Theorem 4.9 is as good as the lower bound (4.16) of Corollary 4.4 for almost all graphs.

4.2.2 Randomized Algorithm

A pseudocode presented in Algorithm 4.3 explicitly describes a randomized algorithm for finding a k-limited packing set, whose size satisfies bound (4.12) with positive probability. Notice that Algorithm 4.3 constructs a (preliminary) k-limited packing X' by recursively removing unwanted vertices from a randomly generated set A. This is different from the probabilistic construction used in the proof of Theorem 4.8. The recursive removal of vertices from the set A may be more effective and efficient, especially if one tries to remove overall as few vertices as possible from A by maximizing intersections of the sets A'_m $(m = k, ..., \Delta)$ and B'_m $(m = k + 1, ..., \Delta)$.

Algorithm 4.3: Randomized k-limited packing.

Input: Graph G and integer k, $1 \leq k \leq \Delta$.
Output: k-Limited packing X in G.

begin

> Compute $p = \left(\frac{1}{\bar{c}_{k+1} \, (1+k)} \right)^{1/k}$;
>
> Initialize $A = \emptyset$;   /* Form a set $A \subseteq V(G)$ */
> **foreach** vertex $v \in V(G)$ **do**
> > with probability p, decide whether $v \in A$ or $v \notin A$;
>
> **end**
> > /* Recursively remove redundant vertices from A
> */
> **foreach** vertex $v \in V(G)$ **do**
> > Compute $r = |N(v) \cap A|$;
> > **if** $v \in A$ and $r \geq k$ **then**
> > > remove any $r - k + 1$ vertices of $N(v) \cap A$ from A;
> >
> > **end**
> > **if** $v \notin A$ and $r > k$ **then**
> > > remove any $r - k$ vertices of $N(v) \cap A$ from A;
> >
> > **end**
>
> **end**
> Set $X' = A$;   /* X' is a k-limited packing */
> Extend X' to a maximal k-limited packing X;
> **return** X;

end

At the final stage, Algorithm 4.3 produces a (greedy) extension of the preliminary k-limited packing X' derived from the randomly generated set A. Experiments with randomly generated problem instances show the following: although the randomized part of Algorithm 4.3 may eventually return a preliminary k-limited packing set slightly smaller than lower bound (4.12), the extension of this set to a maximal k-limited packing always satisfies (4.12). This is of no surprise because the expectation of the size of the randomly formed set A in Algorithm 4.3 is $\mathbb{E}[|A|] = pn$, where

$$p = \left(\binom{\Delta}{k} (\Delta + 1) \right)^{-1/k},$$

whereas the expression for the lower bound in (4.12) yields a smaller value:

$$\frac{kn}{\tilde{c}_{k+1}^{1/k}\,(1+k)^{1+1/k}} = \frac{k}{k+1}pn = \frac{k}{k+1}\mathbb{E}[|A|] < \mathbb{E}[|A|].$$

From the experiments, an initially formed set A may contain only few redundant vertices to be removed to obtain the preliminary k-limited packing X'. As a result, the preliminary k-limited packing X' in many cases satisfies the lower bound (4.12), and the extension of X' to a maximal k-limited packing X seems to always satisfy (4.12). Since the problem of finding a largest-size k-limited packing is NP-hard, Algorithm 4.3 constitutes a simple efficient approach to tackle the problem in practice and, hopefully, can be useful to solve some hard instances of the problem.

Algorithm 4.3 can be implemented to run in $O(n^2)$ time. To compute the probability

$$p = \left(\binom{\Delta}{k}(\Delta+1)\right)^{-1/k},$$

the binomial coefficient $\binom{\Delta}{k}$ can be computed by using the dynamic programming and Pascal's triangle in $O(k\Delta) = O(\Delta^2)$ time using $O(k) = O(\Delta)$ memory. The maximum vertex degree Δ of G can be computed in $O(m)$ time, where m is the number of edges in G. Then, p can be computed in $O(m + \Delta^2) = O(n^2)$ steps. It takes $O(n)$ time to find the initial set A. Computing the intersection numbers $r = |N(v) \cap A|$ and removing unwanted vertices of $N(v) \cap A$ from A can be done in $O(n + m)$ steps. Finally, checking whether X' is maximal and extending X' to a maximal k-limited packing X can be done in $O(n + m)$ time: try adding vertices of $V(G) - X'$ to X' recursively one by one, and check whether the addition of a new vertex $v \in V(G) - X'$ to X' violates the conditions of a k-limited packing for v or its neighbours in G with respect to $X' \cup \{v\}$. Thus, overall Algorithm 4.3 takes $O(n^2)$ time, and since $m = O(n^2)$, it is linear in the graph size $(m + n)$ when $m = \Theta(n^2)$.

Notice that this randomized algorithm for finding k-limited packings in a graph G can be implemented in parallel or as a local distributed algorithm. Such algorithms are especially important, for example in the context of wireless sensor and ad hoc networks. We hope that this approach can also be extended to design self-stabilizing and online algorithms for k-limited packings. For example, a self-stabilizing algorithm searching for maximal 2-packings in a distributed network system is presented in [43]. Notice that self-stabilizing algorithms are distributed and fault tolerant and they use the fact that each node has only a local view/knowledge of the distributed network system. This provides further motivation for efficient distributed algorithms for finding k-limited packings in graphs and networks.

4.2.3 Sharpness of the Lower Bounds

It turns out that the lower bound of Theorem 4.8 is asymptotically best possible for some large values of k. The bound of Theorem 4.8 can be rewritten in the following form for $\Delta \geq k$:

$$L_k(G) \geq \frac{kn}{(k+1)\sqrt[k]{\binom{\Delta}{k}}(\Delta+1)}.$$

Combining this bound with the upper bound of Lemma 8 from [24], we obtain that for any connected graph G of order n with minimum degree $\delta(G) \geq k$,

$$\frac{1}{\sqrt[k]{\binom{\Delta}{k}}(\Delta+1)} \times \frac{k}{k+1} n \leq L_k(G) \leq \frac{k}{k+1} n. \qquad (4.18)$$

Notice that the upper bound in inequality (4.18) is sharp (see [24]), hence these bounds provide an interval of values for $L_k(G)$ in terms of k and Δ when $k \leq \delta$. For regular graphs, $\delta = \Delta$, and when $k = \Delta$, we have

$$\frac{1}{\sqrt[k]{\binom{\Delta}{k}}(\Delta+1)} = \frac{1}{(k+1)^{1/k}} \longrightarrow 1 \quad \text{as} \quad k \to \infty.$$

Therefore, the bound of Theorem 4.8 is asymptotically sharp for regular connected graphs in the case $k = \Delta$. In other words, for large values of n, there are graphs whose k-limited packing number is arbitrarily close to the bound of Theorem 4.8. Thus, the following result holds:

Theorem 4.10 *When n is large, there exist graphs G such that*

$$L_k(G) \leq \frac{kn}{\tilde{c}_{k+1}^{1/k} (1+k)^{1+1/k}} (1+o(1)). \qquad (4.19)$$

As explained above, the graphs satisfying Theorem 4.10 contain regular connected graphs for which $k = \Delta$. This class of graphs can be extended— it is possible to prove that the bound of Theorem 4.8 is asymptotically sharp for connected graphs with $k = \Delta(1 - o(1))$ and $\delta(G) \geq k$.

Notice that, for regular graphs, the condition $k = \Delta \geq 1$ and Lemma 5 from [24] imply $L_k(G) = n - \gamma(G)$. Substituting here the classical upper bound of Theorem 4.1 for $\gamma(G)$, we obtain a weaker lower bound for $L_k(G)$ than the bound of Theorem 4.8.

As shown in Theorem 4.9, in contrast to the situation for relatively 'large' values of k, bound (4.12) of Theorem 4.8 can be improved for distance 2-packings (1-limited packings); that is, when $k = 1$ (see Corollary 4.4). However, this improvement is irrelevant for almost all graphs. A 1-limited packing set X in G has a very strong property that any two vertices in X are at distance at least 3 in G. It is well known that almost every graph has diameter equal to 2 (e.g. see [36]). Therefore, $\rho(G) = L_1(G) = 1$ for almost all graphs. Thus, in the case $k = 1$, Theorem 4.8 yields a lower bound of 1 for almost all graphs, which is similar to Theorem 4.9. Notice that the bound of Theorem 4.9 is sharp, for example for any number of disjoint copies of the Petersen graph. In the other cases, when G has diameter larger than 2, one should use the greedy algorithm and lower bound (4.17) provided by Theorem 4.9 because it improves bound (4.16) of Corollary 4.4 by a factor of $2 + o(1)$.

4.3 Generalizations of Multiple Domination

In this section, we consider Cockayne's and Favaron's general frameworks for $\langle \mathbf{r}, \mathbf{s} \rangle$-domination and parametric domination, respectively. They generalize and unify $\{k\}$-domination, k-domination, total k-domination and k-tuple domination. The classical upper bounds of Theorems 4.1 and 4.3 will be generalized for the $\langle \mathbf{r}, \mathbf{s} \rangle$-domination and parametric domination numbers. The generalizations imply upper bounds for the $\{k\}$-domination and total k-domination numbers. These results are based on the probabilistic method, which is a further development of the proof technique from the previous sections. Moreover, we will study threshold functions, which impose additional restrictions on the minimum vertex degree, and discuss other upper bounds for the aforementioned domination numbers. Those bounds extend some known results for k-tuple domination and total k-domination.

Note that the probabilistic constructions used in the proofs of the theorems on $\langle \mathbf{r}, \mathbf{s} \rangle$-domination imply randomized algorithms for finding s-dominating r-functions, whose weights satisfy the bounds of the corresponding theorems with positive probability. A similar statement is true for the theorems devoted to parametric domination.

The k-domination and k-tuple domination numbers are defined in Section 4.1.1. A vertex set X of a graph G is called a *total k-dominating set* of G if $|N(v) \cap X| \geq k$ for every vertex $v \in V(G)$ and its open neighbourhood $N(v)$. The minimum cardinality of a total k-dominating set of G is the *total k-domination number* $\gamma_k^t(G)$. Note that the total k-domination number is only defined for graphs with $\delta \geq k$ and $\gamma_1^t(G)$ is the well-known total domination number $\gamma_t(G)$. A function $f\colon V(G) \rightarrow \{0, 1, ..., k\}$ is called $\{k\}$-*dominating* if

$$\sum_{u \in N[v]} f(u) \geq k$$

for all $v \in V(G)$. The $\{k\}$-*domination number* of a graph G, denoted by $\gamma_{\{k\}}(G)$, is the smallest weight $\left(\sum_{v \in V(G)} f(v) \right)$ of a $\{k\}$-dominating function f of G. A survey of results devoted to k-domination and k-independence can be found in [9].

4.3.1 Framework for $\langle \mathbf{r}, \mathbf{s} \rangle$-Domination

An interesting framework for domination in graphs was introduced by Cockayne in [13]. Let $V(G) = \{v_1, ..., v_n\}$ denote the vertex set of a graph G, and let $\mathbf{r} = (r_1, ..., r_n)$ and $\mathbf{s} = (s_1, ..., s_n)$ be n-tuples of non-negative integers; that is, $r_i \in \mathbb{N}_0$ and $s_i \in \mathbb{N}_0$. A function $f\colon V(G) \rightarrow \mathbb{N}_0$ is called an $\mathbf{r}$-*function* of G if $f(v_i) \leq r_i$ for all $i = 1, ..., n$. Let us denote $f[v_i] = \sum_{u \in N[v_i]} f(u)$. An r-function f is s-*dominating* if $f[v_i] \geq s_i$ for all $i = 1, ..., n$. The *weight* of a function f is denoted by $|f|$ and defined by $|f| = \sum_{i=1}^{n} f(v_i)$.

The $\langle \mathbf{r}, \mathbf{s} \rangle$-*domination number* of a graph G, denoted by $\gamma \langle \mathbf{r}, \mathbf{s} \rangle (G)$, is the smallest weight of an s-dominating r-function of G. As pointed out in [13], such functions exist if and only if $\sum_{v_j \in N[v_i]} r_j \geq s_i$ for all $i = 1, ..., n$. It is easy to see that $\langle \mathbf{r}, \mathbf{s} \rangle$-domination unifies and generalizes the classical domination,

k-tuple domination and $\{k\}$-domination if we set $r_i = s_i = 1$; $r_i = 1, s_i = k$; and $r_i = s_i = k$ for all $i = 1, ..., n$, respectively.

Let us denote

$$\tau = \min\{r_1, ..., r_n\}, \quad s = \max\{s_1, ..., s_n\}, \quad r = \left\lfloor \frac{s}{\delta + 1} \right\rfloor + 1,$$

$$\theta = (\delta + 1)r - s \quad \text{and} \quad B_t = \binom{(\delta + 1)r}{t}.$$

The following theorem provides an upper bound for the $\langle \mathbf{r}, \mathbf{s} \rangle$-domination number of a graph.

Theorem 4.11 [47] *For any graph G of order n with $\tau \geq r \geq 1$ and $\rho = 1/\theta$,*

$$\gamma \langle \mathbf{r}, \mathbf{s} \rangle (G) \leq \left(1 - \frac{(r\rho)^\rho}{(1 + \rho)^{1+\rho} B_{s-1}^\rho} \right) rn.$$

Proof: For each vertex $v \in V(G)$, we arbitrarily select δ vertices from $N(v)$ and denote the resulting set together with the vertex v by $N'[v]$. Thus, $|N'[v]| = \delta + 1$. For $i = 1, 2, ..., r$, let $a_i(v)$ be a $(0,1)$-function on the set $V(G)$ such that it assigns '1' to every vertex of G independently with probability

$$p = 1 - \left(\frac{r}{(1 + \theta)B_{s-1}} \right)^{1/\theta}.$$

Let us define an **r**-function $a(v)$ as follows: $a(v) = \sum_{i=1}^{r} a_i(v)$.

For $m = 0, 1, ..., s - 1$, we denote

$$C_m = \left\{ v \in V(G) : \sum_{u \in N'[v]} a(u) = m \right\}.$$

Claim 4.1 *For each set C_m, there exists a function $c_m \colon V(G) \to \mathbb{N}_0$ such that*

$$|c_m| \leq (s - m)|C_m| \tag{4.20}$$

and for any vertex $v \in C_m$,

$$a(v) + c_m(v) \leq r \quad \text{and} \quad \sum_{u \in N'[v]} c_m(u) \geq s - m. \tag{4.21}$$

Proof: Let us initially set $c_m(v) = 0$ for all $v \in V(G)$. Then, for each vertex $v \in C_m$, we redefine c_m in the set $N'[v]$ as follows:

Case 1: Suppose that $c_m(u) = 0$ for any vertex $u \in N'[v]$. Note that because $\sum_{u \in N'[v]} a(u) = m$, the 'spare capacity' in $N'[v]$ is

$$(\delta + 1)r - m > s - m;$$

that is, the weight of c_m can be increased in $N'[v]$ by $s - m$ units. Thus, we can obviously redefine c_m in $N'[v]$ in such a way that

$$\sum_{u \in N'[v]} c_m(u) = s - m \tag{4.22}$$

and

$$a(u) + c_m(u) \leq r \quad \text{for any} \quad u \in N'[v]. \tag{4.23}$$

In this case, we increased the weight of c_m in $N'[v]$ by $s - m$ units.

Case 2: Assume that $c_m(u) > 0$ for some $u \in N'[v]$, but

$$\sum_{u \in N'[v]} c_m(u) = \psi < s - m,$$

where $\psi \geq 1$. In this case, we can increase the weight of c_m in $N'[v]$ by $s - m - \psi$ units to make sure that (4.22) and (4.23) hold.

Case 3: Suppose now that $c_m(u) > 0$ for some $u \in N'[v]$ and

$$\sum_{u \in N'[v]} c_m(u) \geq s - m.$$

In this case, we do not change the weight of c_m in $N'[v]$.

Thus, when constructing the function c_m, we increased its weight at most $|C_m|$ times by at most $s - m$ units, and so (4.20) is true. The inequalities (4.21) are also true by construction. ∎

Let us define the function f on the set $V(G)$ as follows:

$$f(v) = a(v) + \max_{0 \leq m \leq s-1} c_m(v).$$

By Claim 4.1, the function f is an $(s, ..., s)$-dominating $(r, ..., r)$-function. Hence, it is also an s-dominating r-function because $s \geq s_i$ and $r \leq \tau \leq r_i$ for all $i = 1, ..., n$. Also,

$$f(v) \leq a(v) + \sum_{m=0}^{s-1} c_m(v).$$

The expectation of $|f|$ is as follows:

$$\mathbb{E}[|f|] \leq \mathbb{E}\left[|a| + \sum_{m=0}^{s-1} |c_m|\right] \leq \sum_{i=1}^{r} \mathbb{E}[|a_i|] + \sum_{m=0}^{s-1} (s-m)\mathbb{E}[|C_m|].$$

We have

$$\mathbb{E}[|C_m|] = \sum_{v \in V(G)} \mathbb{P}[v \in C_m] = \sum_{i=1}^{n} p^m (1-p)^{(\delta+1)r-m} \binom{(\delta+1)r}{m}$$

$$= p^m (1-p)^{(\delta+1)r-m} B_m n.$$

Thus,

$$\mathbb{E}[|f|] \leq pnr + \sum_{m=0}^{s-1} (s-m) p^m (1-p)^{(\delta+1)r-m} B_m n$$

$$= pnr + (1-p)^{(\delta+1)r-s+1} n \sum_{m=0}^{s-1} (s-m) p^m (1-p)^{s-m-1} B_m.$$

Furthermore, for $0 \leq m \leq s-1$,

$$(s-m)B_m = (s-m)\binom{(\delta+1)r}{m} \leq \prod_{j=1}^{\theta+1} \frac{(s-m+j-1)}{j} \binom{(\delta+1)r}{m}$$

$$= \binom{(\delta+1)r-m}{\theta+1}\binom{(\delta+1)r}{m} = \binom{s-1}{m}\binom{(\delta+1)r}{s-1} = \binom{s-1}{m} B_{s-1}.$$

We obtain

$$\gamma \langle \mathbf{r}, \mathbf{s} \rangle (G) \leq \mathbb{E}[|f|]$$

$$\leq pnr + (1-p)^{\theta+1} n B_{s-1} \sum_{m=0}^{s-1} \binom{s-1}{m} p^m (1-p)^{s-m-1}$$

$$= pnr + (1-p)^{\theta+1} n B_{s-1}$$

$$= \left(1 - \frac{(r\rho)^\rho}{(1+\rho)^{1+\rho} B_{s-1}^\rho}\right) rn,$$

as required. The proof of the theorem is complete. ■

The proof of Theorem 4.11 implies a weaker upper bound for the $\langle r, s \rangle$-domination number. This result generalizes the classical bound in Theorem 4.1.

Corollary 4.5 [47] *For any graph G of order n with $r \leq \tau$,*

$$\gamma \langle r, s \rangle (G) \leq \frac{\ln(\theta + 1) + \ln B_{s-1} - \ln r + 1}{\theta + 1} rn.$$

Proof: The proof easily follows if we use the inequality $1 - p \leq e^{-p}$, and then minimize the following upper bound:

$$\gamma \langle r, s \rangle (G) \leq pnr + e^{-p(\theta+1)} n B_{s-1}.$$

∎

It may be pointed out that the bound of Corollary 4.5 can be optimized with respect to r, where r is now any integer between $s/(\delta + 1)$ and τ:

$$\gamma \langle r, s \rangle (G) \leq \min_{s/(\delta+1) \leq r \leq \tau} \left\{ \frac{\ln(\theta + 1) + \ln B_{s-1} - \ln r + 1}{\theta + 1} rn \right\}.$$

$\{k\}$-Domination is a particular case of $\langle r, s \rangle$-domination when $r_i = s_i = k$ for all $i = 1, ..., n$. Theorem 4.11 and Corollary 4.5 imply the following upper bounds for the $\{k\}$-domination number. We have $\tau = s = k$, and hence

$$r = \left\lfloor \frac{k}{\delta + 1} \right\rfloor + 1, \quad \theta = (\delta + 1)r - k \quad \text{and} \quad B_{k-1} = \binom{(\delta + 1)r}{k - 1}.$$

Corollary 4.6 [47] *For any graph G with $\delta > 0$ and $\rho = 1/\theta$,*

$$\gamma_{\{k\}}(G) \leq \left(1 - \frac{(r\rho)^\rho}{(1 + \rho)^{1+\rho} B_{k-1}^\rho} \right) rn \leq \frac{\ln(\theta + 1) + \ln B_{k-1} - \ln r + 1}{\theta + 1} rn.$$

Similar to Corollary 4.5, the latter upper bound in Corollary 4.6 is weaker than the former upper bound, but it has a simpler formula and can be further optimized with respect to r for integers between $k/(\delta + 1)$ and k.

While $\langle r, s \rangle$-domination generalizes the classical domination, k-tuple domination and $\{k\}$-domination, the following definition generalizes total domination by considering open neighbourhoods.

An **r**-function f is called *total* **s**-*dominating* if

$$\sum_{u \in N(v_i)} f(u) \geq s_i \quad \text{for all} \quad i = 1, 2, ..., n.$$

The total $\langle \mathbf{r}, \mathbf{s} \rangle$-*domination number* $\gamma^t \langle \mathbf{r}, \mathbf{s} \rangle (G)$ of a graph G is the smallest weight of a total **s**-dominating **r**-function of G.

Let us denote

$$\tilde{r} = \left\lfloor \frac{s}{\delta} \right\rfloor + 1, \quad \tilde{\theta} = \delta \tilde{r} - s, \quad \tilde{B}_{s-1} = \binom{\delta \tilde{r}}{s-1} \quad \text{and} \quad \tilde{\rho} = 1/\tilde{\theta}.$$

Theorem 4.12 [47] *For any graph G of order n with $\tilde{r} \leq \tau$ and $\delta > 0$,*

$$\gamma^t \langle \mathbf{r}, \mathbf{s} \rangle (G) \leq \left(1 - \frac{(\tilde{r}\tilde{\rho})^{\tilde{\rho}}}{(1+\tilde{\rho})^{1+\tilde{\rho}} \, \tilde{B}_{s-1}^{\tilde{\rho}}} \right) \tilde{r}n \leq \frac{\ln(\tilde{\theta}+1) + \ln \tilde{B}_{s-1} - \ln \tilde{r} + 1}{\tilde{\theta} + 1} \tilde{r}n.$$

Proof: For each vertex $v \in V(G)$, we arbitrarily select δ vertices from $N(v)$ and denote the resulting set by $N'(v)$. Thus, $|N'(v)| = \delta$. The rest of the proof now follows from the proofs of Theorem 4.11 and Corollary 4.5 if $N'[v]$ is replaced by $N'(v)$. ∎

The upper bounds of Corollary 4.7 for the total k-domination number follow from Theorem 4.12 by setting $r_i = 1$ and $s_i = k$ for all $i = 1, 2, ..., n$, in which case $\tilde{r} = 1, k < \delta, \tilde{\theta} = \delta - k$ and $\tilde{B}_{k-1} = b_{k-1} = \binom{\delta}{k-1}$.

Corollary 4.7 [47] *For any graph G with $\bar{\delta} = \delta - k > 0$,*

$$\gamma_k^t(G) \leq \left(1 - \frac{\bar{\delta}}{(1 + \bar{\delta})^{1+1/\bar{\delta}} \, b_{k-1}^{1/\bar{\delta}}} \right) n \leq \frac{\ln(\delta - k + 1) + \ln b_{k-1} + 1}{\delta - k + 1} n.$$

In particular, we obtain an upper bound for the total domination number for any graph G without isolated vertices:

$$\gamma_t(G) \leq \frac{\ln \delta + 1}{\delta} n.$$

4.3.2 Framework for Parametric Domination

While Cockayne's framework is based on functions with prescribed properties, the focus of the generalization considered in this section is on properties of vertex sets called (k, l)-dominating sets. These two frameworks complement each other because the former does not generalize k-domination, while the latter does not include $\{k\}$-domination.

The following definition with minor adaptations is due to Favaron et al. [19]. For integers $k \geq 1$ and $l \geq 1$, a set D is called a (k, l)-dominating set of G if for every vertex $v \notin D$, $|N[v] \cap D| \geq k$, and for every vertex $v \in D$, $|N[v] \cap D| \geq l$. The minimum cardinality of a (k, l)-dominating set of G is the parametric domination number $\gamma_{k,l}(G)$. Note that, using Favaron's terminology [19], the parametric domination number is called $(l-1)$-total k-domination number $\gamma_{l-1,k}(G)$. There is some similarity between this concept and f-domination defined in [44].

The parametric domination number is only defined for graphs with $\delta \geq \max\{k, l-1\}$. Since $V(G)$ is a (k, l)-dominating set of G, the parametric domination is well defined. It is easy to see that $\gamma_{1,1}(G)$ is the domination number $\gamma(G)$, $\gamma_{2,1}(G)$ is the 2-domination number $\gamma_2(G)$, $\gamma_{2,2}(G)$ is the double domination number $\gamma_{\times 2}(G)$ and $\gamma_{1,2}(G)$ is the total domination number $\gamma_{\mathrm{t}}(G)$.

More generally, the parametric domination number unifies the following:

$l = 1$	$\gamma_{k,1}(G)$ is the k-domination number $\gamma_k(G)$
$l = k$	$\gamma_{k,k}(G)$ is the k-tuple domination number $\gamma_{\times k}(G)$
$l = k+1$	$\gamma_{k,k+1}(G)$ is the total k-domination number $\gamma_k^{\mathrm{t}}(G)$

Let us denote $\varphi = \max\{k, l-1\}$ and $b_t = \binom{\delta}{t}$.

Theorem 4.13 [47] *For any graph G with $\bar{\delta} = \delta - \varphi > 0$,*

$$\gamma_{k,l}(G) \leq \left(1 - \frac{\bar{\delta}}{(1 + \bar{\delta})^{1+1/\bar{\delta}}\, b_{\varphi-1}^{1/\bar{\delta}}}\right) n.$$

Proof: For each vertex $v \in V(G)$, we arbitrarily select δ vertices from $N(v)$ and denote the resulting set by $N'(v)$. Let A be a set formed by an independent choice of vertices of G, where each vertex is selected with probability

$$p = 1 - \left(\frac{1}{(1+\bar{\delta})b_{\varphi-1}} \right)^{1/\bar{\delta}}.$$

For $m = 0, 1, ..., k-1$, let us denote

$$B_m = \{v \in V(G) - A : |N'(v) \cap A| = m\}.$$

Also, for $m = 0, 1, ..., l-2$, we denote

$$A_m = \{v \in A : |N'(v) \cap A| = m\}.$$

For each set A_m, we form a set A'_m in the following way. For every vertex $v \in A_m$, we take $l-m-1$ neighbours from $N'(v) - A$ and add them to A'_m. Such neighbours always exist because $\delta \geq l-1$. It is obvious that $|A'_m| \leq (l-m-1)|A_m|$. For each set B_m, we form a set B'_m by taking $k-m$ neighbours from $N'(v) - A$ for every vertex $v \in B_m$. Such neighbours always exist because $\delta \geq k$. We have $|B'_m| \leq (k-m)|B_m|$.

Let us construct the set D as follows:

$$D = A \cup \left(\bigcup_{m=0}^{l-2} A'_m \right) \cup \left(\bigcup_{m=0}^{k-1} B'_m \right).$$

The set D is a (k, l)-dominating set. Indeed, if there is a vertex v which is not (k, l)-dominated by D, then v is not (k, l)-dominated by A. Therefore, v would belong to A_m or B_m for some m, but all such vertices are (k, l)-dominated by the set D by construction.

The expectation of $|D|$ is

$$\mathbb{E}[|D|] \leq \mathbb{E}\left[|A| + \sum_{m=0}^{l-2} |A'_m| + \sum_{m=0}^{k-1} |B'_m| \right]$$

$$\leq \mathbb{E}[|A|] + \sum_{m=0}^{l-2} (l-m-1)\mathbb{E}[|A_m|] + \sum_{m=0}^{k-1} (k-m)\mathbb{E}[|B_m|].$$

We have

$$\mathbb{E}[|A|] = \sum_{i=1}^{n} \mathbb{P}[v_i \in A] = pn.$$

Also,

$$
\begin{aligned}
\mathbb{E}[|A_m|] &= \sum_{i=1}^{n} \mathbb{P}[v_i \in A_m] \\
&= \sum_{i=1}^{n} p \binom{\delta}{m} p^m (1-p)^{\delta-m} = p^{m+1}(1-p)^{\delta-m} b_m n
\end{aligned}
$$

and

$$
\begin{aligned}
\mathbb{E}[|B_m|] &= \sum_{i=1}^{n} \mathbb{P}[v_i \in B_m] \\
&= \sum_{i=1}^{n} (1-p) \binom{\delta}{m} p^m (1-p)^{\delta-m} = p^m (1-p)^{\delta-m+1} b_m n.
\end{aligned}
$$

We obtain

$$
\begin{aligned}
\mathbb{E}[|D|] &\leq pn + \sum_{m=0}^{l-2} (l-m-1) p^{m+1} (1-p)^{\delta-m} b_m n \\
&\quad + \sum_{m=0}^{k-1} (k-m) p^m (1-p)^{\delta-m+1} b_m n \\
&\leq pn + \sum_{m=0}^{\varphi-1} (\varphi-m) p^{m+1} (1-p)^{\delta-m} b_m n \\
&\quad + \sum_{m=0}^{\varphi-1} (\varphi-m) p^m (1-p)^{\delta-m+1} b_m n \\
&= pn + \sum_{m=0}^{\varphi-1} (\varphi-m) b_m n p^m (1-p)^{\delta-m}.
\end{aligned}
$$

Furthermore, for $0 \leq m \leq \varphi - 1$,

$$
\begin{aligned}
(\varphi-m) b_m &= (\varphi-m) \binom{\delta}{m} \leq (\varphi-m) \binom{\delta}{m} \prod_{j=2}^{\varphi-m} \frac{(\delta-\varphi+j)}{j} \\
&= \frac{\delta!}{m!(\varphi-m-1)!(\delta-\varphi+1)!} = \binom{\varphi-1}{m}\binom{\delta}{\varphi-1} \\
&= \binom{\varphi-1}{m} b_{\varphi-1}.
\end{aligned}
$$

Therefore,

$$\mathbb{E}[|D|] \leq pn + nb_{\varphi-1}(1-p)^{\delta-\varphi+1} \sum_{m=0}^{\varphi-1} \binom{\varphi-1}{m} p^m (1-p)^{\varphi-1-m}$$

$$= pn + nb_{\varphi-1}(1-p)^{\delta-\varphi+1}$$

$$= \left(1 - \frac{\bar{\delta}}{(1+\bar{\delta})^{1+1/\bar{\delta}} \, b_{\varphi-1}^{1/\bar{\delta}}}\right) n.$$

Since the expectation is an average value, there exists a particular (k,l)-dominating set of the above order, as required. The proof of Theorem 4.13 is complete. ∎

Corollary 4.8 [47] *For any graph G with $\delta \geq \varphi$,*

$$\gamma_{k,l}(G) \leq \frac{\ln(\delta - \varphi + 1) + \ln b_{\varphi-1} + 1}{\delta - \varphi + 1} n.$$

Proof: Using the inequality $1 - p \leq e^{-p}$, we obtain

$$\mathbb{E}[|D|] \leq pn + nb_{\varphi-1}e^{-p(\delta-\varphi+1)}.$$

The proof easily follows by minimizing the right-hand side in the above inequality. ∎

Note that Theorem 4.13 and Corollary 4.8 imply the result formulated in Corollary 4.7 if we set $l = k + 1$.

The next result is similar to Theorem 4.13 and Corollary 4.8, which provide better bounds if $l \geq k + 1$. However, for small values of l, the bounds of Theorem 4.14 are better. In what follows, we set $b_{-1} = 0$.

Theorem 4.14 [47] *For any graph G with $\widehat{\delta} = \delta - \max\{k,l\} + 1 > 0$,*

$$\gamma_{k,l}(G) \leq \left(1 - \frac{\widehat{\delta}}{(1+\widehat{\delta})^{1+1/\widehat{\delta}} \, (b_{k-1} + b_{l-2})^{1/\widehat{\delta}}}\right) n.$$

Also, for any graph G with $\delta \geq \max\{k, l-1\}$,

$$\gamma_{k,l}(G) \leq \frac{\ln(\widehat{\delta} + 1) + \ln(b_{k-1} + b_{l-2}) + 1}{\widehat{\delta} + 1} n.$$

Proof: Using the same construction as in the proof of Theorem 4.13, we obtain

$$\mathbb{E}[|D|] \leq pn + \sum_{m=0}^{l-2} (l - m - 1)p^{m+1}(1 - p)^{\delta-m}b_m n$$

$$+ \sum_{m=0}^{k-1} (k - m)p^m(1 - p)^{\delta-m+1}b_m n$$

$$= pn + \sum_{m=1}^{l-1} (l - m)p^m(1 - p)^{\delta-m+1}b_{m-1} n$$

$$+ \sum_{m=0}^{k-1} (k - m)p^m(1 - p)^{\delta-m+1}b_m n$$

$$= pn + (1 - p)^{\delta-l+2}n\Theta_1 + (1 - p)^{\delta-k+2}n\Theta_2,$$

where

$$\Theta_1 = \sum_{m=1}^{l-1} (l - m)p^m(1-p)^{l-m-1}b_{m-1}, \quad \Theta_2 = \sum_{m=0}^{k-1} (k - m)p^m(1-p)^{k-m-1}b_m.$$

Now, using an approach similar to that in the proof of Theorem 4.13, we can prove that

$$(l - m)b_{m-1} \leq \binom{l - 1}{m}b_{l-2} \quad \text{and} \quad (k - m)b_m \leq \binom{k - 1}{m}b_{k-1}.$$

Therefore, $\Theta_1 \leq b_{l-2}$ and $\Theta_2 \leq b_{k-1}$. We have

$$\mathbb{E}[|D|] \leq pn + (1 - p)^{\delta-l+2}nb_{l-2} + (1 - p)^{\delta-k+2}nb_{k-1}$$

$$\leq pn + (1 - p)^{\widehat{\delta}+1}n(b_{l-2} + b_{k-1}).$$

The first upper bound of the theorem statement is obtained by minimizing the above function, whereas the second bound is deduced by minimizing the following function:

$$\mathbb{E}[|D|] \leq pn + e^{-p(\widehat{\delta}+1)}n(b_{l-2} + b_{k-1}).$$

∎

The special case $l=1$ in parametric domination is the k-domination number $\gamma_k(G)$, and the case with $l=k$ is the k-tuple domination number $\gamma_{\times k}(G)$. Thus, Theorem 4.14 implies Theorems 4.4, 4.5 and Corollaries 4.1, 4.2.

4.3.3 Threshold Functions for Multiple Domination

The bounds for multiple domination numbers can be improved if we impose additional restrictions on graph parameters, that is, by considering smaller graph classes. Such restrictions on graph parameters are called *threshold functions*. Caro and Roditty [10] and Stracke and Volkmann [44] were the first to consider a threshold function for k-domination in the form $\delta \geq 2k - 1$. For a slightly stronger threshold function, Rautenbach and Volkmann [40] found an interesting upper bound for the k-tuple domination number:

Theorem 4.15 [40] *If $\delta \geq 2k \ln(\delta + 1) - 1$, then*

$$\gamma_{\times k}(G) \leq \left(\frac{k \ln(\delta + 1)}{\delta + 1} + \sum_{i=0}^{k-1} \frac{k - i}{i!(\delta + 1)^{k-i}} \right) n.$$

It may be pointed out that similar threshold bounds for the k-domination number were studied in [26].

In the next theorem, we consider a threshold function of the form $\delta \geq ck - 1$, where $c > 1$ is a constant. Although c is not restricted from above, for given k and δ the constant c should not be taken as large as possible. The best approach would be to optimize c with respect to k and δ in such a way that the bound (4.24) is minimized when $\delta \geq ck - 1$ holds. We will deal with this optimization later.

Theorem 4.16 [47] *For any graph G with $\delta \geq ck - 1$, where $c > 1$ is a constant,*

$$\gamma_{\times k}(G) < \left(\frac{c}{\delta + 1} + \frac{1}{e^{0.5k(c+1/c-2)}} \right) kn. \tag{4.24}$$

Proof: For each vertex $v \in V(G)$, we arbitrarily select δ vertices from $N(v)$ and denote the resulting set together with the vertex v by $N'[v]$. Thus, $|N'[v]| = \delta + 1$. Let A be a set formed by an independent choice of vertices of G, where each vertex is selected with probability $p = \frac{ck}{\delta+1} \leq 1$. For $m = 0, 1, ..., k - 1$, let us denote

$$C_m = \{ v \in V(G) : |N'[v] \cap A| = m \}.$$

For each set C_m, let us form a set C'_m in the following way: for every vertex $v \in C_m$ we take $k - m$ neighbours from $N'[v] - A$ and add them to C'_m. Such neighbours always exist because $\delta \geq k$. It is obvious that

$$|C'_m| \leq (k - m)|C_m| \leq k|C_m|.$$

Let us construct the set D as follows:

$$D = A \cup \left(\bigcup_{m=0}^{k-1} C'_m \right).$$

It is easy to see that D is a k-tuple dominating set. The expectation of $|D|$ is

$$\mathbb{E}[|D|] \leq \mathbb{E}\left[|A| + \sum_{m=0}^{k-1} |C'_m| \right] \leq \mathbb{E}[|A|] + k \sum_{m=0}^{k-1} \mathbb{E}[|C_m|].$$

We have $\mathbb{E}[|A|] = pn$ and

$$\sum_{m=0}^{k-1} \mathbb{E}[|C_m|] = \sum_{m=0}^{k-1} \sum_{i=1}^{n} \mathbb{P}[v_i \in C_m]$$

$$= \sum_{i=1}^{n} \sum_{m=0}^{k-1} \mathbb{P}[|N'[v_i] \cap A| = m]$$

$$= \sum_{i=1}^{n} \mathbb{P}[|N'[v_i] \cap A| < k].$$

Let $X_1, ..., X_t$ be mutually independent random variables with

$$\mathbb{P}[X_i = 1 - p] = p \quad \text{and} \quad \mathbb{P}[X_i = -p] = 1 - p.$$

Also, let $X = X_1 + ... + X_t$; that is, X has distribution $B(t, p) - np$. By Alon–Spencer's theorem,

$$\mathbb{P}[X < -a] < e^{-a^2/2pt},$$

where $a > 0$ (Theorem A.1.13 in [2]). The random set A constructed above can be seen as a set of vertices labelled by 1, where each vertex is assigned label 1 with probability p and label 0 with probability $1 - p$. Let us now subtract p from all the labels; that is, for each vertex v_j we have a random variable τ_j such that

$$\mathbb{P}[\tau_j = 1 - p] = p \quad \text{and} \quad \mathbb{P}[\tau_j = -p] = 1 - p, \quad j = 1, 2, ..., n.$$

For each vertex v_i, $i = 1, 2, ..., n$, we define a random variable

$$\tau_i^* = \sum_{v_j \in N'[v_i]} \tau_j.$$

Taking into account that

$$k - (\delta + 1)p = k(1 - c) < 0,$$

we obtain by Alon–Spencer's theorem:

$$\mathbb{P}[|N'[v_i] \cap A| < k] = \mathbb{P}[\tau_i^* < k - (\delta + 1)p]$$
$$< e^{-\frac{[k-(\delta+1)p]^2}{2p(\delta+1)}} = e^{-\frac{[k-ck]^2}{2ck}} = e^{-0.5k(c+1/c-2)}.$$

Thus,

$$\sum_{m=0}^{k-1} \mathbb{E}[|C_m|] < ne^{-0.5k(c+1/c-2)}$$

and

$$\gamma_{\times k}(G) \leq \mathbb{E}[|D|] < pn + kne^{-0.5k(c+1/c-2)} = \left(\frac{c}{\delta+1} + \frac{1}{e^{0.5k(c+1/c-2)}} \right) kn,$$

as required. The proof of the theorem is complete. ∎

Let us consider a particular case of Theorem 4.16 when $c = 3$, and compare it to Theorem 4.15 for graphs with $\delta \geq 20$. Theorem 4.16 implies

$$\gamma_{\times k}(G) < \left(\frac{3}{\delta+1} + \frac{1}{e^{2k/3}} \right) kn \tag{4.25}$$

for any graph G with $\delta \geq 3k - 1$. This bound is better than the bound of Theorem 4.15 if the former is less than the first term of the latter; that is,

$$\left(\frac{3}{\delta+1} + \frac{1}{e^{2k/3}} \right) k < \frac{k\ln(\delta+1)}{\delta+1}.$$

Taking into account that $\delta \geq 20$, this inequality is equivalent to

$$k > 1.5\ln(\delta+1) - 1.5\ln[\ln(\delta+1) - 3] = 1.5\ln(\delta+1)(1 - o(1)).$$

Since Theorem 4.15 is applicable for $k \leq (\delta + 1)/(2\ln(\delta + 1))$, we conclude that (4.25) provides a better upper bound than Theorem 4.15 if

$$1.5\ln(\delta + 1)(1 - o(1)) < k \leq \frac{\delta + 1}{2\ln(\delta + 1)},$$

which is a larger part of the applicable interval for k.

For example, if $\delta(G) = 1000$, then Theorem 4.15 is applicable for $k \leq 72$, whereas (4.25) is applicable for $k \leq 333$. Since $1.5\ln(1001) - 1.5\ln[\ln(1001) - 3] = 8.3$, the bound (4.25) is stronger than the bound of Theorem 4.15 for $9 \leq k \leq 72$. If $k \leq 8$, then Theorem 4.15 provides a better upper bound than (4.25). However, we can try to optimize the constant c in Theorem 4.16 for given δ and k as follows. The right-hand side of the bound (4.24) is minimized for c that satisfies the following equation:

$$0.5k\left(c + \frac{1}{c} - 2\right) - \ln(0.5k(\delta + 1)) = \ln\left(1 - \frac{1}{c^2}\right).$$

Now, replacing $\ln(1 - \frac{1}{c^2})$ by $-1/c^2$, we obtain the following cubic equation:

$$kc^3 - 2\left(k + \ln[0.5k(\delta + 1)]\right)c^2 + kc + 2 = 0.$$

The real root $c > 1$ of this equation, which satisfies the condition $\delta \geq ck - 1$, can be used in Theorem 4.16. For example, if $k = 5$ and $\delta = 1000$, then the above cubic equation becomes

$$c^3 - 5.13c^2 + c + 0.4 = 0.$$

For this equation, the largest real root is $c = 4.910$ (3 dp). Using this value of c in Theorem 4.16, we obtain $\gamma_{\times 5}(G) < 0.027n$, whereas Theorem 4.15 produces the bound $\gamma_{\times 5}(G) < 0.035n$.

It is possible to generalize Theorem 4.16 for parametric domination and $\langle r, s \rangle$-domination. Let us denote $\mu = \max\{k, l\}$.

Theorem 4.17 [47] *For any graph G with $\delta \geq c\mu - 1$, where $c > 1$ is a constant,*

$$\gamma_{k,l}(G) < \left(\frac{c}{\delta + 1} + \frac{1}{e^{0.5\mu(c+1/c-2)}}\right)\mu n.$$

Theorem 4.18 [47] *For any graph G with $(\delta + 1)\tau \geq cs$, where $c > 1$ is a constant,*

$$\gamma\langle \mathbf{r}, \mathbf{s}\rangle(G) < \left(\frac{c}{\delta + 1} + \frac{1}{e^{0.5s(c+1/c-2)}} \right) sn.$$

Caro and Yuster [12] proved an important asymptotic result that if δ is 'much larger' than k, then the upper bound for the total k-domination number is 'close' to the bound of Theorem 4.1. More precisely, they proved the following:

Theorem 4.19 [12] *If $k < \sqrt{\ln \delta}$, then*

$$\gamma_k^t(G) \leq \frac{\ln \delta}{\delta} n(1 + o_\delta(1)).$$

The same upper bound is therefore true for the k-tuple domination and k-domination numbers by definitions. The threshold function $k < \sqrt{\ln \delta}$ in Theorem 4.19 is indeed very strong, but the corresponding bound is similar to the bound of Theorem 4.1, which is best possible in the class of all graphs. Let us consider a weaker but similar threshold function $k \leq (1 - c) \ln \delta$, where $0 < c < 1$ is a constant. The following explicit and asymptotic bounds can be obtained:

Theorem 4.20 [47] *For any graph G with $k \leq (1 - c) \ln \delta$, where $0 < c < 1$ is a constant,*

$$\gamma_{\times k}(G) < \left(\frac{\ln \delta}{\delta + 1} + \frac{k}{\delta^{0.5c^2}} \right) n \leq \frac{(1 - c) \ln \delta}{\delta^{0.5c^2}} n(1 + o_\delta(1)).$$

Proof: The proof is similar to that of Theorem 4.16 by setting

$$p = \frac{\ln \delta}{\delta + 1} \leq 1.$$

Now,

$$p \geq \frac{k}{(1 - c)(\delta + 1)} > \frac{k}{(\delta + 1)};$$

that is, $k - (\delta + 1)p < 0$ and

$$\mathbb{P}[|N'[v_i] \cap A| < k] = \mathbb{P}[\tau_i^* < k - (\delta + 1)p] < e^{-\frac{[k-(\delta+1)p]^2}{2p(\delta+1)}}$$
$$= e^{-\frac{[k-\ln \delta]^2}{2 \ln \delta}} \leq e^{-0.5c^2 \ln \delta} = \delta^{-0.5c^2}.$$

Thus,

$$\gamma_{\times k}(G) \le \mathbb{E}[|D|] < pn + kn\delta^{-0.5c^2} = \left(\frac{\ln \delta}{\delta + 1} + \frac{k}{\delta^{0.5c^2}} \right) n,$$

as required.   ∎

Theorem 4.20 can be generalized for parametric domination and $\langle r, s \rangle$-domination as follows.

Theorem 4.21 [47] *For any graph G with $\mu \le (1 - c) \ln \delta$, where $0 < c < 1$ is a constant,*

$$\gamma_{k,l}(G) < \left(\frac{\ln \delta}{\delta + 1} + \frac{\mu}{\delta^{0.5c^2}} \right) n.$$

Similar to Theorem 4.18, the upper bound in the next result does not depend on τ. Note also that $s \le (\delta + 1)\tau$ holds because

$$s \le (1 - c) \ln \delta < \ln \delta \le (\delta + 1)\tau;$$

that is, $\gamma \langle r, s \rangle (G)$ is well defined.

Theorem 4.22 [47] *For any graph G with $s \le (1 - c) \ln \delta$, where $0 < c < 1$ is a constant,*

$$\gamma \langle r, s \rangle (G) < \left(\frac{\ln \delta}{\delta + 1} + \frac{s}{\delta^{0.5c^2}} \right) n.$$

4.4 Exercises

4.4.1

Prove the following upper bound for the domination number of a graph G directly (i.e. without using any theorem statements):

$$\gamma(G) \le \frac{\ln(\delta + 1) + 1}{\delta + 1} n.$$

4.4.2

A vertex set X in a graph G is called a *total dominating set* if every vertex in G is adjacent to some vertex of X. The minimum cardinality of a total dominating set in G is the *total domination number* $\gamma_t(G)$.

Prove the following upper bound for the total domination number directly (i.e. without using any theorem statements): for any graph G with $\delta > 0$,

$$\gamma_t(G) \leq \frac{\ln \delta + 1}{\delta} n.$$

4.4.3*

The proof of Theorem 4.4 can be reduced if we consider sets $N'[v]$ instead of sets $N'(v)$. The first part of the simplified proof is as follows:

For each vertex $v \in V(G)$, we arbitrarily select δ vertices from $N(v)$ and denote the resulting set together with the vertex v by $N'[v]$. Thus, $|N'[v]| = \delta + 1$. Let A be a set formed by an independent choice of vertices of G, where each vertex is selected with probability

$$p = 1 - \left(\frac{1}{\tilde{b}_{k-1}(1 + \delta')} \right)^{1/\delta'}.$$

For $m = 0, 1, ..., k - 1$, we denote

$$C_m = \{v \in V(G) : |N'[v] \cap A| = m\}.$$

For each set C_m, let us form a set C'_m in the following way: for every vertex $v \in C_m$ we take $k - m$ neighbours from $N'[v] - A$ and add them to C'_m. Such neighbours always exist because $\delta \geq k$. It is obvious that $|C'_m| \leq (k-m)|C_m|$.

Let us construct the set D as follows:

$$D = A \cup \left(\bigcup_{m=0}^{k-1} C'_m \right).$$

It is easy to see that D is a k-tuple dominating set.

- The reader should complete this proof by implementing the following steps:

1. Find an upper bound for $\mathbb{E}[|D|]$ in terms of p, n, δ, k and $\tilde{b}$. The upper bound must include the term $(k - m)\tilde{b}_m$.
2. Prove that

$$(k - m)\tilde{b}_m \leq \binom{k - 1}{m}\tilde{b}_{k-1}$$

for $0 \leq m \leq k - 1$.
3. Deduce that

$$\mathbb{E}[|D|] \leq \left(1 - \frac{\delta'}{\tilde{b}_{k-1}^{1/\delta'}(1 + \delta')^{1+1/\delta'}}\right) n,$$

as required.

4.4.4

Using the double-counting approach, prove the following bound [5, 35]:

$$L_k(G) \leq \frac{kn}{\delta + 1}.$$

4.4.5

Prove that $L_k(G) \leq k\gamma(G)$ for any graph G [24].

4.4.6

Show that for any graph G with $\delta \geq k - 1$, the following inequality is true [23, 35]:

$$L_k(G) \leq \gamma_{\times k}(G). \tag{4.26}$$

4.4.7

Prove the following lower bound: for any Δ-regular graph G of order n with $\Delta \geq 3$,

$$L_2(G) \geq \frac{2n}{\Delta^2 - \Delta + 2}.$$

4.4.8

A *Roman dominating function* (RDF) of a graph G is defined as a function $f: V(G) \to \{0, 1, 2\}$ satisfying the condition that every vertex u for which $f(u) = 0$ is adjacent to at least one vertex v for which $f(v) = 2$. The *weight* of an RDF is defined as the value $f(V(G)) = \sum_{v \in V(G)} f(v)$. The Roman domination number of a graph G, denoted $\gamma_R(G)$, is equal to the minimum weight of an RDF on G. Roman domination has military and commercial applications.

Prove the following upper bounds for the Roman domination number:

(a) For any graph G with $\delta > 0$,

$$\gamma_R(G) \leq 2\left(1 - \frac{2^{1/\delta}\delta}{(1+\delta)^{1+1/\delta}}\right)n. \tag{4.27}$$

(b) For any graph G with $\delta > 0$,

$$\gamma_R(G) \leq \frac{2\ln(\delta + 1) - \ln 4 + 2}{\delta + 1}n. \tag{4.28}$$

Hint: use part (a) to prove this upper bound.

4.4.9*

Prove that the upper bound (4.28), and therefore bound (4.27), are asymptotically best possible, that is:

When n is large enough, there exists a graph G such that

$$\gamma_R(G) \geq \frac{2\ln(\delta + 1) - \ln 4 + 2}{\delta + 1}n(1 + o(1)).$$

Here is the first part of the proof: Let us consider a complete graph $K_{[\delta \ln \delta]}$, and let F denote its vertex set. Next, let us add a set of new vertices $V = \{v_1, ..., v_\delta\}$, where each vertex v_i is adjacent to δ vertices that are randomly chosen from the set F. The resulting graph is denoted by G and it has $n = [\delta \ln \delta] + \delta$ vertices. We will prove that with positive probability

$$\gamma_R(G) \geq \frac{2 \ln \delta}{\delta} n(1 + o_\delta(1)) = 2 \ln^2 \delta(1 + o_\delta(1)).$$

Let $f = (D_0, D_1, D_2)$ be a γ_R-function of G; that is, f is an RDF and $f(V(G)) = \gamma_R(G)$. Note that $v \in D_{f(v)}$ for any vertex $v \in V(G)$. It is easy to see that we may assume that $D_2 \subseteq F$ and $D_1 \subseteq V$.

Let us consider two cases. If $|D_2| > \ln^2 \delta - \ln \delta \ln \ln^4 \delta$, then $f(V(G)) > 2 \ln^2 \delta(1 + o_\delta(1))$, as required. Suppose now that

$$|D_2| \leq \ln^2 \delta - \ln \delta \ln \ln^4 \delta.$$

- The reader should complete this proof. Hints: the next step is to prove that

$$\mathbb{P}[D_2 \text{ does not dominate } v_i] \geq \frac{\ln^3 \delta}{\delta}.$$

Then, it is necessary to consider the random variable $|N(D_2) \cap V|$ and its expectation.

4.5 Solutions

4.4.1

Form a set of vertices A by selecting each vertex independently with probability p. The optimal value of p will be determined later. Let us denote

$$B = V(G) - N[A].$$

It is easy to see that $S = A \cup B$ is a dominating set.

Now, the expected value of $|A|$ is np, and the probability that a vertex v belongs to B is

$$(1 - p)^{1 + \deg(v)} \leq (1 - p)^{1 + \delta} \leq e^{-p(1+\delta)}.$$

Hence, the expectation of $|S|$ is

$$\mathbb{E}[|S|] = \mathbb{E}[|A|] + \mathbb{E}[|B|] \le np + ne^{-p(1+\delta)} = (p + e^{-p(1+\delta)})n.$$

The last expression is minimized when

$$p = \frac{\ln(\delta+1)}{\delta+1}.$$

Therefore,

$$\mathbb{E}[|S|] \le \frac{\ln(\delta+1)+1}{\delta+1}n.$$

Thus, there is a particular dominating set S in G with at most this cardinality. ∎

4.4.2

Form a set of vertices A, where each vertex is selected independently with probability p. The optimal value of p will be determined later. Let

$$B = V(G) - N(A).$$

In other words, the set B contains isolated vertices in A and also vertices that are not dominated by A. Let us form a set B' as follows. For every vertex $v \in B$, we take a vertex in $N(v)$ and add it to B'. Such a neighbour always exists because $\delta > 0$. It is not difficult to see that $T = A \cup B'$ is a total dominating set.

Now, the expected value of $|A|$ is np, and the probability that a vertex v belongs to B is

$$(1-p)^{\deg(v)} \le (1-p)^{\delta} \le e^{-p\delta}.$$

Hence, the expectation of $|T|$ is

$$\mathbb{E}[|T|] = \mathbb{E}[|A|] + \mathbb{E}[|B'|] \le np + \mathbb{E}[|B|] \le np + ne^{-p\delta} = (p + e^{-p\delta})n.$$

The last expression is minimized when $p = \frac{\ln \delta}{\delta}$. Therefore,

$$\mathbb{E}[|T|] \le \frac{\ln \delta + 1}{\delta}n.$$

Thus, there is a particular total dominating set T with at most this cardinality. ∎

4.4.3

For each vertex $v \in V(G)$, we arbitrarily select δ vertices from $N(v)$ and denote the resulting set together with the vertex v by $N'[v]$. Thus, $|N'[v]| = \delta + 1$. Let A be a set formed by an independent choice of vertices of G, where each vertex is selected with probability

$$p = 1 - \left(\frac{1}{\tilde{b}_{k-1}(1 + \delta')} \right)^{1/\delta'}.$$

For $m = 0, 1, ..., k - 1$, we denote

$$C_m = \{v \in V(G) : |N'[v] \cap A| = m\}.$$

For each set C_m, let us form a set C'_m in the following way: for every vertex $v \in C_m$ we take $k - m$ neighbours from $N'[v] - A$ and add them to C'_m. Such neighbours always exist because $\delta \geq k$. It is obvious that $|C'_m| \leq (k-m)|C_m|$.

Let us construct the set D as follows:

$$D = A \cup \left(\bigcup_{m=0}^{k-1} C'_m \right).$$

It is easy to see that D is a k-tuple dominating set. The expectation of $|D|$ is

$$\mathbb{E}[|D|] \leq \mathbb{E} \left[|A| + \sum_{m=0}^{k-1} |C'_m| \right] \leq \mathbb{E}[|A|] + \sum_{m=0}^{k-1} (k - m)\mathbb{E}[|C_m|].$$

We have

$$\mathbb{E}[|C_m|] = \sum_{v \in V(G)} \mathbb{P}[v \in C_m] = \sum_{i=1}^{n} p^m (1 - p)^{\delta-m+1} \binom{\delta + 1}{m}$$
$$= p^m (1 - p)^{\delta-m+1} \tilde{b}_m n.$$

Thus,

$$\mathbb{E}[|D|] \leq pn + \sum_{m=0}^{k-1} (k - m) p^m (1 - p)^{\delta-m+1} \tilde{b}_m n$$
$$= pn + (1 - p)^{\delta-k+2} n \sum_{m=0}^{k-1} (k - m) p^m (1 - p)^{k-m-1} \tilde{b}_m.$$

Furthermore, for $0 \leq m \leq k - 1$,

$$(k - m)\tilde{b}_m = (k - m)\binom{\delta + 1}{m} \leq \prod_{j=1}^{\delta - k + 2} \frac{(k - m + j - 1)}{j}\binom{\delta + 1}{m}$$

$$= \binom{\delta - m + 1}{\delta - k + 2}\binom{\delta + 1}{m} = \binom{k - 1}{m}\binom{\delta + 1}{k - 1} = \binom{k - 1}{m}\tilde{b}_{k-1}.$$

We obtain

$$\mathbb{E}[|D|] \leq pn + (1 - p)^{\delta'+1}n\tilde{b}_{k-1}\sum_{m=0}^{k-1}\binom{k-1}{m}p^m(1 - p)^{k-m-1}$$

$$= pn + (1 - p)^{\delta'+1}n\tilde{b}_{k-1} = \left(1 - \frac{\delta'}{\tilde{b}_{k-1}^{1/\delta'}(1 + \delta')^{1+1/\delta'}}\right)n,$$

as required. The proof of the theorem is complete. ∎

4.4.4

Let X be a maximum k-limited packing in G of size $L_k(G)$. Each vertex in $V(G) - X$ is adjacent to at most k vertices in X. Hence, the number of edges between $V(G) - X$ and X is as follows:

$$e(V(G) - X, X) \leq k(n - |X|).$$

On the other hand, each vertex $x \in X$ is adjacent to at most $k - 1$ vertices of X, and hence x is adjacent to at least $\delta - k + 1$ vertices in $V(G) - X$. We obtain

$$e(X, V(G) - X) \geq (\delta - k + 1)|X|.$$

Thus,

$$(\delta - k + 1)|X| \leq k(n - |X|),$$

and, therefore,

$$|X| \leq \frac{kn}{\delta + 1}.$$

∎

4.4.5

The following proof is due to Gallant et al. [24]. Let B be a maximum k-limited packing in G, and let D be a minimum dominating set in G. Let U be the set of ordered pairs

$$\{(b, d) : b \in B, \ d \in D, \ b \in N[d]\}.$$

For every $b \in B$, there is at least one $d \in D$ such that $b \in N[d]$ because D is a dominating set for G. Therefore, $|B| \leq |U|$. For each $d \in D$, we know that $|N[d] \cap B| \leq k$, since B is a k-limited packing. Hence, there are at most k vertices $b \in B$ with $(b, d) \in U$, and so $|U| \leq k|D|$. Thus,

$$L_k(G) = |B| \leq |U| \leq k|D| = k\gamma(G).$$

■

4.4.6

We prove inequality (4.26) by contradiction. Let X be a maximum k-limited packing in G of size $L_k(G)$, and let Y be a minimum k-tuple dominating set in G of size $\gamma_{\times k}(G)$. We denote $B = X \cap Y$; that is, $X = A \cup B$ and $Y = B \cup C$, where A and C are disjoint. Assume to the contrary that $L_k(G) > \gamma_{\times k}(G)$, thus $|A| > |C|$.

Since Y is a k-tuple dominating set, each vertex of A is adjacent to at least k vertices of Y. Hence, the number of edges between A and $B \cup C$ is as follows: $e(A, B \cup C) \geq k|A|$. Now, every vertex of C is adjacent to at most k vertices of X because X is a k-limited packing set. Therefore, the number of edges between C and $A \cup B$ satisfies $e(C, A \cup B) \leq k|C|$. We obtain

$$e(C, A \cup B) \leq k|C| < k|A| \leq e(A, B \cup C);$$

that is, $e(C, A \cup B) < e(A, B \cup C)$. By eliminating the edges between A and C, we conclude that $e(C, B) < e(A, B)$.

Now, let us consider an arbitrary vertex $b \in B$ and denote $s = |N(b) \cap A|$. Since $X = A \cup B$ is a k-limited packing set, we obtain $|N(b) \cap X| \leq k - 1$, and hence $|N(b) \cap B| \leq k - s - 1$. On the other hand, $Y = B \cup C$ is a k-tuple dominating set, so $|N(b) \cap Y| \geq k - 1$. Therefore, $|N(b) \cap C| \geq s$. Thus, $|N(b) \cap C| \geq |N(b) \cap A|$ for any vertex $b \in B$. We obtain, $e(C, B) \geq e(A, B)$, a contradiction. We conclude that $L_k(G) \leq \gamma_{\times k}(G)$. ■

4.4.7

Let X be a maximum 2-limited packing in G of size $L_2(G)$. Because G is a Δ-regular graph, the number of edges between X and $Y = V(G) - X$ satisfies

$$e(X, Y) \le \Delta |X|.$$

Now, each vertex $y \in Y$ is adjacent to at most two vertices of X. If y is not adjacent to any vertex of X, then y must be adjacent to a vertex $z \in Y$ which is adjacent to two vertices of X, otherwise X is not maximal. It is not difficult to see that the smallest number of edges between X and Y is achieved when Y is partitioned in $l = \frac{|Y|}{\Delta-1}$ parts, where l is an integer and each part P is of size $\Delta - 1$ and it has a vertex y that is adjacent to two vertices of X and to other $\Delta - 2$ vertices of P. There are just two edges between P and X. Thus, each part contributes two edges to $e(X, Y)$, and hence

$$e(X, Y) \ge \frac{2|Y|}{\Delta - 1}.$$

If l is not an integer, then one of the parts, say P', would have fewer than $\Delta - 1$ vertices. If P' contributes at least two edges to $e(X, Y)$, then

$$e(X, Y) \ge 2 \left\lceil \frac{|Y|}{\Delta - 1} \right\rceil > \frac{2|Y|}{\Delta - 1}.$$

Now, P' contributes at most one edge to $e(X, Y)$ only if P' has just one vertex, which must be adjacent to a vertex in X. In this case,

$$e(X, Y) \ge \frac{2(|Y| - 1)}{\Delta - 1} + 1 = \frac{2|Y| + (\Delta - 3)}{\Delta - 1} \ge \frac{2|Y|}{\Delta - 1}.$$

Thus,

$$\frac{2|Y|}{\Delta - 1} = \frac{2(n - |X|)}{\Delta - 1} \le e(X, Y) \le \Delta |X|$$

or

$$|X| \ge \frac{2n}{\Delta^2 - \Delta + 2},$$

as required.   ∎

4.4.8

(a) Let A be a set formed by an independent choice of vertices of G, where each vertex is selected with probability

$$p = 1 - \left(\frac{2}{1+\delta}\right)^{1/\delta}.$$

We denote $B = N[A] - A$ and $C = V(G) - N[A]$. Let us assume that f is a function $f: V(G) \to \{0, 1, 2\}$ and assign $f(v_i) = 2$ for each $v_i \in A$, $f(v_i) = 0$ for each $v_i \in B$ and $f(v_i) = 1$ for each $v_i \in C$. It is obvious that f is an RDF and $f(V(G)) = 2|A| + |C|$.

It is easy to show that

$$\mathbb{P}[v_i \in C] = (1-p)^{1+\deg(v_i)} \leq (1-p)^{1+\delta}.$$

The expectation of $f(V(G))$ is

$$\mathbb{E}[f(V(G))] \leq 2\mathbb{E}[|A|] + \mathbb{E}[|C|]$$
$$= 2pn + \sum_{i=1}^{n} \mathbb{P}[v_i \in C]$$
$$\leq 2pn + (1-p)^{1+\delta}n \qquad (4.29)$$
$$= 2\left(1 - \frac{\delta\, 2^{1/\delta}}{(1+\delta)^{1+1/\delta}}\right)n.$$

Since the expectation is an average value, there exists a particular RDF of the above order, as required. The proof of the theorem is complete. ∎

(b) Using the inequality $1 - p \leq e^{-p}$, we obtain the following estimation of the expression (4.29):

$$\mathbb{E}[f(V(G))] \leq 2pn + e^{-p(\delta+1)}n.$$

If we set

$$p = \frac{\ln(\delta+1) - \ln 2}{\delta + 1},$$

then

$$\mathbb{E}[f(V(G))] \leq \frac{2\ln(\delta+1) - \ln 4 + 2}{\delta+1} n,$$

as required. ∎

4.4.9

We modify Alon's probabilistic construction [3] as follows. Let us consider a complete graph $K_{[\delta \ln \delta]}$, and let F denote its vertex set. Next, let us add a set of new vertices $V = \{v_1, ..., v_\delta\}$, where each vertex v_i is adjacent to δ vertices that are randomly chosen from the set F. The resulting graph is denoted by G and it has $n = [\delta \ln \delta] + \delta$ vertices. We will prove that with positive probability

$$\gamma_R(G) \geq \frac{2\ln \delta}{\delta} n(1 + o_\delta(1)) = 2\ln^2 \delta(1 + o_\delta(1)).$$

Let $f = (D_0, D_1, D_2)$ be a γ_R-function of G; that is, f is an RDF and $f(V(G)) = \gamma_R(G)$. Note that $v \in D_{f(v)}$ for any vertex $v \in V(G)$. It is easy to see that we may assume that $D_2 \subseteq F$ and $D_1 \subseteq V$.

Let us consider two cases. If $|D_2| > \ln^2 \delta - \ln \delta \ln \ln^4 \delta$, then $f(V(G)) > 2\ln^2 \delta(1 + o_\delta(1))$, as required. Suppose now that

$$|D_2| \leq \ln^2 \delta - \ln \delta \ln \ln^4 \delta.$$

It is not difficult to show that the probability of the set D_2 not dominating a vertex $v_i \in V$ is

$$\mathbb{P}[D_2 \text{ does not dominate } v_i] = \frac{\binom{|F|-|D_2|}{\delta}}{\binom{|F|}{\delta}}$$

$$\geq \left(\frac{|F| - |D_2| - \delta}{|F| - \delta}\right)^\delta$$

$$= \left(1 - \frac{|D_2|}{|F| - \delta}\right)^\delta.$$

Using the inequality $1 - x \geq e^{-x}(1 - x^2)$ if $x < 1$, we obtain the following estimation:

$$\mathbb{P}[D_2 \text{ does not dominate } v_i] \geq e^{-\frac{\ln^2 \delta - \ln \delta \ln \ln^4 \delta}{\delta \ln \delta - \delta} \delta} \left(1 - \left(\frac{\ln^2 \delta - \ln \delta \ln \ln^4 \delta}{\delta \ln \delta - \delta}\right)^2\right)^{\delta}$$

$$= e^{-\frac{\ln \delta + \ln \ln^4 \delta}{1 - 1/\ln \delta}} (1 + o_\delta(1))$$

$$= e^{\ln\left(\frac{\ln^4 \delta}{\delta}\right)(1 + o_\delta(1))} (1 + o_\delta(1))$$

$$= \left(\frac{\ln^4 \delta}{\delta}\right)^{1 + o_\delta(1)} (1 + o_\delta(1))$$

$$\geq \frac{\ln^3 \delta}{\delta}.$$

Therefore,

$$\mathbb{P}[D_2 \text{ dominates } v_i] \leq 1 - \frac{\ln^3 \delta}{\delta}.$$

Let us consider the random variable $|N(D_2) \cap V|$. The expectation of $|N(D_2) \cap V|$ is

$$\mathbb{E}[|N(D_2) \cap V|] = \sum_{i=1}^{\delta} \mathbb{P}[D_2 \text{ dominates } v_i] \leq \delta - \ln^3 \delta.$$

As discussed above, $f(v_i) = 0$ or 1 for any vertex $v_i \in V$. Also, if $f(v_i) = 0$, then v_i must be adjacent to $w \in F$ for which $f(w) = 2$. Thus, we can conclude that there exists a graph G, for which $|D_1| \geq \ln^3 \delta$, that is,

$$f(V(G)) \geq \ln^3 \delta > 2\ln^2 \delta(1 + o_\delta(1)),$$

as required. ∎

Acknowledgements

This chapter is based on the following publications: (A) Reprinted by permission from Elsevier: *Discrete Applied Mathematics*, **161**, A. Gagarin, A. Poghosyan and V. Zverovich, Randomized algorithms and upper bounds for multiple domination in graphs and networks, 604–611, © 2013. *Discrete Applied Mathematics*, **184**, A. Gagarin and V. Zverovich, The probabilistic approach to limited packings in graphs, 146–153, © 2015. *Discrete Mathematics*, **338**, V. Zverovich, On general frameworks and threshold functions for multiple domination, 2095–2104, © 2015. (B) Reprinted by permission from Springer Nature: Springer-Verlag, *Graphs and Combinatorics*, Upper bounds for α-domination parameters, A. Gagarin, A. Poghosyan and V. Zverovich, © 2009. Springer-Verlag, *Graphs and Combinatorics*, On Roman, global and restrained domination in graphs, V. Zverovich and A. Poghosyan, © 2011.

References

[1] J. Alber, N. Betzler and R. Niedermeier, Experiments on data reduction for optimal domination in networks, *Annals of Operations Research*, **146** (2006) 105–117.

[2] N. Alon and J. H. Spencer, *The Probabilistic Method*, 2nd edn, New York, NY: John Wiley & Sons Inc., 2000.

[3] N. Alon, Transversal numbers of uniform hypergraphs, *Graphs and Combinatorics*, **6** (1990) 1–4.

[4] V. I. Arnautov, Estimation of the exterior stability number of a graph by means of the minimal degree of the vertices, *Prikladnaya Mathematika i Programmirovanie*, **11** (1974) 3–8 (in Russian).

[5] P. N. Balister, B. Bollobás and K. Gunderson, Limited packings of closed neighbourhoods in graphs, arXiv:1501.01833 [math.CO], 2015.

[6] B. Bollobás, *Graph Theory: An Introductory Course*, New York, NY: Springer-Verlag, 1979.

[7] A. P. Burger, M. A. Henning and J. H. van Vuuren, On the ratios between packing and domination parameters of a graph, *Discrete Mathematics*, **309** (2009) 2473–2478.

[8] G. J. Chang, The upper bound on k-tuple domination numbers of graphs, *European Journal of Combinatorics*, **29** (2008) 1333–1336.

[9] M. Chellali, O. Favaron, A. Hansberg and L. Volkmann, k-Domination and k-independence in graphs: a survey, *Graphs and Combinatorics*, **28** (1)(2012) 1–55.

[10] Y. Caro and Y. Roditty, A note on the k-domination number of a graph, *International Journal of Mathematics and Mathematical Sciences*, **13** (1990) 205–206.

[11] Y. Caro and Y. Roditty, Improved bounds for the product of the domination and chromatic numbers of a graph, Manuscript, 1997, 6 pages [Theorem 2.3]. (A reduced version of this paper appeared in: *Ars Combinatoria*, **56** (2000) 189–191.)

[12] Y. Caro and R. Yuster, Dominating a family of graphs with small connected subgraphs, *Combinatorics, Probability and Computing*, **9** (2000) 309–313.

[13] E. J. Cockayne, Towards a theory of $\langle r, s \rangle$-domination in graphs, *Utilitas Mathematica*, **80** (2009) 97–113.

[14] C. Cooper, R. Klasing and M. Zito, Lower bounds and algorithms for dominating sets in web graphs, *Internet Mathematics*, **2** (2005) 275–300.

[15] F. Dai and J. Wu, On constructing k-connected k-dominating set in wireless ad hoc and sensor networks, *Journal of Parallel and Distributed Computing*, **66** (2006) 947–958.

[16] M. P. Dobson, V. Leoni and G. Nasini, The multiple domination and limited packing problems in graphs, *Information Processing Letters*, **111** (2011) 1108–1113.

[17] J. E. Dunbar, D. G. Hoffman, R. C. Laskar and L. R. Markus, α-Domination, *Discrete Mathematics*, **211** (2000) 11–26.

[18] O. Favaron, A. Hansberg and L. Volkmann, On k-domination and minimum degree in graphs, *Journal of Graph Theory*, **57** (2008) 33–40.

[19] O. Favaron, M. A. Henning, J. Puech and D. Rautenbach, On domination and annihilation in graphs with claw-free blocks, *Discrete Mathematics*, **231** (2001) 143–151.

[20] A. Gagarin, A. Poghosyan and V. Zverovich, Randomized algorithms and upper bounds for multiple domination in graphs and networks, *Discrete Applied Mathematics*, **161** (2013) 604–611.

[21] A. Gagarin, A. Poghosyan and V. Zverovich, Upper bounds for α-domination parameters, *Graphs and Combinatorics*, **25** (2009) 513–520.

[22] A. Gagarin and V. Zverovich, A generalised upper bound for the k-tuple domination number, *Discrete Mathematics*, **308** (2008) 880–885.

[23] A. Gagarin and V. Zverovich, The probabilistic approach to limited packings in graphs, *Discrete Applied Mathematics*, **184** (2015) 146–153.

[24] R. Gallant, G. Gunther, B. Hartnell and D. Rall, Limited packings in graphs, *Discrete Applied Mathematics*, **158** (12) (2010) 1357–1364.

[25] A. Hansberg, Bounds on the connected k-domination number in graphs, *Discrete Applied Mathematics*, **158** (2010) 1506–1510.

[26] A. Hansberg and L. Volkmann, Upper bounds on the k-domination number and the k-Roman domination number, *Discrete Applied Mathematics*, **157** (7)(2009), 1634–1639.

[27] J. Harant and M. A. Henning, On double domination in graphs, *Discussiones Mathematicae Graph Theory*, **25** (2005) 29–34.

[28] J. Harant and M. A. Henning, A realization algorithm for double domination in graphs, *Utilitas Mathematica*, **76** (2008) 11–24.

[29] T. W. Haynes, S. T. Hedetniemi and P. J. Slater, *Fundamentals of Domination in Graphs*, New York, NY: Marcel Dekker, 1998.

[30] M. A. Henning, C. Löwenstein and D. Rautenbach, Dominating sets, packings, and the maximum degree, *Discrete Mathematics*, **311** (2011) 2031–2036.

[31] D. S. Hochbaum and D. B. Schmoys, A best possible heuristic for the k-center problem, *Mathematics of Operations Research*, **10** (2) (1985) 180–184.

[32] R. Klasing and C. Laforest, Hardness results and approximation algorithms of k-tuple domination in graphs, *Information Processing Letters*, **89** (2004) 75–83.

[33] L. Lovász, On the ratio of optimal integral and fractional covers, *Discrete Mathematics*, **13** (1975) 383–390.

[34] A. Meir and J. W. Moon, Relations between packing and covering numbers of a tree, *Pacific Journal of Mathematics*, **61** (1975) 225–233.

[35] D. A. Mojdeh, B. Samadi and S. M. Hosseini Moghaddam, Limited packing vs. tuple domination in graphs, *Ars Combinatoria*, **133** (2017) 155–161.

[36] J. W. Moon and L. Moser, Almost all (0,1) matrices are primitive, *Studia Scientiarum Mathematicarum Hungarica*, **1** (1966) 153–156.

[37] R. E. Neapolitan and K. Naimipour, *Foundations of Algorithms Using Java Pseudocode*, Burlington, MA: Jones and Bartlett Publishers Inc., 2004.

[38] T. Nieberg, Distributed algorithms in wireless sensor networks, *Electronic Notes in Discrete Mathematics*, **13** (2003) 81–83.

[39] C. Payan, Sur le nombre d'absorption d'un graphe simple, *Cahiers du Centre d'Études de Recherche Opérationnelle*, **17** (1975) 307–317 (in French).

[40] D. Rautenbach and L. Volkmann, New bounds on the k-domination number and the k-tuple domination number, *Applied Mathematics Letters*, **20** (2007) 98–102.

[41] R. R. Rubalcaba, A. Schneider and P. J. Slater, A survey on graphs which have equal domination and closed neighborhood packing numbers, *AKCE International Journal of Graphs and Combinatorics*, **3** (2) (2006) 93–114.

[42] W. Shang, P. Wan, F. Yao and X. Hu, Algorithms for minimum m-connected k-tuple dominating set problem, *Theoretical Computer Science*, **381** (2007) 241–247.

[43] Z. Shi, A self-stabilizing algorithm to maximal 2-packing with improved complexity, *Information Processing Letters*, **112** (2012) 525–531.

[44] C. Stracke and L. Volkmann, A new domination conception, *Journal of Graph Theory*, **17** (1993) 315–323.

[45] M. T. Thai, N. Zhang, R. Tiwari and X. Xu, On approximation algorithms of k-connected m-dominating sets in disk graphs, *Theoretical Computer Science*, **385** (2007) 49–59.

[46] J. Topp and L. Volkmann, On packing and covering numbers of graphs, *Discrete Mathematics*, **96** (1991) 229–238.

[47] V. Zverovich, On general frameworks and threshold functions for multiple domination, *Discrete Mathematics*, **338** (2015) 2095–2104.

[48] V. Zverovich, The k-tuple domination number revisited, *Applied Mathematics Letter*, **21** (2008) 1005–1011.

5

Graph Models for Optimization Problems in Road Networks

V. Zverovich, P. Corcoran and A. Gagarin

In this chapter, we consider two applications of graph theory. The first part is devoted to pedestrian safety, which is one of major issues in modern societies. The focus is on pedestrian safety in urban areas with respect to crashes between pedestrians and vehicles. In particular, we discuss an algorithm for automated construction of a graph model for pavement networks. Then, an algorithm for finding a user-optimal path in a given pavement network is presented. This algorithm is based on three criteria: path safety, distance and path complexity. A user may choose a suitable ranking of these attributes depending on their needs. The approach is validated and tested on pavement networks generated from real road networks. The second part of this chapter is devoted to optimizing the placement of charging stations for electric vehicles in road networks. Electric and hybrid vehicles play an increasing role in road transport networks. Despite their advantages, they have a relatively limited cruising range in comparison to traditional diesel/petrol vehicles and require significant battery charging time. The facility location problem of the placement of charging stations in road networks is modelled as a multiple domination problem on reachability graphs. This model takes into account a threshold for the remaining battery charge and provides some minimal choice for a travel direction to recharge the battery. Experimental evaluation and simulations for the proposed facility location model are given for real road networks of the cities of Boston and Dublin.

5.1 Pedestrian Road Safety in Urban Areas

Pedestrian safety is a significant part of traffic safety and, more generally, a major social problem. Every year, over 250,000 pedestrians worldwide are killed in traffic crashes. Hence, modelling pedestrian safety is a very important

Modern Applications of Graph Theory. Vadim Zverovich, Oxford University Press (2021). © Vadim Zverovich.
DOI: 10.1093/oso/9780198856740.003.0005

research area. A brief overview of the literature devoted to pedestrian safety is given in Section 5.1.3.

A *pavement network* is a graph representing the pavement and walking path structure in a given urban area. Such a network can be used for modelling and analysis of pedestrian movement on pavements and paths in the area. Unfortunately, accurate pavement networks for the UK are not readily available, so they must be generated somehow.

In general, there are four approaches for generation of pavement/pedestrian networks: network buffering, collaborative mapping, remote sensing and crowdsourcing the information. The first method is based on road networks, the second on walking GPS traces, the third on orthoimages and LiDAR point clouds [24], and the last on a web-based tool for collecting sidewalk accessibility data [42]. In this section, the first approach will be explored because of its low cost and high availability of data. More precisely, the generation of a pavement network presented in this section is based on free geographic data from OpenStreetMap (OSM) [31], which provides accurate road networks for many regions in the UK. Such road networks will be used for automated construction of the corresponding pavement networks.

The vertices of a *road network* N correspond to road intersections, dead-ends or designated pedestrian crossings. The type of each designated pedestrian crossing is known: light controlled, pelican, zebra or human controlled. These four types of designated pedestrian crossings were defined by the UK Department for Transport. The edges of N correspond to road segments connecting their end-vertices, and the length and width of each road segment are known. In addition, the OSM road type for each segment is given: trunk, primary, secondary, trunk link, primary link, secondary link, tertiary, tertiary link, residential, service, track, pedestrian, footway, path or steps. It is not too difficult to establish the relationship between the OSM types and the UK classification of roads: motorways, A roads, B roads, slip roads, classified unnumbered road etc.

There are two assumptions when constructing a pavement network: every road segment has a pavement on both sides and a pedestrian crossing (designated, jaywalk, dead-end) at each endpoint. These assumptions are typical in the network buffering approach [24]. Thus, each road intersection includes either designated pedestrian crossings or jaywalk crossings. These assumptions are not necessarily true in real-life scenarios, but they are good approximations to reality as discussed in [21].

5.1.1 Automated Construction of Pavement Networks (PNC-Algorithm)

The pavement network can be constructed using PNC-Algorithm. A pseudo-code of this algorithm was proposed by Hannah et al. [21], and its mathematical version is presented below.

The input of PNC-Algorithm includes the road network N discussed above. We assume that the network N is an undirected connected graph, which is embedded on a plane without edge crossings (i.e. there are no bridges or tunnels). The type of every vertex $v \in V(N)$ is specified: a road intersection, dead-end or designated pedestrian crossing. In the last case, the type of a pedestrian crossing is indicated (i.e. light controlled, pelican, zebra or human controlled), which may also be specified for a road intersection. This information can be formally modelled by the following function:

$$\phi \colon V(N) \to (A, B),$$

where the set A consists of the following elements: road intersection, dead-end, designated pedestrian crossing, and

$$B = \{\text{nil}, \text{light controlled}, \text{pelican}, \text{zebra}, \text{human controlled}\}.$$

For example, (road intersection, nil) and (road intersection, light controlled) mean a junction with jaywalk pedestrian crossings and a junction with light-controlled crossings, respectively.

The input of the algorithm also includes the length, width and type of every road in N. This can be modelled by the following function:

$$\xi \colon E(N) \to (\mathbb{R}^+, \mathbb{R}^+, \text{String}).$$

The function ξ maps every edge of N to a triple containing the length and width of the corresponding road segment, as well as the OSM type of the road. The complete list of the OSM types is given above.

The output of the algorithm is a pavement network represented by an undirected graph G. The vertices of the graph G are endpoints of pavement segments and pedestrian crossings (designated, jaywalk, dead-end), which are the edges of G. One vertex of G may be incident to a few pavement segments and pedestrian crossings. Examples of pavement networks can be

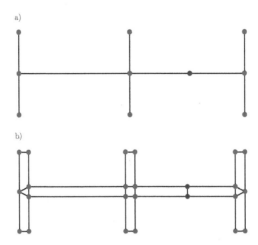

Fig. 5.1 a) Artificial road network N with a designated pedestrian crossing represented by the blue vertex; b) the corresponding pavement network G generated by PNC-Algorithm. (Reproduced from [21] with permission from Elsevier.)

seen in Figures 5.1 and 5.2. For every edge e in G, the following information is available: length of e; type of e if it represents a pedestrian crossing (including a jaywalk crossing and dead-end); OSM type of e if it is a pavement segment or a jaywalk crossing.

In the first step of the algorithm, the vertex and edge sets of the pavement network G are initialized to empty sets. Next, for each vertex $v \in V(N)$ of degree 1 or 2, two copies of v and the edge between them are added to $V(G)$ and $E(G)$, respectively. Note that vertices in N of degree 1 and 2 represent dead-ends and designated pedestrian crossings. In Step 3, for every vertex v of degree at least 3, a cycle of length $k = \deg(v)$ is added to G. In this case, the vertex v represents a road intersection of k roads. For example, if $\deg(v) = 4$, which means that four road segments meet at v, then the cycle C_4 is added to G. Each of the edges of C_4 corresponds to a jaywalk pedestrian crossing, unless otherwise specified. It may be pointed out that it is not difficult to model a mixed situation, where some edges of the cycle represent jaywalk pedestrian crossing and other edges are designated pedestrian crossings.

In Step 4, the algorithm iterates over the edges of N. Every edge of N represents a road segment, for which two pavement segments on both sides of the road are included in $E(G)$. To perform this step correctly, the following function is defined:

$$\text{CyclicOrder} : \{(u, v) \mid uv \in E(N)\} \rightarrow \mathbb{N},$$

which maps every ordered pair (u, v) such that $uv \in E(N)$ to a natural number between 1 and $\deg(v)$. More precisely, CyclicOrder(u, v) is equal to the position of the edge uv in the circular clockwise ordering of edges incident to the vertex u (say, starting at the 12 o'clock position). Such a function is well defined because the graph N is embedded on the plane. Notice that in general CyclicOrder$(u, v) \neq$ CyclicOrder(v, u).

PNC-Algorithm: Pavement network construction.

Input: A road network N represented by an undirected connected graph embedded on a plane.

The type and subtype of every vertex $v \in V(N)$ specified by the function $\phi(v)$.

The length, width and OSM type of every road $e \in E(N)$ given by the function $\xi(e)$.

Output: The undirected graph G representing a pavement network, and parameters of its edges.

(1) Initialize sets $V(G) = \emptyset$ and $E(G) = \emptyset$.

(2) For each vertex $v \in V(N)$ of degree 1 or 2, add two vertices v^0, v^1 to the set $V(G)$, and the edge $v^0 v^1$ to the set $E(G)$.

(3) For each vertex $v \in V(N)$ of degree at least 3, do the following:

Add to $V(G)$ the vertices

$$v^0, v^1, ..., v^{\deg(v)-1}.$$

Add to $E(G)$ the edges

$$v^0 v^1, v^1 v^2, ..., v^{\deg(v)-1} v^0.$$

(4) For each edge $uv \in E(N)$, do the following:

Denote

$$i = \text{CyclicOrder}(u, v), \quad j = \text{CyclicOrder}(v, u).$$

Add to $E(G)$ the edges

$$u^{\Psi(i,u)} v^{(j+1) \bmod \deg(v)}$$

and

$$u^{(i+1) \bmod \deg(u)} v^{\Psi(j,v)},$$

where

$$\Psi(i, u) = \begin{cases} i \text{ if } \deg(u) = 1, \\ i \bmod \deg(u) \text{ if } \deg(u) \geq 2. \end{cases}$$

(5) For each edge $e \in E(G)$, do the following:
   (a) If e is of the form $u^l v^m$, where $u \neq v$, then e represents a pavement segment and its type is the type of the OSM road given by the function $\xi(uv)$. The length of e is equal to the length of the edge uv given by the function $\xi(uv)$.
   (b) If e is of the form $v^l v^m$, then e represents a pedestrian crossing and its type is given by the function $\phi(v)$. The length of e is equal to the width of the corresponding road vw given by the function $\xi(vw)$. If e is a jaywalk crossing, then its OSM type is equal to the OSM type of the corresponding road vw given by the function $\xi(vw)$.
(6) Report the graph G and the information about its edges. Algorithm stops.

Finally, the necessary information about the edges of the graph G must be specified. If $e \in E(G)$ represents a pavement segment, then the length of e is equal to the length of the corresponding road segment. If e represents a pedestrian crossing, then its length is the width of the corresponding road segment. In addition, pedestrian crossing type or OSM type should be specified for every edge of G. Note that if the edge $e = v^l v^m$ is a jaywalk crossing, which happens when $\phi(v) = $ (road intersection, nil), then both OSM type and pedestrian crossing type of e must be specified. This information will be needed for an algorithm presented later.

Figure 5.1 illustrates the application of PNC-Algorithm to an artificial example of a road network with one designated pedestrian crossing [21]. The algorithm was also applied to a road sub-network of the city of York shown in Figure 5.2a. The generated pavement network is illustrated in Figure 5.2b.

5.1.2 Search for and Prioritization of Pedestrian Paths (PPP-Algorithm)

In this section, we present an algorithm for finding a user-optimal path in a given pavement network using three criteria: path safety, distance and

a)

b)

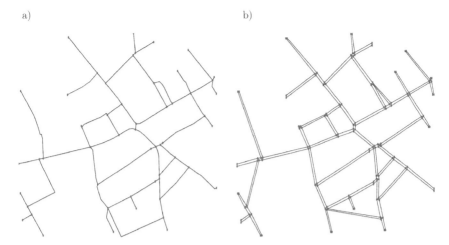

Fig. 5.2 a) Road sub-network of the city of York; b) the corresponding pavement network generated by PNC-Algorithm. (Reproduced from [21] with permission from Elsevier.)

Table 5.1 Risk values (per metre) assigned to designated and dead-end pedestrian crossings [21].

Pedestrian Crossing Type	Risk Value (per metre)
Light controlled	4
Pelican	4
Zebra	3
Human controlled	2
Dead-end	1

path complexity. The user can choose a suitable ranking of these attributes depending on their needs. For instance, an elderly person might want to choose path complexity as the most important criterion to be minimized.

With some adaptations, we adopt risk values of pedestrian crossings and pavement segments from [21]. Because pavement networks are not homogeneous, all risk values in our model are per metre (see Tables 5.1 and 5.2). Moreover, the risk values for jaywalk crossings in Table 5.2 are higher than similar values in [21], and there is some small risk value for the following OSM types: pedestrian, footway, path, steps.

The input of PPP-Algorithm includes a pavement network G discussed in the previous section. For every edge e in G, we know its length, type and the corresponding risk values. The start vertex p and the destination vertex q of the required (p, q)-path are given. In addition, an advanced user may choose

Table 5.2 Risk values (per metre) assigned to pavement segments and jaywalk crossings.

OpenStreetMap Road Type	Pavement Segment Risk Value (per metre)	Jaywalk Crossing Risk Value (per metre)
Trunk, primary, secondary	7	70
Trunk/primary/secondary link	6	60
Tertiary, tertiary link	5	50
Residential, service	4	40
Track	2	20
Pedestrian, footway, path, steps	0.1	0.1

the preferential ranking of the three criteria: path safety (PS), distance (D) and path complexity (PC). We consider seven options: PS > D > PC, PS > PC > D, D > PS > PC, D > PC > PS, PC > D > PS, PC > PS > D and PS = D = PC, where symbol '>' means 'more important'. By default, PS > D > PC. This means that PS is the most important attribute, D is the second most important and PC is the least important.

PPP-Algorithm consists of two parts. In the first part (Steps 1–6), a set R of feasible (p, q)-paths is determined. The set R includes the safest, the shortest and the simplest paths, and also other paths where the aforementioned three attributes are taken into account with different weights. Thus, the set R consists of a number of candidate (p, q)-paths. They form the input for the second part of PPP-Algorithm (Steps 7–14), which represents the stochastic version of the Analytic Hierarchy Process (AHP). The AHP prioritizes the paths from R with respect to the specified preferential ranking of the attributes. Thus, the best (p, q)-path found by the algorithm is the user-optimal path from the set R.

It may be pointed out that PPP-Algorithm can be easily updated if more information about the pavement network is available. For example, risk values could reflect situations when a pavement segment is located extremely close to a road. In a similar vein, complexity numbers could be increased for very narrow pavement segments or when there are other obstacles.

Detailed Description of PPP-Algorithm
In the first step, the algorithm tries to get rid of occasional very long pavement segments. It may be argued that a more homogeneous network would represent a better model, in particular for the path complexity attribute.

In Step 2, the safety is calculated for every edge using the formula

$$S(e) = RV(e) \times D(e),$$

where $D(e)$ is the length of edge e and $RV(e)$ is its risk value defined in Tables 5.1 and 5.2. This formula is well justified because there is a positive correlation between the length of a pedestrian path and likelihood of a pedestrian–vehicle crash along the path.

The edge complexity is calculated in Step 3 for every edge. The first two terms represent an average degree of an edge $e = uv$, that is, $0.5(\deg(u) + \deg(v))$. If the average degree of an edge in a given path is rather high then it is more difficult to choose the right pavement segment to follow and easier to get lost when moving along this edge in the path, compared to a similar situation with a smaller degree. In addition, jaywalk crossings, steps, designated crossings and uphill/downhill movements increase complexity. There is also a term depending on the length of e. Although distance is a separate attribute, it is part of the complexity criterion too. Without the term $0.01D(e)$, we can easily imagine two (p, q)-paths P_1 and P_2 of the same complexity, but of very different lengths, say 100 metres long and 1000 metres long, respectively. Then, the algorithm will not be able to distinguish between the routes from the viewpoint of the complexity attribute and might choose the longest path P_2 as the simplest. This would contradict the common sense perception that the path P_1 is obviously 'simpler' than the path P_2.

Using Dijkstra's algorithm, the safest, shortest and simplest (p, q)-paths are calculated in Step 4. They are stored in the set R, which eventually will consist of (p, q)-paths that are interesting for the user. In Step 5, maximum values of edge safety, length and complexity are found. They will be used in the next Step 6, where a random search for (p, q)-paths is carried out. This search is based on a combined weight $W(e)$, which is a weighted linear combination of normalised edge safety, length and complexity. The normalisation is needed because these attributes use different scales. Also, the three weight coefficients α_i in the combined weight are random numbers between 0 and 1.

The second part of PPP-Algorithm is the application of the stochastic version of the AHP, which is similar to Module 2 of PIR-Algorithm from Section 3.2. Nevertheless, we briefly describe Steps 7–14 of PPP-Algorithm for completeness. It should be pointed out that the stochastic version is needed to guarantee that PPP-Algorithm works in a real or near-real time.

All the (p, q)-paths in the set R are available alternatives and they form the input for the AHP. They are denoted by $P_1, P_2, ..., P_n$ in Step 7. In the next two steps, we calculate path safety, length and complexity for each path in R, followed by computing the necessary statistical parameters.

The hierarchy tree shown in Figure 5.3 represents the initial stage of the AHP. The first hierarchy in this tree consists of three criteria: path safety (PS), distance (D) and path complexity (PC), whereas the lowest three hierarchies comprise the paths in the set R. In Step 10, the comparison matrix M_1 is created for the first hierarchy. This matrix is based on the specified preferential ranking of the criteria. The rows and columns of M_1 correspond to the attributes PS, D and PC, respectively. By default, PS $>$ D $>$ PC, and the corresponding comparison matrix is shown in Step 10. An advanced user can choose one of seven possibilities for the preferential ranking: PS $>$ D $>$ PC, PS $>$ PC $>$ D, D $>$ PS $>$ PC, D $>$ PC $>$ PS, PC $>$ D $>$ PS, PC $>$ PS $>$ D and PS $=$ D $=$ PC. For example, if PS $=$ D $=$ PC, then M_1 consists of 1s; if D $>$ PS $>$ PC, then $M_1[1,2] = 1/2$, $M_1[1,3] = 2$ and $M_1[2,3] = 4$.

In Step 11, three $n \times n$ comparison matrices M_2, M_3 and M_4 are automatically created. The matrices represent the relative importance of the paths in the set R with respect to safety, distance and complexity, respectively. Their construction is based on the corresponding parameters and standard deviations. For example, the (i, j)-element of the matrix M_3 represents the relative importance of the paths P_i and P_j with respect to their lengths in terms of standard deviations. If the difference in lengths is 2.5 standard deviations (i.e. $\beta = (D_j - D_i)/\sigma_D = 2.5$), then one path is weakly more important than the other, which will be reflected by the entry '3' in the comparison matrix.

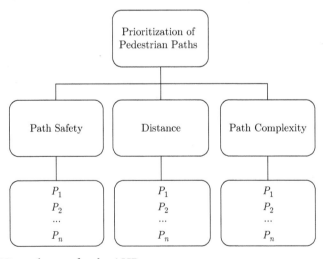

Fig. 5.3 Hierarchy tree for the AHP.

Notice that the use of the exponential function $9^{0.2\beta}$ was justified in Section 3.2 devoted to PIR-Algorithm.

It is easy to see that the matrix M_1 is perfectly consistent. Also, by construction, Consistency Indices and Consistency Ratios of the matrices M_2, M_3 and M_4 are always very small, or equal to zero. Thus, all the matrices are consistent and there is no need to adjust them.

The normalized eigenvector w_i corresponding to the largest eigenvalue λ_i is calculated in Step 12 for each matrix M_i. To achieve this, the standard Direct Iteration Algorithm is used with precision 0.0001. Initially, the matrix is multiplied by the unit vector $(1, 0, ..., 0)^{\mathrm{T}}$. The resulting vector is normalized so that the sum of its entries is equal to 1. This procedure is then repeated with the matrix and the resulting normalized vector. This algorithm quickly converges to the normalized eigenvector corresponding to the largest eigenvalue. For example, for the matrix M_1 in Step 10,

$$w_1 = (0.5714, 0.2857, 0.1429),$$

hence the weight for the path safety attribute is $w_1[1] = 0.5714$.

For each path $P_k \in R$, the aggregate score is computed in Step 13 as follows:

$$AS(P_k) = w_1[1]w_2[k] + w_1[2]w_3[k] + w_1[3]w_4[k].$$

Finally, in Step 14, the path with the highest aggregate score is determined and reported, together with the information about the parameters of this path.

PPP-Algorithm: Prioritization of pedestrian paths in a pavement network.

Input: The undirected graph G representing a pavement network generated by PNC-Algorithm.

Length $D(e)$ of each edge e in the graph G.

Type of every edge e (i.e. type of pedestrian crossing and/or OSM road) and the corresponding risk value $RV(e)$.

The start vertex p and the destination vertex q.

Preferential ranking of path safety (PS), distance (D) and path complexity (PC) (optional, by default PS > D > PC; that is, PS is the most important criterion and D is the second most important one).

Output: User-optimal (p, q)-path and information about its parameters.

(1) For each edge e with $D(e) > 100$ m, replace this edge with a path consisting of $\left\lceil \frac{D(e)}{100} \right\rceil$ edges of equal length. For newly formed edges, specify all the necessary parameters (length, type, risk value and so on).

(2) For each edge e, calculate its safety: $S(e) = RV(e) \times D(e)$.

(3) For each edge $e = uv$, calculate its complexity:

$$C(e) = 0.5 \deg(u) + 0.5 \deg(v) + 0.01 D(e) + \beta,$$

where

$$\beta = \begin{cases} 5 \text{ if } e \text{ is a jaywalk crossing,} \\ 4 \text{ if } e \text{ represents or includes steps,} \\ 3 \text{ if } e \text{ is one of four designated crossings,} \\ 2 \text{ if } e \text{ is going uphill (at least } 10° \text{ on average)} \\ \quad \text{and does not include steps,} \\ 1 \text{ if } e \text{ is going downhill (at least } 10° \text{ on average)} \\ \quad \text{and does not include steps,} \\ 0 \text{ otherwise.} \end{cases}$$

(4) Set $R = \emptyset$, where R is a set of (p, q)-paths. Compute the safest, shortest and simplest (p, q)-paths as follows:

(a) Run Dijkstra's algorithm in the graph G from vertex p with edge weights $S(e)$. It produces the safest available (p, q)-path P. Add this path to the set R.

(b) Run Dijkstra's algorithm in the graph G from vertex p with edge weights $D(e)$. It produces the shortest available (p, q)-path P. Add this path to the set R.

(c) Run Dijkstra's algorithm in the graph G from vertex p with edge weights $C(e)$. It produces the simplest available (p, q)-path P. Add this path to the set R.

(5) Calculate the following parameters:

$$S_m = \max_{e \in E(G)} S(e), \quad D_m = \max_{e \in E(G)} D(e), \quad C_m = \max_{e \in E(G)} C(e).$$

(6) Repeat the following for generation of random (p, q)-paths with different weights of the three attributes:

(a) Generate three random numbers $\alpha_i \in [0, 1]$, $i = 1, 2, 3$.

(b) Run Dijkstra's algorithm from vertex p with edge weights

$$W(e) = \alpha_1 \frac{S(e)}{S_m} + \alpha_2 \frac{D(e)}{D_m} + \alpha_3 \frac{C(e)}{C_m}.$$

(c) Add the resulting (p, q)-path P to the set R if $P \notin R$.
Go to Step 7 if at least one of the stopping criteria is satisfied (a specified number of generated paths or a running time limit).

(7) Denote all the paths in R by $P_1, P_2, ..., P_n$.

(8) For each path $P_k \in R$, calculate the following:
   (a) The safety of P_k: $S_k = \sum_{e \in P_k} S(e)$.
   (b) The length of P_k: $D_k = \sum_{e \in P_k} D(e)$.
   (c) The complexity of P_k: $C_k = \sum_{e \in P_k} C(e)$.

(9) For the paths in R, compute the means and the sample standard deviations of path safeties, distances and complexities: μ_S, σ_S, μ_D, σ_D, μ_C, σ_C.

(10) Create the 3×3 matrix M_1 of relative importance of path safety (PS), distance (D) and path complexity (PC), based on the specified preferential ranking. The rows and columns of M_1 correspond to PS, D and PC, respectively. By default, PS > D > PC; that is,

$$M_1 = \begin{pmatrix} 1 & 2 & 4 \\ 1/2 & 1 & 2 \\ 1/4 & 1/2 & 1 \end{pmatrix}.$$

(11) Create three $n \times n$ comparison matrices M_2, M_3, M_4 for the relative importance of the paths in the set R w.r.t. path safety, distance and complexity as follows:
   - For each entry $M_2[i, j]$, compute $\beta = (S_j - S_i)/\sigma_S$.
     If $\sigma_S = 0$, set $\beta = 0$. Calculate $M_2[i, j] = 9^{0.2\beta}$.
     Set $M_2[i, j] = 1/9$ if $\beta < -5$; set $M_2[i, j] = 9$ if $\beta > 5$.
   - For each entry $M_3[i, j]$, compute $\beta = (D_j - D_i)/\sigma_D$.
     If $\sigma_D = 0$, set $\beta = 0$. Calculate $M_3[i, j] = 9^{0.2\beta}$.
     Set $M_3[i, j] = 1/9$ if $\beta < -5$; set $M_3[i, j] = 9$ if $\beta > 5$.
   - For each entry $M_4[i, j]$, compute $\beta = (C_j - C_i)/\sigma_C$.
     If $\sigma_C = 0$, set $\beta = 0$. Calculate $M_4[i, j] = 9^{0.2\beta}$.
     Set $M_4[i, j] = 1/9$ if $\beta < -5$; set $M_4[i, j] = 9$ if $\beta > 5$.

(12) For each matrix M_i, run the Direct Iteration Algorithm to calculate its largest eigenvalue λ_i and the corresponding normalized eigenvector w_i with precision 0.0001.

(13) For each path $P_k \in R$, calculate the aggregate score:

$$AS(P_k) = w_1[1]w_2[k] + w_1[2]w_3[k] + w_1[3]w_4[k].$$

(14) Determine the path P_m with the highest aggregate score. Report P_m and information about its parameters. Algorithm stops.

In order to evaluate the proposed PPP-Algorithm, we used street network data from OSM. It is a crowdsourcing project for geographical data, which contains accurate street network data for most developed countries and is regularly used in routing applications [31]. Figure 5.4a displays the shortest path with respect to geographical distance between two locations in Cardiff City. This path involves walking beside a busy street and does not consider the possibility of walking through an adjacent park, which would represent a safer path. Figure 5.4b shows the user-optimal path computed using PPP-Algorithm. This path is slightly longer than the shortest path but enters the park and therefore is a safer path (with respect to pedestrian–vehicle crashes). This path represents the optimal trade-off between length, safety and complexity. Figure 5.4c displays the path computed using Google Maps. This

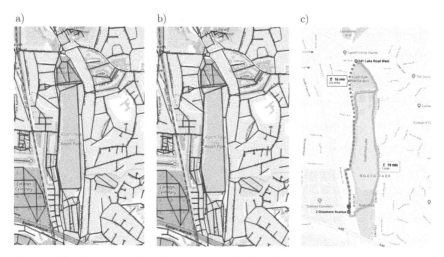

Fig. 5.4 The shortest path and the PPP-path from a location in the top to a location in the bottom are illustrated in a) and b), respectively. In both figures, the pavement network is represented by a set of blue lines and the path in question is represented by a sequence of red lines. The pedestrian path recommended by Google Maps is illustrated in c).

path has equal length to the shortest path above and, similar to this path, it does not consider the possibility of walking through the adjacent park.

5.1.3 Related Work

The most popular models for analysing pedestrian safety with respect to pedestrian–vehicle crashes are various types of regression models. Such models are typically used for finding factors contributing to the severity of pedestrian–vehicle crashes. Sze and Wong [44] developed a binary logistic regression model for identifying associations between pedestrian crashes in Hong Kong and contributing factors such as traffic congestion and pedestrian behaviour. It was found that the factors leading to a higher risk of mortality and severe injury are: victim age above 65, head injury, speed limit above 50 km/h, crash at or within 15 m of a crossing, signalized intersection, and two or more lanes. In a similar vein, Olszewski et al. [33] proposed a binary logit model for studying the effect of different factors on pedestrian fatality risk at zebra crossings in Poland. In this context, the following factors increase the probability of mortality: higher speed limit, darkness, divided road, two-way road, non-built-up area, mid-block crossing and summer time period. Moreover, the risk of fatality is higher for males and it increases with pedestrian age. The relationship between crossing location and injury severity in pedestrian–vehicle crashes in Canada was studied by Rothman et al. [41], who applied binary and multinomial logistic regression models. Their main conclusion was that uncontrolled mid-block crossings have greater injury severity compared with crossings at signalized intersections.

Using a multinomial logit model, Tay et al. [45] investigated factors contributing to pedestrian–vehicle crashes in South Korea. Their findings showed that 'relative to minor crashes, fatal and serious crashes were associated with collisions involving heavy vehicles; drivers who were drunk, male or under the age of 65; pedestrians who were over the age of 65 or female; and pedestrians who were hit in the middle of the road, on high speed roads, in inclement weather conditions, at night, on road links, in tunnels, on bridges, or on wider roads' [45]. Lee and Abdel-Aty [28] identified factors contributing to pedestrian crashes at intersection in Florida using log-linear models. They also applied an ordered probit model for estimating the probability of pedestrian injury severity. Pulugurtha and Sambhara [38] analysed factors that contribute to pedestrian crashes at signalized intersections in Charlotte, North Carolina. They applied a generalized linear model based on negative binomial distribution. Among contributing factors are the pedestrian volume, the number of transit stops, the number of approaches at an intersection and an increase in population.

Socio-economic factors in the context of pedestrian–vehicle crashes were studied by Pour et al. [36] using a boosted regression tree. Based on a dataset of pedestrian–vehicle crashes at mid-blocks in Melbourne, Australia, they concluded that socio-economic factors accounted for 60% of the 20 top contributing factors. In their next paper, Pour et al. [37] carried out a spatial and temporal analysis of pedestrian crashes in Melbourne, Australia. The main conclusion was that spatial and temporal distributions of pedestrian–vehicle crashes vary for different age groups and genders of pedestrians.

Street network structure has a significant impact on pedestrian safety. Marshall and Garrick [30] proposed a negative binomial regression model to study how street network characteristics affect road safety in 24 California cities. Their findings showed that road safety figures are significantly correlated with both street network parameters and street characteristics. In particular, street networks with high intersection densities are associated with fewer crashes for all severity levels. In contrast, increased street connectivity is correlated with more crashes across all severities. The effect of different street designs on crash severity in Calgary, Canada, was investigated by Rifaat et al. [39, 40]. Using logit models of crash severity, they showed that the loops and lollipops design of streets increases the probability of an injury for pedestrian–vehicle crashes. On the other hand, such a pattern reduces the likelihood of fatality and property damage only. Using generalized liner regression and Bayesian techniques, Osama and Sayed [34] studied the relationship between pedestrian–vehicle crashes and various characteristics of sidewalk networks in Vancouver, Canada. One of their conclusions was that lower crash occurrence is associated with higher continuity, linearity, coverage and slope of road pavements. Guo et al. [19] developed a Bayesian Poisson-lognormal model for analysing topological characteristics of road networks affecting pedestrian crashes in Hong Kong. In particular, they found that networks with irregular patterns are safest, whereas networks with grid patterns are the least safe from the viewpoint of pedestrian crash occurrences. Using a random parameter multinomial logit model, Aziz et al. [4] investigated severity levels of pedestrian injuries in New York City. They found that road characteristics, traffic attributes and land use are statistically significant.

Zhang et al. [47] studied the impact of road network structure on pedestrian and bicyclist safety in Alameda County, California. They developed three geographically weighted regression models, which are based on a number of factors and network parameters. The main conclusion of their research was that fewer non-motorist-involved crashes are occurring in more clustered road networks, networks that more highly centred on major roads, or networks with a greater average number of intersections on the shortest path connecting each pair of roads.

The safety of routes with respect to crime was analysed in [18, 25]. Keller and Mazimpaka [25] developed a method for avoiding dangerous urban areas in Los Angeles. Their approach is based on governmental open data about urban infrastructures, historical crime data and volunteered geographic information. The proposed safety index and historical crime hot spots were used for calculating the least dangerous route for vehicle drivers. Galbrun et al. [18] investigated safe urban navigation for pedestrians and developed a method for estimating the relative probability of a crime on any road segment. They considered the problem of safe navigation as a two-objective shortest path problem with two criteria: path length and path risk. The output of the proposed algorithm consists of a small set of paths providing trade-offs between distance and safety. The technique was tested on the street networks of Chicago and Philadelphia.

5.2 Placement of Charging Stations for Electric Vehicles in Road Networks

Zero and low emission electric and hybrid vehicles are playing an ever more important role in road transportation due to increasing concerns about the environment and resulting policies and advances in technology. Despite the advantages of electric vehicles, their relatively limited cruising range in comparison to traditional diesel/petrol vehicles and significant battery charging time often provide major challenges to their usage.

For electric vehicles to be viable, it is necessary to have a sufficient number of charging stations appropriately distributed throughout a road network. Given a particular road network layout, determining appropriate locations and capacities for such charging stations is a challenging multi-objective optimization problem with many constraints. One of the key objectives is to minimize the length of detours from a desired route which are necessary for recharging. On the other hand, constraints in this optimization problem include the requirements that the number of charging stations is reasonably small and the distance between consecutively used stations does not exceed the cruising range of electric vehicles. Also, capacities of charging stations must be sufficient enough to avoid bottlenecks at charging points. The focus of this section is on the problem of optimizing the placement of charging stations such that the length of detours necessary for recharging is minimized, subject to the constraint that the number of charging stations is reasonably small.

In the existing literature (e.g. see [17]), the problem of charging station placement is often modelled as a shortest path vertex cover problem in graphs. In this model, a vehicle is assumed to begin with a fully charged battery and follow a shortest path from an initial point to a final destination without much deviation. However, in many cases this assumption may fail, and hence the model in question is not suitable. For example, mail or groceries delivery drivers are usually concerned about navigating in a way prescribed by delivery options (in time and space) and are not particularly concerned about issues with shortest paths when navigating a certain area. Also, traffic jams, road closures and other temporary or sudden obstacles (e.g. a snow storm) may significantly affect the originally intended shortest path for driving. As a result, it is more natural and plausible to assume that drivers will become concerned about their remaining cruising range and battery charge only after the battery level falls below a certain low threshold, implying the remaining distance they can travel is quite limited.

In this section, we consider a model for placing charging stations in road networks, which is based on multiple domination models for a reachability graph derived from the original road network. A reachability graph models the sets of locations which are reachable from given locations, where a location is reachable if its distance from the location in question is below a certain threshold. The reachability graph appropriately models the situation where a driver becomes concerned about their low battery charge and wishes to make a detour to a recharging station, which is reachable from their current location. By considering multiple domination models on the reachability graph, we can find a set of locations for charging stations such that each node in the network can be served from several charging stations. In other words, multiple charging stations are reachable from each vertex. The driver therefore has several options for selection of a charging station and can in turn select the one that minimizes the necessary detour.

In a practical context, this approach might be used, for example, for electric vehicles that utilize the new emerging ultra-capacitor technology [20]. Electric vehicles of these kinds are known for a shorter cruising range but much faster charging time. For instance, in the case of new electric buses used for public transports in Minsk (Belarus), their driving range is currently about 20 km, and the battery charging time is 5–8 min [6]. Similar electric bus technology using ultra-capacitors has been recently tested in Sofia (Bulgaria), with the battery charging time of 5–6 min ([9] and [46], p.18), and also it is currently used in China. Clearly, the cruising range should increase for electric vehicles of smaller size, and providing decreased battery capacity should respectively help to decrease the charging time.

The approach considered in this section might also be used to decide on efficient placement of fast charging and battery swapping stations. In this context, it may be used, for example, to decide on optimized development of charging infrastructure for urban taxi companies using electric vehicles, where only fast charging (about half an hour) makes sense during taxi service times [3]. The model, which integrates multiple domination and reachability graphs, will guarantee that a taxi driver, observing a low level of battery charge, will be able to reach a fast charging station (ideally at a taxi stand) and have a certain minimal choice of options for driving directions in their service area before serving another customer. Similarly, in the case of battery swapping scenarios [29], a driver would have a guaranteed minimal choice of directions to change the battery at a reachable distance. The same approach may be used to decide on placing portable (mobile) fast charging and battery swapping stations as an extension of the existing network of permanent charging locations [23]. Finally, it can be adapted to decide on optimal locations of refuelling stations for alternative fuel vehicles of other types.

5.2.1 Reachability Graph and Multiple Domination Models

For simplicity, we consider a road network represented by a weighed undirected simple graph $F = (V, E, w : E \rightarrow \mathbb{R})$, where the set of vertices V corresponds to road intersections and dead-ends, while the set of edges E represents road segments connecting these vertices. The weight $w(e)$ of an edge $e \in E$ is the length of the corresponding road segment (in metres). An example of this graph model for the road network of the city of Boston is illustrated in Figure 5.5a.

Given a road network graph $F = (V, E, w)$, its *reachability graph* $G_t^r = (V^r, E_t^r)$ is defined as a simple (unweighted) graph with $V^r = V$ and edges $uv \in E_t^r$ if and only if the length of a shortest path (i.e. distance) between the corresponding vertices u and v in F is less than or equal to a specified *reachability threshold* of t km; that is,

a)

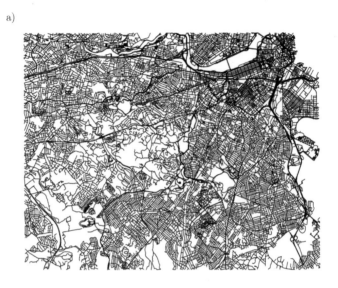

b)

Fig. 5.5 a) The road network for the city of Boston; b) neighbourhood of a vertex in the corresponding reachability graph.

$$w(P_{uv}) \leq 1000\,t,$$

where P_{uv} and $w(P_{uv})$ are a shortest path and the corresponding distance between u and v in F, respectively. The neighbourhood of one vertex in the reachability graph of the Boston road network in Figure 5.5a for $t = 3.0$ km is illustrated in Figure 5.5b. In this figure, red line segments are drawn between a given vertex and each of its neighbouring vertices in $G_{3.0}^r$. The reachability graph G_t^r appropriately models the situation where a driver becomes concerned about their low battery and wishes to make a detour to a recharging station which is reachable from their current location.

Notice that the reachability threshold t used for constructing a reachability graph G_t^r should normally satisfy the following lower bound derived from the road network graph F:

$$1000\,t \geq \max_{u \in V} \min_{v \in N_F(u)} w(uv), \tag{5.1}$$

where $N_F(u)$ is the set of all vertices adjacent to u in F. In other words, from any given vertex $u \in V$, it should be possible to reach at least one of the neighbouring locations $N_F(u)$ using the remaining battery power (to eventually recharge the battery). This would imply that the reachability graph G_t^r has no isolated vertices. Similarly, for better flexibility, more choice and more reliable conditions for reaching possible recharging locations, one may impose the stronger lower bound for the threshold t:

$$1000\,t \geq \max_{e \in E} w(e). \tag{5.2}$$

This would mean that it is possible to reach all the neighbouring locations $N_F(u)$ from any given vertex $u \in V$ using the remaining battery power. The lower bound (5.2) would imply that the vertex degrees of G_t^r are at least the corresponding vertex degrees of F.

However, in the case of a small number of remote locations which are more difficult to reach in the network, it may be too demanding and expensive to satisfy the lower bound (5.2) or even (5.1) for the whole network. Therefore, when conditions of the lower bound (5.2) or (5.1) are not satisfied, the remote locations ('outliers' of the road network) should be treated separately.

Having constructed a road network graph F and the corresponding reachability graph G_t^r, the problem of placing charging stations in the road network becomes a facility location problem, which can be modelled on the graphs F and G_t^r as described below.

In general, if G is a graph of order n, then $V(G) = \{v_1, v_2, ..., v_n\}$ is the set of vertices of G, the degree of vertex v_i is denoted by $\deg(v_i)$, $i = 1, 2, ..., n$, and the minimum vertex degree of G is denoted by $\delta = \delta(G)$. The neighbourhood of a vertex v in G is denoted by $N_G(v)$ or $N(v)$; and $N[v] = N(v) \cup \{v\}$ is the closed neighbourhood of v. A subset $X \subseteq V(G)$ is called a *dominating set* of G if every vertex not in X is adjacent to at least one vertex in X. The minimum cardinality of a dominating set of G is called the *domination number* of G and denoted by $\gamma(G)$. Dominating sets in graphs are natural general models for facility location problems in networks.

Given an integer $k \geq 1$, a set $X \subseteq V(G)$ is called a *k-dominating set* of G if every vertex $v \in V(G) - X$ has at least k neighbours in X. The minimum cardinality of a k-dominating set of G is the *k-domination number* $\gamma_k(G)$. Clearly, $\gamma_1(G) = \gamma(G)$, and $\gamma_{k_1}(G) \leq \gamma_{k_2}(G)$ when $k_1 \leq k_2$. Given a real number α, $0 < \alpha \leq 1$, a set $Y \subseteq V(G)$ is called an *α-dominating set* of G if for every vertex $v \in V(G) - Y$, $|N(v) \cap Y| \geq \alpha \deg(v)$. This means that v has at least $\lceil \alpha \deg(v) \rceil$ (i.e. $\alpha \times 100\%$) neighbours in Y. The minimum cardinality of an α-dominating set of G is called the *α-domination number* $\gamma_\alpha(G)$. It is easy to see that for graphs without isolated vertices $\gamma(G) \leq \gamma_\alpha(G)$, and $\gamma_{\alpha_1}(G) \leq \gamma_{\alpha_2}(G)$ for $\alpha_1 < \alpha_2$. Also, $\gamma(G) = \gamma_\alpha(G)$ when α is sufficiently close to 0.

The k- and α-domination are two types of multiple domination in graphs. The concept of α-domination differs from the k-domination in that a vertex must be dominated by a certain percentage ($\alpha \times 100\%$) of the vertices in its neighbourhood instead of a fixed number k of its neighbours. Either of these two types of multiple domination can be used to model the situation when an electric vehicle driver starts to look for a conveniently located battery charging station, ideally having several options where to recharge the battery. The focus of this section is on k-domination, which means that in any location (vertex) of the network (graph) the driver can use one out of at least k possible options, $k = 1, 2, ..., \delta$. Clearly, in the case $k > \delta$, this model suggests that the vertices of degree less than k are all included into a k-dominating set or ignored (i.e. treated separately). Hence, without loss of generality, we may assume that $k \leq \delta$, and the graph has no isolated vertices.

The problems of finding exact values of $\gamma_k(G)$ and $\gamma_\alpha(G)$ and the corresponding smallest k-dominating and α-dominating sets of vertices in graphs are known to be NP-hard [13, 27]. Therefore, it is important to have efficient heuristic algorithms and methods for finding some reasonably small k- and α-dominating sets in graphs. Also, good theoretical bounds for $\gamma_k(G)$ and $\gamma_\alpha(G)$ would be helpful for estimating quality of a given solution set. Two general upper bounds for the k- and α-domination numbers are given in Chapter 4 (Theorems 4.5 and 4.6). These results generalize a classical upper bound for the domination number and imply randomized algorithms for finding k- and α-dominating sets, whose expected sizes satisfy the corresponding upper bounds.

Clearly, given a reachability graph G_t^r, increasing the reachability threshold t can only extend the neighbourhoods of vertices in G_t^r to obtain G_q^r for $q > t$; that is, G_t^r is a spanning subgraph of G_q^r. Therefore, given a k-dominating set $X \subseteq V^r$ in G_t^r, one can deduce some properties about this set X in the reachability graph G_q^r, where $q \geq t$. If k and the set X are fixed, then $k \leq |X|$ and every vertex $v \in V^r - X$ in the graph G_q^r is dominated by at least k vertices in X. As the reachability threshold q increases, keeping the set X fixed and considering k as a parameter, the multiplicity of domination (k) can eventually be increased. When the reachability threshold q is at least the diameter of the network graph F, the reachability graph G_q^r becomes a complete graph, and every vertex $v \in V^r - X$ in the graph G_q^r is dominated by all the vertices in X, so that we can set $k = |X|$.

If $q \leq t$, one cannot infer any domination properties of the k-dominating set X of G_t^r in the reachability graph G_q^r, which is a spanning subgraph of G_t^r in this case. However, as q approaches t, the set X will start to behave like a k-dominating set in the reachability graph G_q^r, and eventually the reachability threshold t might be lowered. Some of the above properties are illustrated in the subsection devoted to experimental results.

5.2.2 Basic Algorithms and Heuristics

In this subsection, we describe the basic algorithmic ideas for computing reachability graphs and finding k-dominating sets. A k-dominating set in a reachability graph provides facility location points for charging stations in the corresponding road network.

The following procedure is used to compute the reachability graph $G_t^r = (V^r, E_t^r)$. First, the vertices of the road network graph $F = (V, E)$ are copied into V^r. Next, for each vertex v in G_t^r, we add an edge between v and all the vertices in G_t^r which are within the distance t from v in F. Here the distance between two vertices in F is measured as the length of the shortest path between those vertices. The distances are computed by applying Dijkstra's algorithm from the source vertex v in F and terminating the search when all vertices within the distance t from v have been found. In our simulations, the graph F is sparse. Therefore, to minimize running time, we implemented Dijkstra's algorithm using a binary heap-based priority queue. This gives a running time of $O\big((|V| + |E|) \log |V|\big)$ for each call of this algorithm [5]. The algorithm is called for each node $v \in V$ as the source vertex, giving a total running time of

$$O\Big(|V|(|V| + |E|) \log |V|\Big)$$

for computing the reachability graph.

Algorithm 5.1 below is a randomized heuristic to compute a small and minimal-by-inclusion k-dominating set in G_t^τ and it is an adjustment of Algorithm 4.2 from Chapter 4. Its input is a reachability graph G_t^τ and a positive integer k, $k \leq \delta(G_t^\tau)$. The algorithm returns a minimal-by-inclusion k-dominating set D' in G_t^τ. The set D' provides a collection of locations for charging stations in the road network graph F such that, from any given vertex in F, a driver has at least k different feasible options to reach a charging station when the remaining battery charge is enough for t kilometres. The expected size of the set D' in G_t^τ returned by Algorithm 5.1 satisfies the upper bound of Theorem 4.5.

The upper bound of Theorem 4.5 is known to be asymptotically best possible for general graphs on n vertices in the case of 1-dominating sets [1]. However, it turned out that this bound is not sharp enough in the case of particular reachability graphs of road networks for Boston and Dublin. Hence, instead of using the minimum vertex degree $\delta(G_t^\tau)$ of the reachability graphs G_t^τ to compute the probability p and the parameters δ' and b_{k-1} in Algorithm 5.1, we use in our experiments the average vertex degree of G_t^τ; that is,

$$\bar{d}(G_t^\tau) = \frac{1}{n} \sum_{i=1}^n \deg(v_i).$$

Notice that

$$k \leq \delta(G_t^\tau) \leq \bar{d}(G_t^\tau).$$

In general, using $\bar{d}(G_t^\tau)$ instead of $\delta(G_t^\tau)$ in Algorithm 5.1 does not guarantee obtaining a k-dominating set that satisfies the upper bound of Theorem 4.5 with positive probability. However, in the particular cases of road networks of Boston and Dublin, using $\bar{d}(G_t^\tau)$ in Algorithm 5.1 provides better computational results, which also satisfy this upper bound. In contrast to Algorithm 4.2, the final part of Algorithm 5.1 includes construction of a minimal (by inclusion) k-dominating set D' from the set D, which can be found by Algorithm 5.3. Also, in our experiments, Algorithm 5.1 is run several times to obtain k-dominating sets in G_t^τ of smaller size.

In the experiments, the results obtained by the randomized approach of Algorithm 5.1 were compared with those returned by a simple recursive greedy method described in Algorithm 5.2, and the results returned by Algorithm 5.2

Algorithm 5.1: Randomized k-dominating set.

Input: A reachability graph $G_t^r = (V^r, E_t^r)$ and an integer k, $k \leq \delta$.
Output: A minimal k-dominating set D' of G_t^r.

begin

   Compute the probability $p = 1 - \dfrac{1}{\sqrt[\delta']{b_{k-1}(1 + \delta')}}$,

   where $b_{k-1} = \begin{pmatrix} \delta \\ k-1 \end{pmatrix}$ and $\delta' = \delta - k + 1$;

   Initialize set $A = \emptyset$;

   foreach vertex $v \in V^r$ **do**

      with probability p, decide whether $v \in A$,
      otherwise $v \notin A$;   /* This forms a subset $A \subseteq V^r$ */;

   end

   Initialize $B = \emptyset$;

   foreach vertex $v \in V^r - A$ **do**

      if $|N(v) \cap A| < k$ **then**

         /* v is dominated by less than k vertices of A
         */

         add v into B; /* This forms a subset $B \subseteq V^r - A$ */;

      end

   end

   Set $D = A \cup B$;   /* D is a k-dominating set in G_t^r */

   If possible, remove some vertices from D to have a minimal
     k-dominating set D' in G_t^r;

   return D';

end

Algorithm 5.2: Greedy k-dominating set.

Input: A reachability graph $G_t^r = (V^r, E_t^r)$ and an integer k.
Output: A k-dominating set D of G_t^r.

Initialize $D = \emptyset$;

while $|\{v \in V^r - D : |N(v) \cap D| < k\}| > 0$ **do**

   Set $U = \{v \in V^r - D : |N(v) \cap D| < k\}$;

   Find $u \in V^r - D$ such that $|N[u] \cap U|$ is maximum possible;

   Set $D = D \cup \{u\}$;

end

---

were used as a benchmark to run Algorithm 5.1 several times for obtaining better results. Notice that when $k = 1$, Algorithm 5.2 is a simple deterministic (greedy) approach de-randomizing Algorithm 5.1 [2].

The k-dominating sets D constructed by Algorithms 5.1 and 5.2 are normally not minimal (by inclusion). Hence, a simple greedy procedure was used to reduce them to minimal k-dominating sets. A pseudocode for this elimination of redundancy is presented in Algorithm 5.3. In general, sets that are minimal by inclusion with respect to a given property may have their cardinality significantly larger than the smallest sets possessing the property (the latter must be minimal by definition). In other words, the cardinality of minimal k-dominating sets returned by Algorithm 5.1 may be larger than the k-domination number $\gamma_k(G)$ of the graph in question; the same statement is true for Algorithm 5.2 whose k-dominating sets D are reduced to minimal ones. As mentioned above, it is an NP-hard problem to find $\gamma_k(G)$ and corresponding sets.

Algorithm 5.3: Minimal k-dominating set.

Input: A reachability graph G_t^r and a k-dominating set D of G_t^r.
Output: A minimal k-dominating set D of G_t^r.
Order the vertices in D as
$L = (v_1, ..., v_{|D|}) : \ v_i \in D, \ |N(v_i) - D| \leq |N(v_{i+1}) - D|$;
for $i = 1$ to n **do**
    if $D - \{v_i\}$ is a k-dominating set of G_t^r **then**
     | set $D = D - \{v_i\}$;
    end
end

5.2.3 Experimental Evaluation

To evaluate the proposed methodology, we describe the experiments with multiple domination models and corresponding algorithms in the case of two road network graphs F for the cities of Boston in the United States and Dublin in Ireland. Computer codes implementing the algorithms were written using the programming language Python and executed on a laptop containing a 2.4 GHz Intel core i7-5500u processor, 16 GB 1600 MHz DDR3 RAM and running Ubuntu 17.04.

Data

The two aforementioned road networks were obtained from OpenStreetMap [12]. They are illustrated in Figures 5.5a and 5.6a. The graph F corresponding to Boston consists of 21,542 vertices and 31,112 edges. It is contained within a rectangular region of width 15.5 km and height 12.1 km. The graph F corresponding to Dublin consists of 55,162 vertices and 64,437 edges. It is contained within a rectangular region of width 29.5 km and height 24.6 km. Notice that both road network graphs are either planar or 'almost planar': when considering them embedded in the plane as road maps, the edge crossings are only possible in the case of road bridges and tunnels. Moreover, these two graphs are sparse because the number of edges m satisfies the following linear upper bound in terms of the number of vertices n for planar graphs: $m \leq 3n - 6$, as opposed to the general worst-case quadratic upper bound $m \leq 0.5n(n-1) \in \Theta(n^2)$.

An appropriate reachability threshold t for the reachability graph G_t^r is a function of a number of parameters. This includes the number of electric vehicles which require charging, the number of charging stations which can be installed, the number of charging options one wishes to offer and the cost of installing a charging station. Determining this threshold would probably best be done by consultation with city planners. In this section, we assume that the most appropriate reachability threshold for both cities' road networks and electric vehicles is 3.0 km.

For each road network graph F, the corresponding reachability graph $G_{3.0}^r$ was computed. These graphs are illustrated in Figures 5.5b and 5.6b. The reachability graph $G_{3.0}^r$ corresponding to Boston contains 21,542 vertices and 23,052,466 edges (9.94%). The reachability graph $G_{3.0}^r$ corresponding to Dublin contains 55,162 vertices and 54,306,700 edges (3.57%). The CPU time required to compute these reachability graphs was 464 and 1054 minutes, respectively. All k-dominating sets were computed using these two reachability graphs.

Computing k-Dominating Sets

For each of the reachability graphs $G_{3.0}^r$, one k-dominating set was computed using the greedy algorithm, and ten k-dominating sets were computed using the randomized algorithm for $k = 1, 2, 4$. Table 5.3 displays the cardinalities of the k-dominating sets computed using the greedy algorithm and the cardinalities of the smallest k-dominating sets computed by the randomized algorithm for each of the cities and each value of $k = 1, 2, 4$. In four out of the six cases, the randomized algorithm computed a smaller dominating set than the

a)

b)

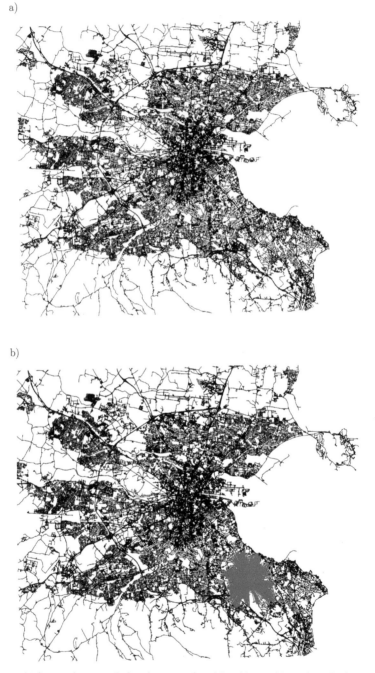

Fig. 5.6 a) The road network for the city of Dublin; b) neighbourhood of a vertex in the corresponding reachability graph.

Table 5.3 Cardinalities of the k-dominating sets computed using the greedy algorithm, and the size parameters (smallest, average, standard deviation) of k-dominating sets computed by the randomized algorithm for each city and each value of k.

Network:	Boston				Dublin			
Algorithm:	Greedy	Randomized			Greedy	Randomized		
		Min	Mean	Std		Min	Mean	Std
$k = 1$	32	31	33.2	1.6	110	111	114.4	2.0
$k = 2$	64	56	61.2	2.8	214	215	220.7	3.7
$k = 4$	122	115	120.3	3.2	413	411	418.0	3.4

Table 5.4 CPU time of the greedy algorithm, and mean and standard deviation of the CPU time of the randomized algorithm for finding k-dominating sets (in minutes).

Network:	Boston			Dublin		
Algorithm:	Greedy	Randomized		Greedy	Randomized	
		Mean	Std		Mean	Std
$k = 1$	24	53	5	87	145	10
$k = 2$	33	89	7	121	275	14
$k = 4$	55	107	7	192	516	18

same multiplicity dominating set returned by the greedy algorithm. The two 2-dominating sets of sizes 64 and 56 for the city of Boston are displayed in Figure 5.7. The two 4-dominating sets of sizes 413 and 411 for the city of Dublin are shown in Figure 5.8.

A visual inspection of Figures 5.7 and 5.8 reveals that spatial locations of the elements in the dominating sets tend to be more spatially clustered when computed using the greedy algorithm. This can be attributed to the greedy nature of the approach: vertices of high degree in the corresponding reachability graphs tend to be spatially clustered, and the greedy algorithm will add these high degree vertices to the dominating set first. On the other hand, the randomized algorithm initially adds a random set of vertices to the future dominating set, and these vertices are likely to be spatially distributed in a more uniform way.

The CPU time used by each algorithm to compute the k-dominating sets in the reachability graphs for the cities of Boston and Dublin are reported in Table 5.4. For the randomized algorithm, the mean and standard deviation of

a)

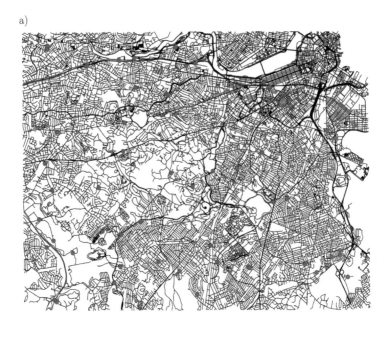

b)

Fig. 5.7 a) The 2-dominating set of 64 vertices computed by the greedy algorithm; b) the smallest 2-dominating set of 56 vertices computed by the randomized algorithm. (Both for the city of Boston; the vertices in the 2-dominating sets are in red.)

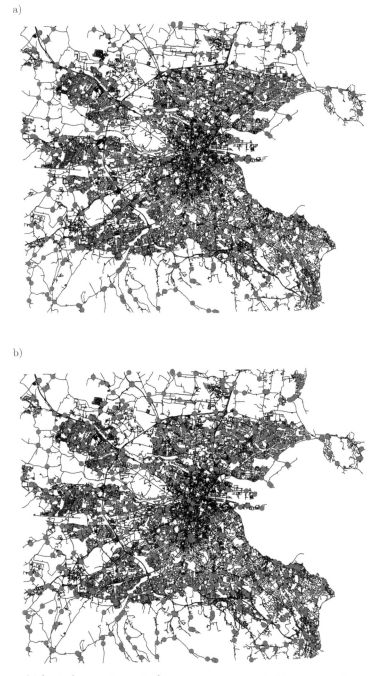

Fig. 5.8 a) The 4-dominating set of 413 vertices computed by the greedy algorithm; b) the smallest 4-dominating set of 411 vertices computed by the randomized algorithm. (Both for the city of Dublin; red dots indicate the vertices in the 4-dominating sets.)

the times for ten runs are reported instead of showing them individually. An interesting point is that the run time for the randomized algorithm is generally greater than that of the greedy algorithm, which can be attributed to a more subtle and sophisticated nature of the randomized approach.

Reachability of Charging Stations

Given a fixed k-dominating set X in a reachability graph G_t^r corresponding to a road network graph $F = (V, E)$, the number of elements in X reachable from a given vertex in F is a non-decreasing function of distance. To examine this phenomenon, we consider the smallest 2-dominating sets for Boston and Dublin computed by the randomized algorithm. These two sets contain 56 and 215 elements, respectively. The set corresponding to Boston is illustrated in Figure 5.7b.

For a 2-dominating set X, the mean and standard deviation were computed for the number of vertices in X reachable from a vertex in $V - X$ as a function of distance. These values for the road networks of Boston and Dublin are displayed in Table 5.5. An analysis of this table reveals the following. Despite the fact that the dominating sets were computed for a reachability graph with the reachability threshold of 3 km, the mean number of vertices in X within the distance of 1 km from a vertex in $V - X$ for each of the cities is 0.5. Furthermore, for both cities, the mean number of elements in X within the distance of 3 km from a vertex in $V - X$ is significantly larger than 2. This means more support and flexibility, than only two a priori guaranteed options, for recharging electric vehicles in many vertices of these road networks.

On the other hand, increasing the reachability threshold to some $q > t = 3$ km and keeping the set of vertices X fixed in G_q^r may eventually

Table 5.5 Three statistics for the number of charging stations reachable from a vertex outside the charging station locations (represented by a 2-dominating set) for the road networks of Boston and Dublin computed as a function of distance.

Network:	Boston			Dublin		
Stats:	Mean	Std	Min	Mean	Std	Min
1 km	0.5	0.7	0	0.5	0.7	0
2 km	1.9	1.1	0	2.0	1.1	0
3 km	4.3	1.4	2	4.6	1.5	2
4 km	7.3	1.9	2	8.5	2.3	2
5 km	10.9	2.6	2	13.7	3.1	2
6 km	14.9	3.7	2	20.0	4.2	3

increase the minimum multiplicity of coverage of each vertex $v \in V - X$ by the vertices in X; that is, X may become a k-dominating set in G_q^r with $k > 2$. The minimum multiplicity of coverage by the same 2-dominating set X in the corresponding reachability graph G_q^r was computed with the reachability threshold q increasing to 4, 5 and 6 km. As shown in Table 5.5, this increase of the reachability threshold did not increase the minimum number of options for the city of Boston, but it did turn the 2-dominating set X of $G_{3.0}^r$ into a 3-dominating set in $G_{6.0}^r$ for the city of Dublin. In other words, the drivers in Dublin will have at least three options available within the distance of 6 km for recharging their batteries when using the same 2-dominating set X from $G_{3.0}^r$.

Shortest Detours

The number of options available for recharging electric vehicles in a road network F increases as a function of the multiplicity value k in the corresponding k-dominating set. In turn, this may reduce the length of detours required for recharging electric vehicles. To quantify this phenomenon, let us consider the situation where a driver of an electric vehicle wishes to travel from a source location to a destination but first needs to have their vehicle recharged. The driver considers all charging stations within the distance of 3 km from the source and charges their vehicle at a charging station which minimizes the detour. Here the detour is the sum of distances from the source to the charging station and from the charging station to the destination, minus the distance from the source to the destination.

To illustrate this, let us consider the situation where the source and the destination are represented by red dots in the left side and the upper right corner of Figure 5.9, respectively. For the smallest 56-vertex 2-dominating set computed by the randomized algorithm (see Figure 5.7b), there are five charging stations within the distance of 3 km from the source. These five charging stations are marked by green dots in Figure 5.9. The route that minimizes the detour is represented by the blue line in the figure, and the detour in question is 92 m.

For each of the cities of Boston and Dublin, 200 random pairs of source and destination locations were selected, and for each pair of the locations the corresponding detour for recharging was calculated. For the smallest k-dominating sets computed by the randomized algorithm for different values of $k = 1, 2, 4$, the corresponding mean and standard deviation of detours required for recharging are displayed in Table 5.6. As expected, for both cities

Fig. 5.9 A detour through a charging station for the city of Boston.

Table 5.6 Statistics of detours (in metres) required for recharging batteries for random pairs of source and destination locations in the road networks for different values of k.

Network:	Boston		Dublin	
Stats:	Mean	Std	Mean	Std
$k = 1$	769	777	747	863
$k = 2$	436	541	501	578
$k = 4$	316	415	298	465

the mean and standard deviation values decrease as the multiplicity k of the domination parameter increases.

Realistically Constrained Scenarios

To demonstrate applicability of the proposed general approach, let us consider the following constrained real-world scenario. Assume that there exists a set of already installed charging stations, which we wish to transform into a k-dominating set through addition of new charging stations, where the additional charging stations may only be placed at a specified subset of locations.

To make the randomized and greedy algorithms applicable to this scenario, each of them can be adapted in the following way. First, instead of initializing respectively the sets A and D in the randomized and greedy algorithms to be the empty set, we initialize each of these sets to be the set of already existing charging stations. Second, for adding new charging stations, we only consider a specified subset of locations where additional charging stations may be installed. Finally, when reducing a k-dominating set to be minimal, we do not remove any elements belonging to the set of already installed charging stations.

To evaluate the new more constrained model and the adapted randomized and greedy algorithms, we consider the actually installed set of charging stations in the city of Dublin, whose locations can be obtained from the Irish state-owned electricity company Electricity Supply Board (ESB). The data in question are freely available from the ESB website in Keyhole Markup Language (KML) format [14]. The set of installed charging stations has cardinality 90 and it is illustrated in Figure 5.10a. For this set of charging stations, the minimum number of charging stations within reach of 3 km from any vertex is 0. Therefore, this set of charging stations is not a k-dominating set for any value of k with respect to the reachability threshold of 3 km considered here. We specify the subset of other locations, where additional charging stations may be installed, to be 5000 randomly chosen vertices in the road network graph with currently no charging stations installed. Note that this represents less than 10% of the vertices in the graph, which contains 55,162 vertices. These locations are illustrated in Figure 5.10b.

Table 5.7 displays the cardinalities of the k-dominating sets computed by the adapted greedy algorithm and the cardinalities of the smallest of ten k-dominating sets computed by the adapted randomized algorithm for each

Table 5.7 Cardinalities of the k-dominating sets computed by the adapted greedy algorithm and the smallest k-dominating sets computed by the adapted randomized algorithm for each value of k.

Network:	Dublin	
Algorithm:	Greedy	Randomized
$k = 1$	178	177
$k = 2$	277	271
$k = 4$	456	460

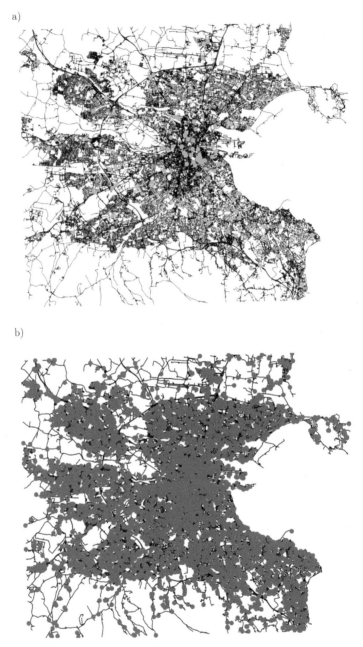

Fig. 5.10 a) Current set of charging stations (green dots) installed in the city of Dublin [14]; b) 5000 locations (red dots) where new charging stations are allowed for installation.

a)

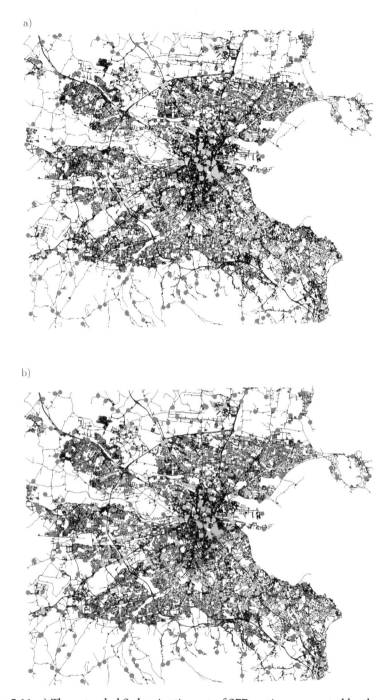

b)

Fig. 5.11 a) The extended 2-dominating set of 277 vertices computed by the adapted greedy algorithm for the city of Dublin; b) the smallest extended 2-dominating set of 271 vertices computed by the adapted randomized algorithm for the city of Dublin. (Green and red dots represent currently installed and new charging stations, respectively.)

value of $k = 1, 2, 4$. The generated 2-dominating sets are illustrated in Figure 5.11. By comparing the corresponding values in Tables 5.3 and 5.7, one can see that the dominating sets are larger in this constrained case scenario than those computed in the unconstrained case. Clearly, this increase is a consequence of the fact that the introduction of the constraints reduces the number of possible feasible solutions: the resulting sets must contain the original 90 vertices, and their extension is only possible by using the specified 9.1% of other locations.

5.2.4 Mixed Integer Linear Programming (MILP) Formulation

As an example of dealing with generalizations of the main model of reachability graphs and k-dominating sets, we show how to incorporate information about the demand and capacities into an MILP formulation of the problem. Assume that the demand $d_i \geq 0$ and the potential capacity $c_i \geq 0$ for installing a charging station at each vertex v_i of the n-vertex reachability graph $G_t^r = (V^r, E_t^r)$ of a road network are known, $i = 1, 2, ..., n$. Also, suppose that a decision-maker wants to minimize the total capacity of actually installed charging stations in the road network while preserving the property of k-dominating sets in the reachability graph G_t^r and satisfying the demand for charging in the network. We use Boolean (0–1) decision variables x_i to indicate whether the charging station of a given capacity c_i is installed at vertex v_i of G_t^r, $i = 1, 2, ..., n$, and real decision variables y_{ij} to indicate the proportion of the demand d_i at vertex v_i that is supposed to be satisfied by a charging station at vertex v_j of capacity c_j, $i, j = 1, 2, ..., n$.

Adapting the MILP formulation of the classical transportation problem, the MILP model for our optimization problem can be written as follows:

$$\min \ \sum_{i=1}^{n} c_i x_i \tag{5.3}$$

subject to

$$\sum_{v_i \in N(v_j)} x_i \geq k(1 - x_j), \quad j = 1, 2, ..., n \tag{5.4}$$

$$\sum_{v_i \in N[v_j]} d_i y_{ij} \leq c_j x_j, \quad j = 1, 2, ..., n \tag{5.5}$$

$$\sum_{v_j \in N[v_i]} y_{ij} = 1, \quad i = 1, 2, ..., n \tag{5.6}$$

$$y_{ij} = 0, \quad v_i v_j \notin E_t^r, \quad i \neq j, \quad i, j = 1, 2, ..., n \tag{5.7}$$

$$y_{ii} = x_i \min\left\{1, \frac{c_i}{d_i}\right\}, \qquad i = 1, 2, ..., n \tag{5.8}$$

$$x_i \in \{0, 1\}, \quad i = 1, 2, ..., n \tag{5.9}$$

$$y_{ij} \in [0, 1], \quad i, j = 1, 2, ..., n, \tag{5.10}$$

where

- (5.3) is the objective function to minimize the total capacity of installed charging stations in the network;
- constraints (5.4) guarantee that either a vertex v_j is in the k-dominating set ($x_j = 1$), or else ($x_j = 0$) the vertex v_j is dominated by at least k neighbours in the k-dominating set;
- constraints (5.5) represent the requirement that the total demand satisfied at vertex v_j does not exceed the capacity c_j of v_j in case a charging station is installed at v_j (i.e. when $x_j = 1$);
- constraints (5.6), (5.7) and (5.8) guarantee that the whole demand d_i at vertex v_i is satisfied at some charging stations installed in the vertices v_j of its closed neighbourhood $N[v_i]$. Also, in case a charging station is installed at v_i, the whole demand d_i must be satisfied locally at v_i if $d_i \leq c_i$; if $d_i > c_i$, then maximal part of demand d_i (i.e. c_i) is satisfied locally.

It is not difficult to see that the smallest-size k-dominating set problem in reachability graphs can be formulated as a particular case of the MILP problem (5.3)–(5.10), resulting in an integer programming formulation. Such a formulation can be used for finding the smallest k-dominating sets in graphs; that is, exact solutions of the problem considered in the previous subsections.

To be sure that the greedy solutions used as a benchmark in the experimental evaluation in the previous subsection are meaningful, we compared greedy solutions with exact solutions for reachability graphs G_t^T corresponding to small-size areas of Boston using reasonably chosen values for the threshold parameter t. The largest reachability graphs, for which it was possible to obtain exact solutions for $k = 1, 2$ and 4, correspond to a road network area of about 0.5 km by 0.5 km with the threshold t equal to 175 m and 200 m. These reachability graphs contain 50 vertices. For these graph instances, the greedy solution either coincides with an exact solution or is at most three vertices larger, which is within 8.9% of the optimal exact solutions. For appropriate larger reachability graph instances (up to 146 vertices, corresponding to an area of about 1 km $\times$ 1 km with the threshold of 300 m), it was possible to compute exact solutions only for 1-dominating sets (i.e. when $k = 1$): all the corresponding greedy solutions turned out to be within 22.3% of the optimal solutions. We believe these experimental results with exact solutions justify well the choice of greedy heuristic solutions as a benchmark for the large-scale optimization considered in the previous subsection.

5.2.5 Related Work

There exists quite a large volume of recent literature related to electric vehicles and optimization in road networks focusing on different aspects of problem modelling and corresponding solution methods. For example, Poghosyan et al. [35] discussed possible scenarios of distribution of loads in the power grids and their dependence on temporal, spatial and behavioural charging patterns for electric vehicles.

Given a set of charging stations and their locations fixed in the network, Storandt and Funke [43] proposed a method for computing all locations that are reachable from a given initial location, assuming a specified number of battery recharges can be done. In their work, the locations of charging stations are assumed to be fixed, and there is no attempt to optimize the placement of charging stations in the network.

Lam et al. [26] considered a specific type of the general facility location problem called the problem of electric vehicle charging station placement. In their work, the authors try to minimize construction costs for placement of charging stations in few preselected locations subject to a set of constraints. The problem is modelled using MILP with some non-linear constraints. The authors showed that the problem is NP-hard and proposed several solution methods by reduction to MILP problems and using heuristics. An experimental evaluation is first done with randomly generated small-size synthetic instances using MATLAB and generic MILP solvers. Then, the model and methods are evaluated for possible scenarios of building charging stations in Hong Kong by considering 18 preselected locations for potential construction of charging stations in different districts of the country. Notice that in this model the sites for potential construction of charging stations are preselected, and the average cruising distance of fully charged electric vehicles is used to select the sites minimizing the total construction costs.

Funke et al. [17] modelled the problem of placement of charging stations as a shortest path cover problem in a road network graph $F = (V, E)$. One needs to find the smallest subset of vertices $L \subseteq V$ such that every minimal shortest path in F that exceeds the electric vehicle battery capacity contains a vertex of the set L with a charging station. The problem is then modelled as a special type of the hitting set problem: the collection of subsets of V to be hit by the charging stations corresponds to the minimal shortest paths in F that exceed the battery capacity. An adaptation of the standard greedy approach provides an $O(\log |V|)$-approximation algorithm to solve this problem. The instance construction and representation were described as the main challenges with respect to limited computational memory and time resources. As a result, using different representations and searching for minimal shortest paths turned out to be quite a complicated task, which is involved with many details. Overall, the problem does not seem to scale well, and the heuristic improvements for the implementation would be very challenging to reproduce.

A good description of optimization problems and different practical problem scenarios, mostly related to car-sharing systems employing electric vehicles, was presented in the overview paper by Brandstätter et al. [7]. Notice that the models and optimization scenarios, which are based on the integration of multiple domination and reachability graphs, roughly correspond to the strategic and tactical problems described in [7]. He et al. [22] considered a mathematical programming model that incorporates details of customer adoption behaviour and fleet management in car-sharing systems, including repositioning and charging electric vehicles under imbalanced travel plans.

Asamer et al. [3] estimated the potential charging demand by areas for a taxi company operating fossil-fuelled vehicles in the city of Vienna, Austria, and proposed a method for placing in an optimal way a predefined fixed number of charging stations to maximize the coverage of the estimated charging demand. The problem is formulated as an MILP problem and solved by using a generic MILP solver CPLEX. The authors point out that, in this case, only fast (Level 3) charging stations can be used at taxi stands to recharge taxis quickly during their operational service time while waiting for the next customer. This is opposed to currently prevailing slow (Level 1) and standard (Level 2) charging of vehicles when they are not in use. Their solution is supposed to be further refined for each region, depending on taxi stand locations and other real-life constraints.

Chen et al. [10] used parking and personal trip information for a downtown area of Seattle, United States, to determine possible non-residential public parking locations for installing standard (Level 2) charging stations. In their behavioural models, they first predict (at least 15 min) parking demand for different areas. Then, using different parking demand variables, the authors formulate an MILP problem to determine optimal locations (by areas) for placement of charging stations at parking lots. The MILP problem ensures that charging stations are not too clustered and have good accessibility by users. Also, some limitations of the model and optimization are discussed.

Frade et al. [16] presented a study on possible efficient location of slow and standard (Levels 1 and 2) charging stations in public parking lots of a downtown area of Lisbon, Portugal. The area is characterized by a mixed high usage for both residential and workplace/business parking, with a low number of private parking spots. Therefore, many vehicles are parked for a long time in public parking lots, which can be used 24 hours per day. First, the authors estimate the recharging demand during the day and night time for smaller regions (census blocks) of the area by means of numerical and other characteristics of households and the volume of employment and type of buildings. Then, an MILP formulation of the maximal coverage problem is presented to decide at which parking lots a limited predefined number of charging stations should be installed, and with what number of supply points, to maximize the total demand coverage. Four different scenarios are considered and discussed.

Given estimated demand for charging at specific locations, Cavadas et al. [8] proposed three MILP problem formulations to decide on locations for slow and standard charging stations (for long-term parking). The first formulation is to maximize the satisfied demand coverage with respect to a fixed budget and taking into account the distance between the demand sites and actual charging station locations and the stations' capacities. The second MILP problem is an extension of the first one by allowing a transfer of the demand from one site to another in case a driver can charge at either of the two locations. Finally, the first two MILP models are refined to take into consideration variations of the demand for charging during different time intervals of the day. Simulations with the MILP models and optimization are described for the city of Coimbra, Portugal. To estimate the demand, the authors used data from a mobility survey for the city and represented the relevant demand areas by square grid cells and, in the case of high demand, by their subgrid cells (129 cells in total). The theoretical optimization results were compared to the actually implemented placement of nine charging stations in the city.

General MILP models for stochastic problems of refuelling station locations for fast-fill stations (Level 3 charging or battery swapping in the case of electric vehicles) were considered and described by Hosseini and MirHassani [23].

The authors proposed two-stage uncapacitated and capacitated MILP models for alternative-fuel vehicles, where the first stage decides on locations for permanent refuelling stations, and the second stage places portable (mobile) refuelling stations in a road network. Uncertain traffic flows depending on a number of time-dependent traffic scenarios in a road network are used as input parameters for the models. The resulting models are quite complicated, involved with details and computationally intractable. Hence, heuristic methods are proposed to solve the problems, and simulations are presented in the case of an intercity road network of the state of Arizona having 50 candidate facility nodes.

Different models, business scenarios and studies for development of a network of battery swapping facilities were considered by Mak et al. [29]. Zhao and Ma [48] presented agent-based simulations to show how different layouts of a limited number of refuelling stations for alternative fuel vehicles can influence adoption rates of the corresponding vehicles. These simulations are based on randomly generated traffic flows for the city of Shanghai, China, and layouts of alternative fuel stations are optimized in different scenarios by using a genetic algorithm with the main (largest) road network graph consisting of 532 nodes. Nie and Ghamami [32] considered a conceptual optimization model related to the development of fast refuelling facilities in the case of medium- and long-distance travels by electric vehicles along a 'corridor'.

5.3 Exercises

5.3.1

Apply PNC-Algorithm for constructing a pavement network G for the road network N of Figure 5.12, where the vertex d represents a zebra crossing, the vertex f is a road intersection with light-controlled pedestrian crossings, and g is a road intersection with jaywalk pedestrian crossings.

The road (c, d, f, g, h) is a primary road of width 40 m. All road sections cd, df, fg and gh are 1 km long. The roads af, bg and fl are residential, each is 10 m wide and 100 m long.

5.3.2

In Step 5 of PNC-Algorithm, if the edge $e = v^l v^m$ is a jaywalk crossing, then the OSM type of e must be specified. How can we determine the corresponding

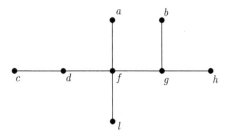

Fig. 5.12 Artificial road network N.

OSM road for e using the vertex labellings of the graph G; that is, without the graphical illustration of G?

5.3.3

(a) For a road network, propose a method for modelling a road intersection which has different types of pedestrian crossings, including situations when pedestrian crossings are forbidden for some roads at the intersection. The method must be suitable for constructing the corresponding pavement network. Hint: the necessary information may be kept in the form of a matrix.

(b) Propose a method for modelling bridges when constructing a pavement network, including the case when a given road network is not planar; that is, it cannot be drawn on a plane without edge crossings.

5.3.4

(a) For the pavement network G constructed in Exercise 5.3.1, find the shortest, safest and simplest (c^1, h^0)-paths using Dijkstra's algorithm or otherwise. In terms of the road network N, the start vertex of those paths is the north part of the intersection c, and the destination vertex is the north part of the junction h.

(b) There were some improvements in the pavement network G, resulting in the construction of short uphill steps on pavement segments $f^3 g^0$ and $g^0 h^1$. In terms of the road network N, these segments are located on the south parts of the road segments fg and gh. Find the shortest, safest and simplest (c^1, h^0)-paths in the improved pavement network

G'. In addition, determine the simplest (c^1, h^0)-path that does not include a jaywalk crossing.

Hint: the pavement segments $f^3 g^0$ and $g^0 h^1$ should be classified as before ('primary'), however their complexities must be increased by 4 because of the introduced steps.

5.3.5

(i) Briefly describe research devoted to pedestrian safety in urban areas with respect to pedestrian–vehicle crashes in terms of
   (a) Entities (objects) for which safety is modelled;
   (b) Types of models;
   (c) Independent variables or data used;
   (d) Outputs.
(ii) Briefly describe research devoted to pedestrian safety in urban areas with respect to crime in terms of (a)–(d) from part (i).

5.3.6

The following questions are devoted to the k-domination number $\gamma_k(G)$:

(a) Find $\gamma_k(C_n)$ for the cycle C_n of order $n \geq 3$.
(b) Show that for any graph G of order n with maximum degree Δ, the following inequality is true [15]:

$$\gamma_k(G) \geq \frac{kn}{k + \Delta}.$$

Hint: use the double-counting approach.

(c)* Prove that $\gamma_k(G) \leq \frac{kn}{k+1}$ for any graph G with $\delta \geq k$ [11].

The first part of the proof is as follows:

Assume to the contrary that there exists a graph G with $\delta \geq k$ such that $\gamma_k(G) > \frac{kn}{k+1}$ and G has the smallest possible number of edges. Let us denote $l = n - \gamma_k(G)$. Then

$$\gamma_k(G) > \frac{k(l + \gamma_k(G))}{k + 1}.$$

Hence, $\gamma_k(G) > kl$ or $\gamma_k(G) \geq kl + 1$.

- The reader should complete the proof by implementing the following steps:
  (1) Consider the set $W = \{w \in V(G) : \deg(w) > k\}$, which might be empty. Show that the set W is independent; that is, this set does not contain edges.
  (2) Let us consider the maximal independent set W^+ containing the set W. Explain why the set $D = V(G) - W^+$ is k-dominating. State a lower bound for $|D|$, and find an upper bound for $|D \cap N[x]|$ for a vertex $x \in D$.
  (3) Construct an independent set of vertices $X = \{x_1, x_2, ..., x_{l+1}\}$ by selecting vertices from the set D. Show that your construction is well defined; that is, all $l + 1$ vertices x_i can be found in D.
  (4) In the graph G, find a k-dominating set of size at most $\gamma_k(G) - 1$, which is a contradiction.

5.3.7

In this exercise, we will explore 2-dominating sets in the cube Q_3, which is shown in Figure 5.13.

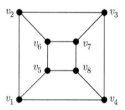

Fig. 5.13 The cube Q_3.

(a) Apply Algorithm 5.2 to the cube Q_3 assuming that $k = 2$. If there is a choice when selecting a vertex $v_i \in V^r - D$ such that $|N[v_i] \cap U|$ is maximum possible, choose a vertex v_i with the smallest subscript i. This represents a deterministic version of the algorithm.

(b) Apply Algorithm 5.1 to the cube Q_3 three times for $k = 2$. To construct the random set A in this algorithm, you will need to generate three sets of random numbers between 0 and 1. Alternatively, the following sets of random numbers can be used:

$$0.497, 0.897, 0.595, 0.942, 0.665, 0.463, 0.829, 0.130;$$

$$0.767, 0.891, 0.352, 0.759, 0.187, 0.498, 0.034, 0.806;$$

$$0.612, 0.620, 0.798, 0.220, 0.708, 0.307, 0.382, 0.101.$$

(c) Find $\gamma_2(Q_3)$ and justify your answer.

5.3.8

The following questions are devoted to the α-domination number $\gamma_\alpha(G)$ of a graph G without isolated vertices. They are based on results from [13]:

(a) Let P_n, C_n and K_n denote the path, the cycle and the complete graph of order $n \geq 3$, respectively. Find $\gamma_\alpha(P_n)$, $\gamma_\alpha(C_n)$ and $\gamma_\alpha(K_n)$ for $\alpha = 0.8$.
(b) Prove that for aiy graph G of order n with minimum degree δ and maximum degree Δ,

$$\gamma_\alpha(G) \geq \frac{\alpha \delta n}{\Delta + \alpha \delta}.$$

Hint: use the double-counting approach.
(c) Prove the following upper bound:

$$\gamma_\alpha(G) \leq \frac{\Delta n}{\Delta + (1 - \alpha)\delta}.$$

Hint: use the known bound $\gamma_\alpha(G) + \gamma_{1-\alpha}(G) \leq n$ for $0 < \alpha < 1$.

5.3.9

(a) Carry out the complexity analysis of Algorithm 5.1, assuming that the average vertex degree $\bar{d}$ is used instead of the minimum vertex degree δ. You may use the ideas of a similar analysis from Section 4.1.5.
(b) Carry out the complexity analysis of Algorithm 5.1, assuming that the maximum vertex degree Δ is used instead of the minimum vertex degree δ.

5.3.10

The following questions are devoted to the MILP problem formulation (5.3)–(5.10):

(a) Explain how the smallest-size k-dominating set problem in reachability graphs can be formulated as a particular case of the MILP problem.

(b) Assume that $d_i \leq c_i$ for $i = 1, 2, ..., n$. Now, let us suppose that it is necessary to guarantee that the whole demand of the closed neighbourhood of a charging station installed at vertex v_j can eventually be satisfied at v_j; that is, all the drivers who cannot charge their batteries locally and are at the reachability distance from v_j decide to go to charge at v_j. Formulate an additional constraint that should be added to the MILP problem in this case.

(c) Explain why the total demand in the network is satisfied; that is,

$$\sum_{j=1}^{n} d_j \leq \sum_{j=1}^{n} c_j x_j.$$

5.4 Solutions

5.3.1

The function ϕ is defined as follows:
$\phi(x) = (\text{dead-end, nil})$ if $x \in \{a, b, c, h, l\}$;
$\phi(d) = (\text{designated pedestrian crossing, zebra})$;
$\phi(f) = (\text{road intersection, light controlled})$;
$\phi(g) = (\text{road intersection, nil})$.

Also,
$\xi(e) = (1000, 40, \text{primary})$ if $e \in \{cd, df, fg, gh\}$ and
$\xi(e) = (100, 10, \text{residential})$ if $e \in \{af, bg, fl\}$.

Steps 1–3 of PNC-Algorithm result in the following vertex and edge sets of the pavement network G:

$$V(G) = \{a^0, a^1, b^0, b^1, c^0, c^1, d^0, d^1, f^0, f^1, f^2, f^3, g^0, g^1, g^2, h^0, h^1, l^0, l^1\}$$

and

$$E(G) = \{a^0 a^1, b^0 b^1, c^0 c^1, d^0 d^1, f^0 f^1, f^1 f^2, f^2 f^3,$$

$$f^3 f^0, g^0 g^1, g^1 g^2, g^2 g^0, h^0 h^1, l^0 l^1\}.$$

All the necessary cyclic clockwise orderings are found in Step 4:

$$
\begin{aligned}
&\text{CyclicOrder}(a,f) = 1, && \text{CyclicOrder}(f, a) = 1, \\
&\text{CyclicOrder}(b, g) = 1, && \text{CyclicOrder}(g, b) = 1, \\
&\text{CyclicOrder}(c, d) = 1, && \text{CyclicOrder}(d, c) = 2, \\
&\text{CyclicOrder}(d, f) = 1, && \text{CyclicOrder}(f, d) = 4, \\
&\text{CyclicOrder}(f, g) = 2, && \text{CyclicOrder}(g, f) = 3, \\
&\text{CyclicOrder}(g, h) = 2, && \text{CyclicOrder}(h, g) = 1, \\
&\text{CyclicOrder}(f, l) = 3, && \text{CyclicOrder}(l, f) = 1.
\end{aligned}
$$

Moreover, in Step 4, for every edge of N we generate two edges in the graph G. For example, for $af \in E(N)$, we have

$$\Psi(i, a) = i = \text{CyclicOrder}(a, f) = 1$$

and

$$\Psi(j, f) = j \bmod \deg(f) = \text{CyclicOrder}(f, a) \bmod 4 = 1.$$

Hence, the following edges of G are generated:

$$a^{\Psi(i,a)} f^{(j+1)\bmod \deg(f)} = a^1 f^2$$

and

$$a^{(i+1)\bmod \deg(a)} f^{\Psi(j,f)} = a^0 f^1.$$

In addition, the following edges are added to the graph G:

$$c^1 d^1, c^0 d^0, d^1 f^1, d^0 f^0, f^2 g^1, f^3 g^0, f^3 l^0, f^0 l^1, g^1 b^0, g^2 b^1, g^2 h^0, g^0 h^1.$$

The resulting pavement network G is shown in Figure 5.14.

In Step 5, the necessary information about the edges of G should be specified. This information is given in Table 5.8. ∎

5.3.2

Let us consider two examples, which are based on Exercise 5.3.1. Suppose we want to determine the OSM type of the jaywalk crossing $g^0 g^1$, but the picture

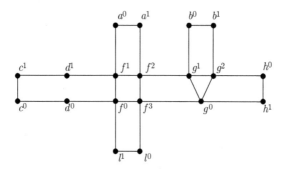

Fig. 5.14 Pavement network G.

Table 5.8 Information about edges of the pavement network G.

Edges	Segment Type	OSM Type	Length (m)
$a^0 f^1, a^1 f^2, b^0 g^1, b^1 g^2,$ $f^0 l^1, f^3 l^0$ $c^0 d^0, d^0 f^0, f^3 g^0, g^0 h^1,$	pavement segment	residential	100
$c^1 d^1, d^1 f^1, f^2 g^1, g^2 h^0$	pavement segment	primary	1000
$a^0 a^1, b^0 b^1, l^0 l^1$	dead-end		10
$c^0 c^1, h^0 h^1$	dead-end		40
$d^0 d^1$	zebra crossing		40
$f^0 f^1, f^2 f^3$	light-controlled crossing		40
$f^1 f^2, f^3 f^0$	light-controlled crossing		10
$g^0 g^1, g^2 g^0$	jaywalk crossing	primary	40
$g^1 g^2$	jaywalk crossing	residential	10

of the graph G is not available. First, we list all vertices that are adjacent to either g^0 or g^1:

$$g^1, g^2, h^1, f^3, b^0, g^2, g^0, f^2.$$

Next, identify the vertex u such that $u \neq g$ and u is listed twice in the list with different superscripts. In our case, such a vertex is f, which is listed twice: f^3 and f^2. Therefore, the OSM type of the jaywalk crossing is equal to the OSM type of the road segment gf (i.e. primary).

Now, suppose we wish to find out the OSM type of the jaywalk crossing $g^1 g^2$. Here is the neighbourhood of g^1 and g^2:

$$b^0, b^1, h^0, g^0, f^2, g^1, g^2.$$

The vertex b is the required vertex. Hence, the OSM type of g^1g^2 is the same as for gb (i.e. residential).   ∎

5.3.3

(a) The method is illustrated using the road network N from Exercise 5.3.1. Suppose that at the intersection f the road segment fa has a zebra pedestrian crossing, the road fg has a pelican pedestrian crossing, the road fl has a light-controlled crossing, and the road fd has a jaywalk crossing.

   In general, the information about pedestrian crossings in a road network can be arranged in an $n \times n$ matrix PC, where n is the number of vertices in the network and the rows and columns correspond to the vertices with the same ordering. If $u \neq v$, then the entry in Row u and Column v provides information about a pedestrian crossing (if it exists) of the road uv at the intersection u. If u represents a designated pedestrian crossing or a dead-end, then this information can be coded in the entry of Row u and Column u.

   In our example, the fifth row of the 8×8 matrix PC is as follows:

$$(\mathrm{z}, -, -, \mathrm{j}, -, \mathrm{p}, -, \mathrm{lc}),$$

where 'z' means 'zebra', 'j' stands for 'jaywalk', 'p' denotes 'pelican' and 'lc' represents 'light controlled'. If a pedestrian crossing is forbidden for a particular road at an intersection (e.g. by means of a physical obstruction), then this can be marked by 'f' in the corresponding place of the matrix. The information kept in the matrix PC can be easily passed on to the corresponding pavement network. For example, if a pedestrian crossing is forbidden for a particular road at an intersection, then the corresponding edge representing such a crossing should be removed from the generated pavement network.   ∎

(b) Let us suppose that in a given road network N an edge ab represents a road segment with a bridge going over a road segment cd. Graphically this means that the edges ab and cd intersect each other, but their intersection point is not a vertex (junction) in the network N.

   We will briefly outline two ideas, the details are left to the reader. The first approach is to introduce an artificial junction (vertex) w at the intersection point and repeat the same for all bridges in N. Then, PNC-Algorithm is applied, and in the generated pavement network all artificial vertices are removed and the necessary edges are added. In our two-edge example, the artificial vertices w^0, w^1, w^2, w^3 should be deleted from the pavement network, together with incident edges. Also, two edges between the sets $\{a^0, a^1\}$ and $\{b^0, b^1\}$ should be introduced making sure that they do not intersect each other. In a similar way, two edges between the sets $\{c^0, c^1\}$ and $\{d^0, d^1\}$ should be introduced making sure that they do not intersect each other.

Another approach would be to remove the middle part of the edge ab; that is, ab is replaced by two edges: aa_1 and b_1b, where a_1 and b_1 are artificial dead-ends. The same should be done for all bridges before PNC-Algorithm is applied. Then, all artificial dead-ends must be removed and the necessary edges should be correctly introduced in the generated pavement network.   ■

5.3.4

(a) The shortest (c^1, h^0)-path is $(c^1, d^1, f^1, f^2, g^1, g^2, h^0)$ with length 4020 m. The safest (c^1, h^0)-path is $(c^1, c^0, d^0, f^0, f^3, g^0, h^1, h^0)$ with total risk value 28,120. The latter is also the simplest path with total complexity 64.9.   ■

(b) The shortest and safest paths are the same as in part (a). The simplest (c^1, h^0)-path is $(c^1, d^1, f^1, f^2, g^1, g^2, h^0)$ with total complexity 69.2. The second simplest (c^1, h^0)-path $(c^1, d^1, f^1, f^2, g^1, b^0, b^1, g^2, h^0)$ has total complexity 70.2 and it does not include a jaywalk crossing, as required.   ■

5.3.5

(i) (a) Crash locations, intersections, signalized intersections, mid-block crossings, unsignalized zebra crossings, road networks, road patterns, boroughs, paths.

(b) Logistic regression models, statistical tests, regression trees, kernel density estimation, Bayesian techniques, negative binomial regression models, graph-theoretic methods.

(c) Temporal, spatial, demographic, socio-economic, road network characteristics, street designs, topological characteristics, crossing locations.

(d) Risk factors, fatality factors, severity factors, injury severity likelihood, pedestrian crash likelihood, crash patterns, crash severity, likelihood of injury and fatality.   ∎

(ii) (a) Path, route.

(b) Density estimation, path planning.

(c) Crime activity spatial density, historical crime data.

(d) Safest path/route.   ∎

5.3.6

(a) It is straightforward to see that $\gamma_1(C_n) = \lceil \frac{n}{3} \rceil$, $\gamma_2(C_n) = \lceil \frac{n}{2} \rceil$ and $\gamma_k(C_n) = n$ if $k \geq 3$.   ∎

(b) Let A be a k-dominating set in G such that $|A| = \gamma_k(G)$. Consider the set of edges L between A and $V(G) - A$. Counting the edges L from the set A, we obtain

$$|L| \leq \sum_{u \in A} \deg(u) \leq \Delta|A| = \Delta\gamma_k(G).$$

On the other hand, counting the edges from $V(G) - A$, we have

$$|L| \geq \sum_{u \in V(G)-A} k = k\,|V(G) - A| = k(n - \gamma_k(G)).$$

Thus,

$$\Delta\gamma_k(G) \geq k(n - \gamma_k(G))$$

or

$$\gamma_k(G) \geq \frac{kn}{k + \Delta},$$

as required.   ∎

(c) Assume to the contrary that there exists a graph G with $\delta \geq k$ such that $\gamma_k(G) > \frac{kn}{k+1}$ and G has the smallest possible number of edges. Let us denote $l = n - \gamma_k(G)$. Then

$$\gamma_k(G) > \frac{k(l + \gamma_k(G))}{k + 1}.$$

Hence, $\gamma_k(G) > kl$ or $\gamma_k(G) \geq kl + 1$.

Consider the set $W = \{w \in V(G) : \deg(w) > k\}$, which might be empty. This set cannot have edges. Indeed, if there is an edge $e = uv \in E(G)$ such that $u, v \in W$, then $\gamma_k(G-e) \geq \gamma_k(G)$ and $\delta(G-e) \geq k$; that is, the graph G is not a minimal counterexample. Therefore, the set W is independent.

Let us consider the maximal independent set W^+ containing the set W. It is easy to see that the set $D = V(G) - W^+$ is k-dominating, and hence

$$|D| \geq kl + 1. \tag{5.11}$$

Now, because W^+ is maximal, each vertex $x \in D$ is adjacent to a vertex of W^+. Taking into account that $\deg(x) = k$, we obtain

$$|D \cap N[x]| \leq k. \tag{5.12}$$

Let us construct an independent set of vertices $X = \{x_1, x_2, ..., x_{l+1}\}$ by selecting vertices from the set D as follows. First, take any vertex $x_1 \in D$. Then, select any vertex x_2 in the set $D - N[x_1]$. Further, select any vertex x_3 in the set $D - N[x_1] - N[x_2]$, and so on. Finally, $x_{l+1} \in D - N[x_1] - ... - N[x_l]$. This procedure is well defined because from (5.11) and (5.12), we obtain

$$|D - N[x_1] - ... - N[x_l]| \geq kl + 1 - kl = 1.$$

The set X is independent by construction. Hence, $V(G) - X$ is a k-dominating set of size

$$n - (l+1) \leq n - (n - \gamma_k(G) + 1) = \gamma_k(G) - 1,$$

a contradiction.   ∎

5.3.7

(a) Let us apply the greedy algorithm to the cube Q_3. In the first step, $U = V(Q_3)$ and $|N[v_i] \cap U| = 4$ for any vertex. We set $D = \{v_1\}$ because v_1 has the smallest subscript. In the second step, $U = V(Q_3) - \{v_1\}$ and v_3 has the smallest subscript with $|N[v_3] \cap U| = 4$. Hence, $D = \{v_1, v_3\}$. In the next step, $U = \{v_5, v_6, v_7, v_8\}$ and the maximum value of $|N[v_i] \cap U| = 3$ is achieved for all the vertices belonging to

U. The vertex with the smallest subscript is v_5, so $D = \{v_1, v_3, v_5\}$. Further, $U = \{v_6, v_7, v_8\}$ and the maximum value of $|N[v_i] \cap U| = 3$ is achieved for one vertex v_7. Hence, $D = \{v_1, v_3, v_5, v_7\}$. The algorithm stops because $U = \emptyset$. It is easy to see that the set D is a minimal 2-dominating set of size 4. ∎

(b) Now, let us apply Algorithm 5.1. We have $b_1 = 3$ and $\delta' = 2$. Hence, $p = 1 - \frac{1}{3} = \frac{2}{3}$. We will apply Algorithm 5.1 three times because it is a randomized algorithm. Here are three sets of random numbers:

$$0.497, 0.897, 0.595, 0.942, 0.665, 0.463, 0.829, 0.130;$$
$$0.767, 0.891, 0.352, 0.759, 0.187, 0.498, 0.034, 0.806;$$
$$0.612, 0.620, 0.798, 0.220, 0.708, 0.307, 0.382, 0.101.$$

Note that eight random numbers in each set correspond to the vertices $v_1, ..., v_8$ and determine the vertices that belong to the set A. For example, the first number 0.497 is less than the probability $p = 2/3$, and hence $v_1 \in A$.

The first set of random numbers results in the set $A = \{v_1, v_3, v_5, v_6, v_8\}$, which is a 2-dominating set. Hence, $B = \emptyset$, $D = A$ and the vertex v_5 can be removed from D to generate D'. Thus, the minimal 2-dominating set D' has size 4.

The second set of random numbers results in the set $A = \{v_3, v_5, v_6, v_7\}$, which is not a 2-dominating set because the vertices v_1 and v_4 are not 2-dominated. If v_1 is added to B, then the redundant vertex v_6 should be removed from D to generate D'. If v_4 is added to B, then the redundant vertex v_7 should be removed from D to generate D'. In any case, the minimal 2-dominating set D' has size 4.

Finally, the third set of random numbers results in the set $A = \{v_1, v_2, v_4, v_6, v_7, v_8\}$, which is a 2-dominating set. Hence, $B = \emptyset$, $D = A$ and the vertices v_2 and v_8 can be removed from D to generate D'. Thus, the minimal 2-dominating set D' has size 4. ∎

(c) Using the result of the greedy algorithm from (a), we obtain $\gamma_2(Q_3) \leq 4$. On the other hand, using the lower bound from the previous exercise, $\gamma_2(Q_3) \geq \frac{2n}{2+3} = 0.4n = 3.2$. Therefore, $\gamma_2(Q_3) = 4$. ∎

5.3.8

(a) It is not difficult to see that $\gamma_{0.8}(P_n) = \lfloor n/2 \rfloor$, $\gamma_{0.8}(C_n) = \lceil n/2 \rceil$ and $\gamma_{0.8}(K_n) = \lceil 0.8(n-1) \rceil$. ∎

(b) Let A be an α-dominating set in G such that $|A| = \gamma_\alpha(G)$. Consider the set of edges between A and $V(G) - A$, denoted by L. Counting the edges L from the set A, we obtain

$$|L| \leq \sum_{u \in A} \deg(u).$$

On the other hand, counting the edges from $V(G) - A$, we have

$$|L| \geq \sum_{u \in V(G) - A} \alpha \deg(u).$$

Thus,

$$\Delta|A| \geq \sum_{u \in A} \deg(u) \geq |L| \geq \sum_{u \in V(G) - A} \alpha \deg(u) \geq \alpha\delta \, |V(G) - A|.$$

Hence,

$$\Delta\gamma_\alpha(G) \geq \alpha\delta(n - \gamma_\alpha(G)) \quad \text{or} \quad \gamma_\alpha(G) \geq \frac{\alpha\delta n}{\Delta + \alpha\delta},$$

as required. ∎

(c) We have for $0 < \alpha < 1$: $\gamma_\alpha(G) \leq n - \gamma_{1-\alpha}(G)$. Now, using the lower bound from (b), we obtain

$$\gamma_\alpha(G) \leq n - \frac{(1-\alpha)\delta n}{\Delta + (1-\alpha)\delta} = \frac{\Delta n}{\Delta + (1-\alpha)\delta}$$

for $0 < \alpha < 1$. If $\alpha = 1$, the inequality is obviously true. ∎

5.3.9

(a) The binomial coefficient b_{k-1} in Algorithm 5.1 can be computed by using the dynamic programming and Pascal's triangle, which require $O(\bar{d}^2)$ time and space in this case. The average vertex degree $\bar{d}$ of G_t^r can be computed in $O(m + n) = O(m)$ time, where m is the number of edges and n is the number of vertices in G_t^r. Because

$$\bar{d} = \frac{1}{n} \sum_{i=1}^{n} d_i = \frac{2m}{n}$$

and

$$\bar{d}^2 = \frac{4m^2}{n^2} \leq \frac{4m \times n(n-1)}{n^2 \times 2} < 2m,$$

we see that $O(\bar{d}^2)$ does not exceed $O(m)$. Therefore, computing the probability p can be done in $O(m)$ time. The remaining parts of the algorithm stay the same as in the case of usage of the minimum vertex degree. It takes $O(n)$ time to find the set A. The numbers $z = |N(v) \cap A|$ for each vertex $v \in V^r - A$ can be computed separately or when finding the set A. We need to keep track of them only for $z < k$. Since we may need to browse through all the neighbours of vertices in A, it can take in total $O(m)$ steps to calculate all the necessary values z for all $v \in V^r - A$. Hence, the set B can be found in $O(m+n)$ steps. Removing unnecessary vertices from the set $D = A \cup B$ to have it minimal k-dominating can be done in $O(m+n)$ time. Thus, when using $\bar{d}$, Algorithm 5.1 runs in $O(m+n)$ time, similar to the case when the minimum vertex degree δ is used to compute the probability p. ∎

(b) Let the maximum vertex degree Δ of G_t^r be used to compute the probability p. Notice that Δ can be linear in n (i.e. $\Theta(n)$), whereas m can be sub-quadratic in n (i.e. $o(n^2)$). Then again, the binomial coefficient b_{k-1} in Algorithm 5.1 can be computed by using the dynamic programming and Pascal's triangle, which require $O(\Delta^2)$ time and space in this case, and Δ can be computed in $O(m)$ time. However, because m is $o(n^2)$ and Δ^2 is $\Theta(n^2)$ in our case, $O(\Delta^2)$ cannot be $O(m)$. Therefore, in the worst case, it requires $O(n^2)$ time and space to compute the probability p. The remaining parts of the algorithm and their analysis stay the same. Thus, Algorithm 5.1 runs in $O(n^2)$ time in the case when the maximum vertex degree is used to compute the probability p, which would be more time and space consuming than when using δ or $\bar{d}$ of G_t^r. ∎

5.3.10

(a) In view of the MILP problem formulation (5.3)–(5.10), the smallest-size k-dominating set problem in reachability graphs can be formulated as a particular case of this formulation by setting $c_i = 1$ for $i = 1, 2, ..., n$ in the objective function (5.3) and considering only constraints (5.4) and the decision variables $x_i \in \{0, 1\}$ in (5.9):

$$\min \ \sum_{i=1}^{n} x_i$$

subject to

$$\sum_{v_i \in N(v_j)} x_i \geq k(1 - x_j), \quad j = 1, 2, ..., n$$

$$x_i \in \{0, 1\}, \quad i = 1, 2, ..., n$$

This provides an integer programming formulation, which can be used for finding the smallest k-dominating sets in graphs. ∎

(b) To guarantee that the whole demand of the closed neighbourhood of a charging station installed at vertex v_j can be eventually satisfied at v_j, the following additional constraint should be added to the above MILP problem:

$$d_j + \sum_{v_i \in N(v_j)} d_i(1 - x_i) \leq c_j.$$

This constraint guarantees that the sum of the demand at the vertex v_j and the demands at its neighbours, where a charging station is not installed, is less than or equal to the capacity c_j of the charging station at v_j. ∎

(c) The satisfaction of the total demand in the network, that is,

$$\sum_{j=1}^{n} d_j \leq \sum_{j=1}^{n} c_j x_j$$

is guaranteed by the constraints (5.5) taking into account (5.6). Indeed, (5.5) implies

$$\sum_{j=1}^{n} \sum_{v_i \in N[v_j]} d_i y_{ij} \leq \sum_{j=1}^{n} c_j x_j.$$

Taking into account (5.6), we obtain

$$\sum_{j=1}^{n} \sum_{v_i \in N[v_j]} d_i y_{ij} = d_1 \left(\sum_{v_j \in N[v_1]} y_{1j} \right) + ... + d_n \left(\sum_{v_j \in N[v_n]} y_{nj} \right)$$

$$= \sum_{j=1}^{n} d_j.$$

∎

Acknowledgements

Section 5.2 is based on the article in *Computers and Operations Research*, **96**, A. Gagarin and P. Corcoran, Multiple domination models for placement of electric vehicle charging stations in road networks, 69–79, © 2018, with permission from Elsevier.

References

[1] N. Alon, Transversal numbers of uniform hypergraphs, *Graphs and Combinatorics*, **6** (1990), 1–4.

[2] N. Alon and J. H. Spencer, *The Probabilistic Method*, New York, NY: John Wiley & Sons Inc., 1992.

[3] J. Asamer, M. Reinthaler, M. Ruthmair, M. Straub and J. Puchinger, Optimizing charging station locations for urban taxi providers, *Transportation Research*, Part A **85** (2016), 233–246.

[4] H. A. Aziz, S. V. Ukkusuri and S. Hasan, Exploring the determinants of pedestrian–vehicle crash severity in New York city, *Accident Analysis and Prevention*, **50** (2013), 1298–1309.

[5] H. Bast, D. Delling, A. Goldberg, M. Müller-Hannemann, T. Pajor, P. Sanders, D. Wagner and R. Werneck, Route planning in transportation networks, *Algorithm Engineering: Selected Results and Surveys. Lecture Notes in Computer Science*, **9220** (2016), 19–80.

[6] K. Belous, First electrobuses in the streets of Minsk, 'Minsk-News' Agency, 16 May 2017 (in Russian), http://minsknews.by/blog/2017/ 05/16/page/2/ (accessed 25 September 2017).

[7] G. Brandstätter, C. Gambella, M. Leitner, E. Malaguti, F. Masini, J. Puchinger, M. Ruthmair and D. Vigo, Overview of optimization problems in electric car-sharing system design and management, in H. Dawid, K. Doerner, G. Feichtinger, P. Kort and A. Seidl (eds), *Dynamic Perspectives on Managerial Decision Making, Dynamic Modeling and Econometrics in Economics and Finance*, **22**, Berlin: Springer, 2016, 441–471.

[8] J. Cavadas, G. H. Correia and J. Gouveia, A MIP model for locating slow-charging stations for electric vehicles in urban areas accounting for driver tours, *Transportation Research*, Part E **75** (2015), 188–201.

[9] Chariot e-bus product description, Chariot Motors Group, http://www.chariot-electricbus.com/products/chariot-e-bus/ (accessed 25 September 2017).

[10] T. D. Chen, K. M. Kockelman and M. Khan, Locating electric vehicle charging stations: parking-based assignment method for Seattle, Washington, *Transportation Research Record*, (2385)(2013), 28–36.

[11] E. J. Cockayne, B. Gamble and B. Shepherd, An upper bound for the k-domination number of a graph, *Journal of Graph Theory*, **9** (1985), 533–534.

[12] P. Corcoran, P. Mooney and M. Bertolotto, Analysing the growth of Open-StreetMap networks, *Spatial Statistics*, **3** (2013), 21–32.

[13] J. E. Dunbar, D. G. Hoffman, R. C. Laskar and L. R. Markus, α-Domination, *Discrete Mathematics*, **211** (2000), 11–26.

[14] Electricity Supply Board (ESB) web-site, https://www.esb.ie/electric-cars/kml/charging-locations.kml (accessed 25 September 2017).

[15] J. F. Fink and M. S. Jacobson, n-Domination in graphs, in *Graph Theory with Applications to Algorithms and Computer Science*, New York, NY: Wiley, 1985, 283–300.

[16] I. Frade, A. Ribeiro, G. Gonçalves and A. P. Antunes, Optimal location of charging stations for electric vehicles in a neighborhood in Lisbon, Portugal, *Transportation Research Record*, (2252)(2011), 91–98.

[17] S. Funke, A. Nusser and S. Storandt, Placement of loading stations for electric vehicles: no detours necessary! *Journal of Artificial Intelligence Research*, **53** (2015), 633–658.

[18] E. Galbrun, K. Pelechrinis and E. Terzi, Urban navigation beyond shortest route: the case of safe paths, *Information Systems*, **57** (2016), 160–171.

[19] Q. Guo, P. Xu, X. Pei, S. Wong and D. Yao, The effect of road network patterns on pedestrian safety: a zone-based Bayesian spatial modeling approach, *Accident Analysis and Prevention*, **99** (2017), 114–124.

[20] T. Hamilton, Next stop: ultracapacitor buses, MIT Technology Review, 19 October 2009, https://www.technologyreview.com/s/415773/next-stop-ultracapacitor-buses/ (accessed 25 September 2017).

[21] C. Hannah, I. Spasić and P. Corcoran, A computational model of pedestrian road safety: The long way round is the safe way home, *Accident Analysis and Prevention*, **121** (2018), 347–357.

[22] L. He, H.-Y. Mak, Y. Rong and Z.-J. M. Shen, Service region design for urban electric vehicle sharing systems, *Manufacturing & Service Operations Management*, **19** (2)(2017), 309–327.

[23] M. Hosseini and S. A. MirHassani, Refueling-station location problem under uncertainty, *Transportation Research*, Part E **84** (2015), 101–116.

[24] H. A. Karimi and P. Kasemsuppakorn, Pedestrian network map generation approaches and recommendation, *International Journal of Geographical Information Science*, **27** (5)(2013), 947–962.

[25] A. Keler and J. D. Mazimpaka, Safety-aware routing for motorised tourists based on open data and VGI, *Journal of Location Based Services*, **10** (1)(2016), 64–77.

[26] A. Y. S. Lam, Y.-W. Leung and X. Chu, Electric vehicle charging station placement: formulation, complexity, and solutions, *IEEE Transactions on Smart Grid*, **5** (2014), 2846–2856.

[27] J. K. Lan and G. J. Chang, Algorithmic aspects of the k-domination problem in graphs, *Discrete Applied Mathematics*, **161** (2013), 1513–1520.

[28] C. Lee and M. Abdel-Aty, Comprehensive analysis of vehicle–pedestrian crashes at intersections in Florida, *Accident Analysis and Prevention*, **37** (4)(2005), 775–786.

[29] H.-Y. Mak, Y. Rong and Z.-J. M. Shen, Infrastructure planning for electric vehicles with battery swapping, *Management Science*, **59** (7)(2013), 1557–1575.

[30] W. E. Marshall and N. W. Garrick, Does street network design affect traffic safety? *Accident Analysis and Prevention*, **43** (3)(2011), 769–781.

[31] P. Mooney and P. Corcoran, Using OSM for LBS—an analysis of changes to attributes of spatial objects, in G. Gartner and F. Ortag (eds), *Advances in Location-Based Services*, London: Springer, 2012, 165–179.

[32] Y. M. Nie and M. Ghamami, A corridor-centric approach to planning electric vehicle charging infrastructure, *Transportation Research*, Part B **57** (2013), 172–190.

[33] P. Olszewski, P. Szagała, M. Wolański and A. Zielińska, Pedestrian fatality risk in accidents at unsignalized zebra crosswalks in Poland, *Accident Analysis and Prevention*, **84** (2015), 83–91.

[34] A. Osama and T. Sayed, Evaluating the impact of connectivity, continuity, and topography of sidewalk network on pedestrian safety, *Accident Analysis and Prevention*, **107** (2017), 117–125.

[35] A. Poghosyan, D. V. Greetham, S. Haben and T. Lee, Long term individual load forecast under different electrical vehicles uptake scenarios, *Applied Energy*, **157** (2015), 699–709.

[36] A. T. Pour, S. Moridpour, R. Tay and A. Rajabifard, Neighborhood influences on vehicle–pedestrian crash severity, *Journal of Urban Health*, (2017), 1–14.

[37] A. T. Pour, S. Moridpour, R. Tay and A. Rajabifard, Influence of pedestrian age and gender on spatial and temporal distribution of pedestrian crashes, *Traffic Injury Prevention*, **19** (1)(2018), 81–87.

[38] S. S. Pulugurtha and V. R. Sambhara, Pedestrian crash estimation models for signalized intersections, *Accident Analysis and Prevention*, **43** (1)(2011), 439–446.

[39] S. M. Rifaat, R. Tay and A. de Barros, Effect of street pattern on the severity of crashes involving vulnerable road users, *Accident Analysis and Prevention*, **43** (1)(2011), 276–283.

[40] S. M. Rifaat, R. Tay and A. de Barros, Urban street pattern and pedestrian traffic safety, *Journal of Urban Design*, **17** (3)(2012), 337–352.

[41] L. Rothman, A. W. Howard, A. Camden and C. Macarthur, Pedestrian crossing location influences injury severity in urban areas, *Injury Prevention*, **18** (6)(2012), 365–370.

[42] M. Saha, M. Saugstad, H. T. Maddali, A. Zeng, R. Holland, S. Bower, A. Dash, S. Chen, A. Li, K. Hara and J. Froehlich, Project Sidewalk: a web-based crowdsourcing tool for collecting sidewalk accessibility data at scale, *Proceedings of the 2019 CHI Conference on Human Factors in Computing Systems*, ACM, 4–9 May, Glasgow, 2019.

[43] S. Storandt and S. Funke, Cruising with a battery-powered vehicle and not getting stranded, *Proceedings of the 26th AAAI Conference on Artificial Intelligence*, **3** (2012), 1628–1634.

[44] N.-N. Sze and S. Wong, Diagnostic analysis of the logistic model for pedestrian injury severity in traffic crashes, *Accident Analysis and Prevention*, **39** (6)(2007), 1267–1278.

[45] R. Tay, J. Choi, L. Kattan and A. Khan, A multinomial logit model of pedestrian–vehicle crash severity, *International Journal of Sustainable Transportation*, **5** (4)(2011), 233–249.

[46] Zero emission urban bus system (ZeEUS) project, ZeEUS eBus Report #1 (2016), http://zeeus.eu/uploads/publications/documents/zeeus-ebus-report-internet.pdf (accessed 25 September 2017).

[47] Y. Zhang, J. Bigham, D. Ragland and X. Chen, Investigating the associations between road network structure and non-motorist accidents, *Journal of Transport Geography*, **42** (2015), 34–47.

[48] J. Zhao and T. Ma, Optimizing layouts of initial AFV refueling stations targeting different drivers, and experiments with agent-based simulations, *European Journal of Operational Research*, **249** (2)(2016), 706–716.

6

Graphs in Molecular Epidemiology

P. Skums and V. Zverovich

Graphs and networks are used in molecular epidemiology to model the evolution of viruses and their spread during outbreaks and epidemics. They are instrumental at different stages of the computational pipelines. This includes the inference of transmission networks using viral sequences sampled from infected individuals, studies of selection and accumulation of mutations in viral populations and their interactions with hosts' immune systems. In this chapter, we will describe some algorithmic and graph-theoretic problems associated with these stages to illustrate the relevance of the concepts of graph theory to molecular epidemiology of viral infections. We will demonstrate how graph-theoretic methods combined with the machinery of differential equations, the Bayesian inference and computational genomics allow us to uncover hidden biological and epidemiological patterns of virus evolution and transmission.

6.1 A Brief Overview of Graph Models

Molecular epidemiology is a scientific discipline at the intersection of biology and public health. Its goal is to study the spread of infectious diseases using an analysis of the genomes of pathogens. In the last decade, discrete algorithms and mathematical modelling became the major tools of molecular epidemiology. From a computational point of view, the main objects studied in molecular epidemiology are DNA or RNA *sequences* of viruses; that is, words over the four-letter alphabet of *nucleotides*:

$$\mathcal{A} = \{A, C, U, G\} \text{ for RNA}$$

or

$$\mathcal{A}' = \{A, C, T, G\} \text{ for DNA}.$$

Modern Applications of Graph Theory. Vadim Zverovich, Oxford University Press (2021). © Vadim Zverovich.
DOI: 10.1093/oso/9780198856740.003.0006

Each sequence is a genome of a particular *genomic variant*, and the set of these variants forms a *viral population*. In what follows, we will use the words 'sequence' and 'variant' interchangeably. A viral population can be considered as a metric space with the distance between sequences being either Hamming distance or some variant of edit distance (e.g. Levenshtein distance). In evolutionary biology, this space is usually referred to as a *sequence space* [20]. Note that viruses mutate extremely rapidly, and hence each infected individual usually carries a viral subpopulation rather than a single genomic variant.

Graphs are widely used as models and data structures, which are highly instrumental for our understanding of pathogen evolution and the spread of diseases in populations of susceptible individuals. Graph-theoretic models in molecular epidemiology include several examples, which can informally be described as follows:

- A *genetic network* is a graph-based representation of a sequence space, which is another model that represents the structure of a viral population. It is defined as an undirected graph with vertices corresponding to viral variants, and edges connecting variants at a distance below a certain threshold, for example at Hamming distance 1 (i.e. different by a single mutation).

- *Phylogenetic trees* are the most typical and widely used biological models, going back to the time of Charles Darwin. Rooted phylogenies are mostly analysed in molecular epidemiology, for example rooted binary trees. The leaves in such trees correspond to observed viral variants, and internal nodes represent the most recent common ancestor of those variants.

- A *host network* represents epidemiological relationships between infected individuals. Its vertices are infected hosts, and two vertices are adjacent whenever they are epidemiologically linked; that is, when there is evidence that one of the hosts directly infected the other or they are linked by a sufficiently short chain of viral transmissions between observed and, possibly, unobserved individuals. Connected components of a host network are called *transmission clusters* or *outbreaks*.

- A *transmission network* indicates who infected whom. It is a directed graph whose vertices are infected hosts, and each link uv indicates that the host u directly infected the host v.

- A *cross-immunoreactivity network* is a directed graph with vertices corresponding to viral variants. Two variants a and b are connected by the link ab if a elicits antibodies that can bind to b. This network represents

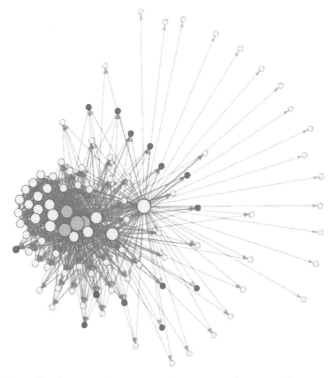

Fig. 6.1 Example of a cross-immunoreactivity network at equilibrium. Selfish and altruistic variants are shown in red and green, respectively; other variants are shown in yellow.

interactions between viral variants and the host's immune system. In some models, there are altruistic and selfish variants, which will be discussed later in this chapter.

Note that graphs have been used in traditional epidemiology for at least a century [23]. Many classical mathematical epidemiological models (e.g. susceptible–infected–recovered [29]) belong to the class of the so-called compartmental models. In such models, the population of infected individuals is separated into disjoint subsets or compartments, for example susceptible individuals, infected individuals and recovered individuals. The associated differential equation model (or stochastic model) describes the movement of individuals between groups. The scheme of flows between compartments is a graph, and therefore the model basically describes a dynamical process on the graph. Such models work on the level of infected individuals and do not consider genomes and associated biological features of spreading pathogens.

In contrast, computational molecular epidemiology specifically takes these factors into account and strives to make use of them. This has become a standard tool of biomedical research over only the last decade and was caused by the emergence of the so-called next-generation sequencing technologies. These technologies have revolutionized practically all fields of biology by allowing for fast, relatively inexpensive and accurate determination of DNA or RNA sequences of different life forms and organisms. For epidemiology, this means the possibility to study millions of viral sequences sampled from thousands of infected individuals. The availability of such data provides numerous research opportunities, but also brings to life many algorithmic and mathematical problems. Graphs are highly instrumental in addressing many of such problems.

In some cases, the aforementioned graphs and networks are major objects to be inferred and analysed. Other graphs or networks are usually used as auxiliary structures that are instrumental at different stages of the computational pipelines for algorithmic and mathematical problems. Some of the problems, such as the construction of phylogenetic trees from DNA or RNA sequences and their analysis, are typical for many areas of biology and belong to the oldest and best-studied branches of computational biology and bioinformatics. There is a significant body of literature dedicated to classical phylogenetics, and therefore the associated problems are not discussed here. Instead, our intention is to focus on the problems that are specific for computational molecular epidemiology with a particular emphasis on graph models and algorithms used at different stages of the modelling pipeline. This includes the modelling of viral evolution, virus interactions with the immune systems of infected hosts and the reconstruction of viral transmission networks.

6.2 Graphs and the Molecular Evolution of Viruses

Concepts of graph theory have broad applications in many scientific and engineering disciplines. Viral evolution is one of the most complex scientific disciplines that started embracing these concepts, and this has had substantial benefits for the field. In this section, we show the relevance of the concepts of graph theory to the molecular evolution of viral infections and how crucial molecular evolutionary models can be viewed in light of graph theory. In particular, the state of the population of selectively neutral viral variants converges to the distribution fundamentally determined by the graph-theoretic properties of its genetic network.

Interestingly, there is some connection with applications in computer science because the basic mathematical model of viral evolution can be considered in relation to the concept of PageRank [3]. It is one of the most influential concepts used by internet search engines, which appears to be a special case of the mathematical evolutionary model. Notice that PageRank and the dynamics of selectively neutral viral variants are determined by the spectral properties of the graph's normalized Laplacian matrix and the adjacency matrix, respectively. Methods of graph theory are used in studies of PageRank, indicating their applicability to problems of viral evolution.

6.2.1 Graphs and the Quasispecies Model

The classical quasispecies model was introduced by Eigen to describe the evolution of self-replicating chemical molecules as part of his general model of the origins of life [14, 15]. It was noticed later that the quasispecies model is applicable to the evolution of RNA viruses, such as Hepatitis C (HCV) and Human Immunodeficiency Virus (HIV) [12, 13]. The model found many proponents who used it widely to study viral evolution. The most intriguing predictions of the model are *survival of the flattest* and *error catastrophe*. The latter refers to the phenomenon whereby the effects of having a too high mutation rate (exceeding the so-called error threshold) outweigh the effects of natural selection and cause the population to lose its structural coherence and eventually die out. More detailed accounts of the quasispecies model can be found in [30, 39]. In this chapter, we will describe the first of the aforementioned phenomena and its relation to the graph-theoretic properties of the sequence space.

Consider the sequence space S consisting of n different viral RNA sequences of fixed length L:

$$S \subseteq \{A, C, U, G\}^L.$$

Let t denote time. Suppose that the sequences have time-dependent frequencies

$$c_1(t), c_2(t), ..., c_n(t),$$

where

$$\sum_{i=1}^{n} c_i(t) = 1 \text{ for every } t \geq 0.$$

In addition, the sequences have replication rates (*fitnesses*)

$$f_1, f_2, ..., f_n$$

that are time independent. Let

$$\phi(t) = \sum_{i=1}^{n} f_i c_i(t)$$

be the average fitness of the population at time t. Let $q_{i,j}$ be the probability of a mutation from the i-th to the j-th sequence. In other words, $q_{i,j}$ is the probability that replication of sequence i produces sequence j. We may assume that $q_{i,j} = q_{j,i}$.

The standard quasispecies model can be written as the following system of differential equations [15]:

$$\frac{dc_i(t)}{dt} = \sum_{j=1}^{n} f_j\, q_{j,i}\, c_j(t) - \phi(t)\, c_i(t), \quad i = 1, 2, ..., n, \qquad (6.1)$$

or in matrix form

$$\frac{d\mathbf{c}(t)}{dt} = W^{\mathrm{T}}\mathbf{c}(t) - \phi(t)\,\mathbf{c}(t), \qquad (6.2)$$

where W is an $n \times n$ matrix with

$$W_{i,j} = f_i q_{i,j}.$$

The probability $q_{j,i}$ can be calculated as follows:

$$q_{j,i} = \left(\frac{\epsilon}{3}\right)^{h_{i,j}} (1 - \epsilon)^{L - h_{i,j}},$$

where ϵ is the *error rate* (i.e. the probability of mutation at a single position) and $h_{i,j}$ is the Hamming distance between sequences i and j. For example, the error rate of HCV replication is estimated to be 10^{-5} to 10^{-4} [12]. Since the probability of multiple mutations is low, for simplicity it may be assumed that $q_{j,i} = 0$ if $h_{j,i}$ is very large.

In reality, the overwhelming majority of sequences from the set

$$\{A, C, U, G\}^L$$

contain deleterious mutations and cannot replicate; that is, $f_j = 0$ for such sequences. For example, if the whole sequence contains only one open reading frame, which is an accurate assumption for many RNA viruses, mutations that generates the in-phase stop codons are deleterious and should be prohibited. As shown in [36], this results in a reduction of the number of viable sequences to

$$61^{L/3} = o(4^L).$$

The above number is actually much lower, since many more mutations negatively affect viral replication. The variants that cannot replicate, however, do not influence the dynamics of a viral population and hence they can be excluded from evaluation.

With this consideration in mind, the quasispecies model can be simplified. Let

$$D = \left\{ i \in \{1, 2, ..., n\} : f_i = 0 \right\}.$$

All quasispecies from D can be replaced with one 'generalized' variant x such that $f_x = 0$ and

$$q_{i,x} = \sum_{j \in D} q_{i,j}, \quad q_{x,i} = 0 \ \text{ for every } \ i \notin D.$$

After relabelling, the replication-competent sequences are labelled by $1, 2, ..., n$, and the 'generalized' non-replicating variant is labelled by $n + 1$. The removal of this variant from (6.1) results in the system that describes the evolution of relative frequencies only for replicating variants and it can be achieved via the following substitution:

$$b_i(t) = \frac{c_i(t)}{C(t)}, \quad C(t) = \sum_{i=1}^{n} c_i(t).$$

Taking into account that

$$C(t) = 1 - c_{n+1}(t) \quad \text{and} \quad \frac{dC(t)}{dt} = -\frac{dc_{n+1}(t)}{dt},$$

we obtain the system

$$\frac{db_i(t)}{dt} = \sum_{j=1}^{n} f_j \, q_{j,i} \, b_j(t) - \psi(t) \, b_i(t), \quad i = 1, 2, ..., n, \tag{6.3}$$

where

$$\psi(t) = \sum_{i=1}^{n} f_i \, q_i \, b_i(t) \quad \text{and} \quad q_i = \sum_{j=1}^{n} q_{i,j}.$$

Now, the sequence space can be considered as the following genetic network, which is a vertex- and edge-weighted graph:

$$G = (V(G), E(G), f, q),$$

where

$$V(G) = \{1, 2, ..., n\} \quad \text{and} \quad E(G) = \{ij : q_{i,j} \neq 0\}.$$

Assuming that the replication rates of all variants are equal, we obtain without loss of generality that $f_i = 1$ for all $i = 1, 2, ..., n$ (in accordance with the neutral theory of molecular evolution [24]). Further, considering single-nucleotide mutations only, which occur with fixed probability q, the sequence space can be viewed as an induced subgraph $G = (V(G), E(G))$ of the generalized hypercube

$$(K_4)^n.$$

The system (6.3) can be rewritten as

$$\frac{db_i(t)}{dt} = q \times \sum_{ji \in E(G)} b_j(t) - \psi(t) \, b_i(t), \quad i = 1, 2, ..., n, \qquad (6.4)$$

or in matrix form

$$\frac{d\boldsymbol{b}(t)}{dt} = A\boldsymbol{b}(t) - \psi(t) \, \boldsymbol{b}(t), \qquad (6.5)$$

where A is the adjacency matrix of G,

$$\psi(t) = q \sum_{i=1}^{n} d_i \, b_i(t)$$

and d_i is the degree of vertex i in G.

We will assume that the graph G is connected; that is, the matrix A is irreducible. Suppose also that A is an aperiodic matrix; the following conclusions hold true without this assumption but it simplifies the presentation. The substitution

$$g_i(t) = b_i(t) \, e^{\int_0^t \psi(t)dt}$$

transforms the system (6.5) into a linear system with constant coefficients:

$$\frac{d\boldsymbol{g}(t)}{dt} = A\boldsymbol{g}(t).$$

The general solution can be written in the form

$$\boldsymbol{g}(t) = \sum_{i=1}^{r} e^{\lambda_i t} \sum_{j=1}^{k_i} P_i^{j-1}(t) \, \boldsymbol{u}_i^j,$$

where

$$\lambda_1 > \lambda_2 > ... > \lambda_r$$

are different eigenvalues of the matrix A with multiplicities

$$k_1, k_2, ..., k_r.$$

Also,

$$(\boldsymbol{u}_i^1, \boldsymbol{u}_i^2, ..., \boldsymbol{u}_i^{k_i})$$

are Jordan chains of vectors corresponding to the eigenvalues λ_i, and P_i^{j-1} are polynomials of degree $(j-1)$, $i = 1, 2, ..., r$. Since A is a non-negative aperiodic irreducible matrix, the Perron–Frobenius Theorem implies that $k_1 = 1$, $\boldsymbol{u}_1^1$ is a non-negative eigenvector corresponding to the eigenvalue λ_1 and $|\lambda_i| < \lambda_1$ for every $i = 2, 3, ..., r$. Without loss of generality, we may assume that

$$\mathbf{1}^{\mathsf{T}}\boldsymbol{u}_1^1 = 1,$$

where $\mathbf{1}$ is the column vector consisting of ones. Since

$$\boldsymbol{b}(t) = \frac{1}{\sum_{s=1}^{n} g_s(t)} \boldsymbol{g}(t),$$

we have

$$\boldsymbol{b}(t) \to \boldsymbol{u}_1^1 \quad \text{as} \ t \to \infty.$$

Thus, the distribution of the frequencies of viral variants converges to the eigenvector corresponding to the largest eigenvalue of the graph G; this parameter is referred to as the vector of *eigenvalue centralities* of G. A similar conclusion was achieved in [37] using a different type of argument. Therefore, the state of the population of selectively neutral viral variants converges to the distribution that is completely determined by the graph-theoretic properties of the sequence space. This observation is the foundation of the 'survival of the flattest' phenomenon. According to this concept, the highest population size is achieved by the genomic variants with the highest centrality. Thus, under high mutation rates, natural selection favours variants that are robust against deleterious mutational effects.

There are currently no simple rules describing sequences without deleterious mutations, and the derivation of such rules in the near future is not warranted. With this limitation in mind, it is reasonable to assume for mathematical modelling that the sequence space is a random subgraph of the generalized hypercube

$$(K_4)^n,$$

where each vertex is chosen independently with fixed probability. Thus, the spectral properties of such random subgraphs are particularly interesting, and the connectivity and metric properties are important too.

Random subgraphs of the hypercube $(K_2)^n$ were studied in a number of papers [1, 2, 6, 16, 25, 26, 35]. All these studies evaluated the connectivity and metric properties of random subgraphs of the hypercube. The model of random subgraphs, where each edge is chosen with certain probability, was considered most frequently. The spectral properties of random edge-deleted subgraphs of the hypercube were studied in [35], and the estimation for the largest eigenvalue of the adjacency matrices of those subgraphs was obtained. However, random edge-deleted subgraphs of the hypercube, unlike vertex-deleted subgraphs, have a limited biological significance. Random vertex-deleted subgraphs of the hypercube are significantly less studied. This type of random subgraph was considered in [25, 26].

The most widely used graph invariant determined by its spectral properties is PageRank. The concept of PageRank was introduced by Brin and Page in [3] and it forms the basis for Google web search algorithms. For each web page, PageRank assigns a value that represents its 'importance'. The 'importance' of a page is determined by the total importance of web pages that refer to it. As will be shown below, PageRank can be determined as the stationary distribution in a particular case of the quasispecies model.

Formally, PageRank is defined as follows. Consider the web digraph

$$\mathcal{W} = (V(\mathcal{W}), E(\mathcal{W})),$$

where

$$V(\mathcal{W}) = \{1, 2, ..., n\}$$

is the set of web pages and $ij \in E(\mathcal{W})$ if and only if there exists a hyperlink from web page i to web page j. Let d_i^+ and d_i^- denote the out-degree and the in-degree of vertex i, respectively. The $n \times n$ edge-weight matrix $R = (r_{i,j})_{i,j=1}^n$ is defined as follows:

$$r_{i,j} = \begin{cases} \frac{1}{d_i^+} & \text{if } ij \in E(\mathcal{W}), \\ 0 & \text{if } ij \notin E(\mathcal{W}). \end{cases}$$

These weights are usually interpreted as the probabilities of moving between edges. Let $\alpha \in (0, 1)$ and

$$\boldsymbol{s} = (s_1, s_2, ..., s_n)^\mathrm{T},$$

where

$$s_i > 0 \text{ and } \sum_{i=1}^n s_i = 1.$$

The vector

$$\boldsymbol{p} = (p_1, p_2, ..., p_n)^\mathrm{T}$$

is the solution to the following system of linear equations:

$$\boldsymbol{p} = (1 - \alpha)\boldsymbol{s} + \alpha R \boldsymbol{p}. \tag{6.6}$$

This vector represents the *PageRanks of the web pages*. The term $(1 - \alpha)\boldsymbol{s}$ in (6.6) is used to avoid problems associated with the fact that the matrix R is not always simple and irreducible. The introduction of this term is equivalent to the addition of low-weight edges to $\mathcal{W}$ to make it strongly connected.

Let us show that PageRank is a special case of the quasispecies model. Indeed, let

$$G = (V(G), E(G))$$

be a digraph with the vertex set

$$V(G) = \{1, 2, ..., n\}.$$

Consider the quasispecies equations (6.1) with $f_i = 1$ for every $i = 1, 2, ..., n$ and

$$q_{j,i} = \alpha r_{j,i} + (1 - \alpha)s_i.$$

Then, the equation (6.2) takes the form

$$\frac{d\mathbf{c}(t)}{dt} = (Q^{\mathrm{T}} - I)\mathbf{c}(t).$$

It is a linear system of differential equations with constant coefficients. Therefore, the general solution can be written in the form

$$\mathbf{c}(t) = \sum_{i=1}^{r} e^{(\lambda_i - 1)t} \sum_{j=1}^{k_i} P_i^{j-1}(t)\, \mathbf{u}_i^j,$$

where

$$\lambda_1 > \lambda_2 > ... > \lambda_r$$

are different eigenvalues of the matrix Q with multiplicities $k_1, k_2, ..., k_r$. Also, $(\mathbf{u}_i^1, \mathbf{u}_i^2, ..., \mathbf{u}_i^{k_i})$ are Jordan chains of vectors corresponding to the eigenvalues λ_i, and P_i^{j-1} are polynomials of degree $(j-1)$, $i = 1, 2, ..., r$.

Because Q is a strictly positive stochastic matrix, we have $\lambda_1 = 1$, $k_1 = 1$, $\mathbf{u}_1^1$ is a non-negative eigenvector corresponding to the eigenvalue λ_1 and $\lambda_i - 1 < 0$ for every $i = 2, 3, ..., r$. Therefore,

$$\mathbf{c}(t) \to \mathbf{p} = C\mathbf{u}_1^1 \text{ as } t \to \infty,$$

where $C = P_1^0$. With the proper normalization, we can assume that $\sum_{j=1}^{n} p_j = 1$. Taking into account this property, it follows from the equation

$$\mathbf{p} = Q^{\mathrm{T}}\mathbf{p}$$

that $\mathbf{p}$ satisfies (6.6). Thus, $\mathbf{p}$ is the PageRank of G.

Both PageRank and quasispecies models can fundamentally be described as Markov processes. The powerful methods of the theory of random walks on graphs and spectral graph theory have been applied to study PageRank [9, 10]. Hence, the same methods of graph theory may be applicable to the problems of viral evolution.

6.2.2 Graphs and Immune Adaptation

The quasispecies model is a high-level model that basically describes the emergence of mutation–selection balance in viral populations, without detailed consideration of the nature of the selection pressures affecting the virus. The major evolutionary factor influencing viral evolution is, of course, the impact of the host's immune system. Despite extensive immunological studies, interactions between immune systems and viruses, such as HCV, are still not well understood. For example, for a very long time, continuous immune escape by genetic diversification was assumed to be the major mechanism allowing a virus to evade the host's immune system and establish a chronic infection. However, new biotechnologies have produced several observations that put this common knowledge into question. One such observation was the long-term survival of certain viral variants that existed in infected hosts for years. Thus, mathematical modelling was employed to find an alternative mechanism of viral adaptation to the host's immune system [33].

There are a number of mathematical models describing the interactions between viruses and the immune system. Graphs usually arise in the models that take into account cross-immunoreactivity—the phenomenon when antibodies specific to one antigen may recognize another antigen. These are called *cross-immunoreactive antigens*. We will describe one such model, as studied in [5, 33].

The model considers a population of n viral genomic variants x_i with different antigenic properties. This means that the variants induce n immune responses r_i. Thus, the value $x_i(t)$ is the number of viral particles with the i-th genome at time t, and $r_i(t)$ is the number of antibodies that specifically recognize viral particles with the i-th genome at time t.

Viral variants exhibit cross-immunoreactivity, which is reflected in the *cross-immunoreactivity network*—it is a directed weighted graph $G = (V, E)$ with vertices corresponding to viral variants and links connecting cross-immunoreactive variants. In immunology, it is known that activation of the immune response and neutralization of antigens by antibodies are different biochemical processes. The model takes this into account by assuming that every edge ij of the cross-immunoreactivity graph has two weights, $u_{i,j}^-$ and

$u_{i,j}^+$. The first weight represents the rate of neutralization of the j-th variant by the immune response r_i, and the second weight is the rate of activation of the response r_i by the j-th variant. We assume that $u_{i,i}^- = u_{i,i}^+ = 1$; in other words, the immune response r_i against the variant x_i is fully neutralizing. The weight functions are represented by immune neutralization and immune stimulation matrices

$$U^- = (u_{i,j}^-)_{i,j=1}^n \quad \text{and} \quad U^+ = (u_{i,j}^+)_{i,j=1}^n.$$

Furthermore, the model assumes that variants x_i replicate at rates $f_i > 0$ and are eliminated by immune responses r_j at rates

$$p\, u_{j,i}^- r_j,$$

where $p > 0$ is a constant. It is known that any antigen x_j preferentially stimulates already existing immune responses capable of binding to x_j. This feature is modelled by setting the rate of stimulation of the immune response r_i by the variant x_j as proportional to the weighted relative frequency of r_i calculated as

$$g_{j,i} = \frac{u_{j,i}^+ r_i}{\sum_{k=1}^n u_{j,k}^+ r_k}.$$

Finally, when the immune response r_i is not stimulated by antigens, it decays at rate b. All these assumptions are summarized in the following system of differential equations, which describes the mutual dynamics of the viral population and the immune system:

$$\dot{x}_i = f_i x_i - p\, x_i \sum_{j=1}^n u_{j,i}^- r_j, \quad i = 1, 2, ..., n, \tag{6.7}$$

$$\dot{r}_i = c \sum_{j=1}^n \frac{u_{j,i}^+ r_i}{\sum_{k=1}^n u_{j,k}^+ r_k} x_j - b\, r_i, \quad i = 1, 2, ..., n, \tag{6.8}$$

where $c > 0$ is a constant. Note that without cross-immunoreactivity, we have

$$U^- = U^+ = I,$$

where I is an identity matrix. In this case, the system (6.7)–(6.8) reduces to the simpler model known in the literature [31]. The system (6.7)–(6.8) is non-linear and rather complex, which makes its analytical solution rather impossible to find even for simple cross-immunoreactivity networks. Therefore, it can be studied either by finding the solutions numerically or by analysing equilibrium (or steady-state) solutions, for example solutions with $\dot{x}_i = \dot{r}_i = 0$, $i = 1, 2, ..., n$. The results obtained by these approaches are described below.

In what follows, we will focus on the effects of the topological structure of the cross-immunoreactivity network G. Thus, we consider a special case of the model (6.7)–(6.8), where all viral variants have the same replication rate f. Also, for every pair of cross-immunoreactive variants x_i and x_j, the corresponding immune stimulation and neutralization weights are equal to the constants α and β, respectively, $0 \leq \alpha, \beta \leq 1$. Considering that the neutralization of viruses usually requires binding of a viral particle by K antibody molecules, we can suppose that $\beta \leq \alpha$ or, more specifically, $\beta = \alpha^K$. In this case, immune neutralization and stimulation matrices are determined by the adjacency matrix of G:

$$U^- = I + \beta A^{\mathrm{T}}, \quad U^+ = I + \alpha A.$$

The major property of the system (6.7)–(6.8) is the emergence of the so-called mechanism of *viral adaptation by antigenic cooperation* and the phenomena of *local immunodeficiency* and *population transition via indirect variant interactions*. To better understand the aforementioned mechanism and phenomena, we will examine the 2-vertex and 3-vertex graphs shown in Figure 6.2.

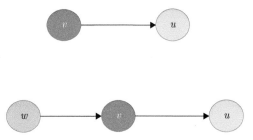

Fig. 6.2 Toy examples of cross-immunoreactivity networks with two and three variants.

Antigenic Cooperation and Local Immunodeficiency
Let us consider the 2-vertex graph shown in Figure 6.2 and assume that $f = f_u = f_v > 0$. For this example, the system (6.7)–(6.8) takes the following form:

$$\dot{x}_u = f x_u - p\, x_u r_u, \tag{6.9}$$

$$\dot{x}_v = f x_v - p\, x_v (\beta r_u + r_v), \tag{6.10}$$

$$\dot{r}_u = c x_u + c \frac{\alpha r_u}{\alpha r_u + r_v} x_v - b r_u, \tag{6.11}$$

$$\dot{r}_v = c \frac{r_v}{\alpha r_u + r_v} x_v - b r_v. \tag{6.12}$$

Without cross-reactivity (i.e. $\alpha = \beta = 0$), the system converges to the equilibrium solution

$$x_u^\circ = \frac{bf}{cp}, \quad x_v^\circ = \frac{bf}{cp}, \quad r_u^\circ = \frac{f}{p}, \quad r_v^\circ = \frac{f}{p}. \tag{6.13}$$

We will use this solution as a benchmark for comparison with other solutions. It will help us to understand better how the behaviour of the population changes when cross-immunoreactivity is involved. Let us assume that

$$0 < \beta \le \alpha \le 1.$$

The Jacobian of the system is given by

$$J(x_u, x_v, r_u, r_v)$$

$$= \begin{pmatrix}
f - p r_u & 0 & -p x_u & 0 \\
0 & f - p(\beta r_u + r_v) & -p\beta x_v & -p x_v \\
c & \frac{c\alpha r_u}{\alpha r_u + r_v} & \frac{c\alpha r_v x_v}{(\alpha r_u + r_v)^2} - b & -\frac{c\alpha r_u x_v}{(\alpha r_u + r_v)^2} \\
0 & \frac{c r_v}{\alpha r_u + r_v} & \frac{-c\alpha r_v x_v}{(\alpha r_u + r_v)^2} & \frac{c\alpha r_u x_v}{(\alpha r_u + r_v)^2} - b
\end{pmatrix}$$

Recall that the solution of the system of differential equations is asymptotically stable if the real parts $\Re(\lambda)$ of all eigenvalues λ of its Jacobian for this solution are negative; and unstable if some eigenvalue has a positive real part. The system (6.9)–(6.12) has several stationary solutions, and we will consider them in turn.

Solution 1: The most illustrative is the solution

$$x_u^* = \frac{b\,(1-\alpha)f}{c} \frac{f}{p}, \quad x_v^* = \frac{b\,(\alpha - \beta)f + f}{c} \frac{}{p}, \quad r_u^* = \frac{f}{p}, \quad r_v^* = \frac{f - \beta f}{p}. \tag{6.14}$$

We have

$$x_v^* - x_v^\circ = \frac{b(\alpha - \beta)}{cp},$$

hence v converges to the higher population size than in solution (6.13), whereas the equilibrium size of its specific immune response r_v^* is lower:

$$r_v^* - r_v^\circ = -\frac{\beta f}{p}.$$

This is what we call the phenomenon of *local immunodeficiency*—the reduction in the specific immune responses against v, whereas the response against u stays the same in both solutions. However, the beneficial effect of cross-immunoreactivity for the variant v is accompanied by its detrimental effect on the variant u, whose equilibrium population size declines with respect to solution (6.13). Thus, the model suggests differential roles for the variants u and v during the viral evolution—this mechanism is called *antigenic cooperation* between u and v. The variant v acts selfishly and stimulates immune responses against the variant u. This diminishes the capability of the immune responses to produce the v-specific antibodies and allows the variant v to survive. In contrast, the variant u acts altruistically by maintaining its own specific immune response on a level sufficiently high to allow the variant v to survive, even by decreasing its own survival chances. Thus, the viral population as a whole exhibits quasi-social behaviour by the mechanism that can be called *immune escape by antigenic cooperation*.

Proposition 6.1 *The solution (6.14) is stable.*

Proof: To calculate the characteristic polynomial $P(\lambda)$ of the Jacobian $J(x_u^*, x_v^*, r_u^*, r_v^*)$, we will use the following formula [19]:

$$P(\lambda) = \lambda^4 - S_1\lambda^3 + S_2\lambda^2 - S_3\lambda + S_4,$$

where S_i is the sum of all principal i-minors of $J(x_u^*, x_v^*, r_u^*, r_v^*)$. We obtain

$$S_1 = b\frac{2(\beta - 1) - \alpha\beta}{\alpha - \beta + 1},$$

$$S_2 = bf(2 + \alpha\beta - \alpha - \beta) + b^2\frac{(1 - \alpha)(1 - \beta)}{\alpha - \beta + 1},$$

$$S_3 = -b^2 f\frac{(1 - \alpha)(1 - \beta)(2 + \alpha - \beta)}{\alpha - \beta + 1},$$

$$S_4 = b^2 f^2(1 - \alpha)(1 - \beta).$$

By the Routh–Hurwitz criterion [19], the solution (6.14) is stable if the following conditions hold:

$$S_2, S_4 > 0, \quad S_1, S_3 < 0,$$
$$\Delta_1 = -S_1 S_2 + S_3 > 0,$$
$$\Delta_2 = S_1 S_2 S_3 - S_1^2 S_4 - S_3^2 > 0.$$

For Δ_1, we have

$$\Delta_1 = \frac{b^2 f}{\alpha - \beta + 1} g_1(\alpha, \beta) + g_2(\alpha, \beta),$$

where

$$g_1(\alpha, \beta) = \alpha^2 \beta^2 + (1 - \beta)(2\alpha\beta + 2 - \beta) - \alpha(1 - \alpha)(1 - 2\beta)$$

and

$$g_2(\alpha, \beta) = b^3 \frac{(\alpha\beta + 2(1 - \beta))(1 - \alpha)(1 - \beta)}{(\alpha - \beta + 1)^2}.$$

It is obvious that $g_2(\alpha, \beta) \geq 0$, so it remains to show that $g_1(\alpha, \beta) > 0$. We have

$$g_1(\alpha, \beta) \geq (1 - \beta)(2\alpha\beta + 2 - \beta) - \alpha(1 - \alpha)(1 - \beta) \quad (6.15)$$
$$= (1 - \beta)(2\alpha\beta + 2 - \beta - \alpha(1 - \alpha)) \geq 0. \quad (6.16)$$

The inequality (6.16) achieves equality if and only if $\beta = 1$. Therefore, in this case $g_1(\alpha, \beta) = \alpha$. Now, the assumption $\alpha \geq \beta$ implies $g_1(\alpha, \beta) > 0$, and hence $\Delta_1 > 0$.

In a similar way, we can show that $\Delta_2 > 0$. This is left to the reader as an exercise. Thus, the solution (6.14) is stable. ∎

In addition to the solution (6.14), the system (6.9)–(6.12) has several other equilibrium solutions. The majority of them are unstable, but the few stable solutions also have biological interpretations. Below we list those solutions and prove their stability or instability.

Solution 2:

$$x_u^* = 0, \quad x_v^* = \frac{b}{c}\left(\frac{f}{p} + (1 - \beta)h\right), \quad r_u^* = h, \quad r_v^* = \frac{f}{p} - \beta h.$$

This family of solutions is parameterized by $h > 0$, which depends on the initial conditions and the parameters of (6.9)–(6.12). These solutions exist only when $\alpha = 1$; that is, cross-immunoreactivity between the v-specific antibodies and the variant u is so strong that these antibodies are fully neutralizing. In this case, the stronger form of antigenic cooperation emerges. In the evolutionary scenario leading to these solutions, the variant u is completely eliminated; however, with the same initial conditions, the population of v achieves a higher equilibrium level with respect to solution (6.14).

If $\beta > 0$ then $h \leq f/(\beta p)$, and hence the size of the equilibrium population of the variant v is bounded by

$$\frac{b}{c}\frac{f}{\beta p}.$$

This means that the immune system is able to control the growth of the variant v at some predefined level. For $\beta = 0$, the equilibrium solution has the following form:

$$x_u^* = 0, \quad x_v^* = \frac{b}{c}\left(\frac{f}{p} + h\right), \quad r_u^* = h, \quad r_v^* = \frac{f}{p}.$$

In this case, the equilibrium population size x_v^* for the variant v can be arbitrarily high.

Simple calculations demonstrate that the Jacobian $J(x_u^*, x_v^*, r_u^*, r_v^*)$ has the following eigenvalues:

$$\lambda_1 = f - ph,$$
$$\lambda_2 = \frac{1}{2}\left(-b - \sqrt{b^2 - 2cp\beta x_v - 2p(1-\beta)br_v}\right),$$
$$\lambda_3 = \frac{1}{2}\left(-b + \sqrt{b^2 - 2cp\beta x_v - 2p(1-\beta)br_v}\right),$$
$$\lambda_4 = 0.$$

It follows that the solution is unstable if $h < f/p$. Therefore, for the solution to be stable, the concentration of the u-specific antibodies should be supported at a sufficiently high level. Since the variant u dies out, this task should be carried out by the variant v.

Solution 3:

$$x_u^{\bullet} = \frac{b}{c}\frac{f}{p}, \quad x_v^{\bullet} = 0, \quad r_u^{\bullet} = \frac{f}{p}, \quad r_v^{\bullet} = 0.$$

This solution describes a situation where the mechanism of antigenic cooperation is not involved, and the variant u persists, whereas v is eliminated. Although the model allows for such a situation, this solution turns out to be unstable. Indeed, the Jacobian $J(x_u^\bullet, x_v^\bullet, r_u^\bullet, r_v^\bullet)$ has the following eigenvalues:

$$\lambda_1 = \frac{1}{2}\left(-b - \sqrt{b^2 - 4bf}\right),$$
$$\lambda_2 = \frac{1}{2}\left(-b + \sqrt{b^2 - 4bf}\right),$$
$$\lambda_3 = -b,$$
$$\lambda_4 = f(1 - \beta).$$

The instability of this solution for $\beta < 1$ follows from the fact that $\lambda_4 > 0$.

Solution 4:

$$x_u^\circ = 0, \quad x_v^\circ = \frac{b\,f}{c\,p}, \quad r_u^\circ = 0, \quad r_v^\circ = \frac{f}{p}.$$

We can show that Solution 4 is unstable. This is left to the reader as an exercise.

Solution 5:

$$x_u^\circ = h, \quad x_v^\circ = \frac{b\,f}{c\,p} - h, \quad r_u^\circ = \frac{f}{p}, \quad r_v^\circ = 0, \tag{6.17}$$

where h is a parameter depending on the initial conditions and the parameters of (6.9)–(6.12), and

$$0 < h < \frac{b\,f}{c\,p}.$$

The solution (6.17) exists only if $\beta = 1$. It describes the outcome when the populations of both variants u and v can be controlled at the equilibrium level solely by the u-specific immune response. This is due to its ability to neutralize the variant v just as efficiently as the v-specific immune response.

The eigenvalues of $J(x_u^\circ, x_v^\circ, r_u^\circ, r_v^\circ)$ are as follows:

$$\lambda_1 = 0,$$
$$\lambda_2 = \frac{1}{2}\left(-b - \sqrt{b^2 - 4(\beta bf + (1 - \beta)phc)}\right),$$
$$\lambda_3 = \frac{1}{2}\left(-b + \sqrt{b^2 - 4(\beta bf + (1 - \beta)phc)}\right),$$
$$\lambda_4 = \frac{(1 - \alpha)bf - cph}{\alpha f}.$$

We have

$$\Re(\lambda_2) < 0 \quad \text{and} \quad \Re(\lambda_3) < 0.$$

Thus, the solution is unstable if

$$h < (1 - \alpha)\frac{b}{c}\frac{f}{p}.$$

Solution 6:

$$x_u^\diamond = 0, \quad x_v^\diamond = \frac{b}{c}\frac{f}{\beta p}, \quad r_u^\diamond = \frac{f}{\beta p}, \quad r_v^\diamond = 0,$$

where $\beta > 0$. The Jacobian $J(x_u^\diamond, x_v^\diamond, r_u^\diamond, r_v^\diamond)$ has the following eigenvalues:

$$\lambda_1 = f - f/\beta,$$
$$\lambda_2 = b/\alpha - b,$$
$$\lambda_3 = \frac{1}{2}\left(-b - \sqrt{b^2 - 4bf}\right),$$
$$\lambda_4 = \frac{1}{2}\left(-b + \sqrt{b^2 - 4bf}\right).$$

This solution is unstable for $\alpha < 1$ because the eigenvalue λ_2 is strictly positive.

Population Transition via Indirect Variant Interactions

The antigenic cooperation is just one example of the result of interactions between viral variants in a cross-immunoreactivity network. Such interactions occur between adjacent vertices in the network. In general, the interactions are complex, and the viral variants can influence other variants even if they are not adjacent. The model suggests that a new vertex joining the network can affect even distant variants and either trigger their extinction or, in contrast, boost their frequencies. Such effects are called *population transition via indirect variant interactions*.

To see how these effects emerge from the properties of the system (6.9)–(6.12), let us consider its equilibrium solutions x^* and r^*. The following sets will be needed:

$$\mathcal{I} = \left\{i \in \{1, 2, ..., n\} : r_i^* + \beta \sum_{ij \in E} r_j^* \neq f/p\right\},$$
$$\bar{\mathcal{I}} = \{1, 2, ..., n\} - \mathcal{I},$$
$$\mathcal{J} = \left\{i \in \{1, 2, ..., n\} : r_i^* = 0\right\},$$
$$\bar{\mathcal{J}} = \{1, 2, ..., n\} - \mathcal{J}.$$

Clearly, for every $i \in \mathcal{I}$ it is true that $x_i^* = 0$. Furthermore, the equilibrium solutions r^* and x^* satisfy the following system of equations:

$$r_i^* + \beta \sum_{ij \in E} r_j^* = \frac{f}{p}, \quad i \in \bar{\mathcal{I}}, \tag{6.18}$$

$$\delta_i x_i^* + \alpha \sum_{\substack{ji \in E \\ j \in \bar{\mathcal{I}}}} \delta_j x_j^* = \frac{b}{c}, \quad i \in \bar{\mathcal{J}}, \tag{6.19}$$

where

$$\delta_i = \left(r_i^* + \alpha \sum_{ik \in E} r_k^* \right)^{-1}.$$

Let us take a closer look at the constraints imposed by this system. According to the sub-system (6.19), for each vertex $i \in V$, the weighted sum of the equilibrium populations of its neighbours and the vertex itself is fixed and should be equal to the constant $\frac{b}{c}$. Thus, adding a new vertex or a new edge to the network may cause an increase or decline in population size for some variants from the same closed neighbourhood. This, in turn, may change the population sizes of the neighbours of the affected vertices and so on. Consequently, the changes propagate through the cross-immunoreactivity network to other closed neighbourhoods. Thus, the system (6.18)–(6.19) suggests the inter-dependency between viral variants that belong to the same connected component of the cross-immunoreactivity network. This inter-dependency is determined by the whole spectra of paths in the network and reveals how the survival of viral variants depends on the global structure of the cross-immunoreactivity network rather than just the properties of their neighbours.

As before, these effects can be more easily understood when small graph examples are examined. Let us consider two cross-immunoreactivity networks shown in Figure 6.2. Let us also assume for simplicity that $\alpha = 1$. Then, for both networks, there exist equilibrium solutions with $\mathcal{J} = \emptyset$. In this case, the system (6.19) has the following forms for the two graphs, respectively:

$$\delta_u x_u^* + \delta_v x_v^* = \frac{b}{c},$$

$$\delta_v x_v^* = \frac{b}{c},$$

and

$$\delta_u x_u^* + \delta_v x_v^* = \frac{b}{c},$$

$$\delta_v x_v^* + \delta_w x_w^* = \frac{b}{c},$$

$$\delta_w x_w^* = \frac{b}{c}.$$

For the first cross-immunoreactivity network, the solution is

$$x_u^* = 0, \quad x_v^* = \frac{b}{c\delta_v} > 0.$$

Thus, the variant u goes extinct, whereas the variant v survives. For the second network, the situation is the opposite. We have

$$x_u^* = \frac{b}{c\delta_u}, \quad x_v^* = 0, \quad x_w^* = \frac{b}{c\delta_w}.$$

Therefore, the variant u survives, but the variant v is eliminated. This change is mediated by the addition of the variant w to the network, such that w is able to stimulate immune pressure on the variant v, thus relieving the pressure on the variant u that was stimulated by v. Therefore, the survival of u is due to the effect of the variant w—this phenomenon is called an *enhancement* of u by w. It can be viewed as another form of cooperation among viral variants, in this case between w as a 'benefactor' and u as a 'beneficiary'. However, unlike the antigenic cooperation described above, the interaction between w and u is indirect and is being mediated by the variant v.

Numerical Analysis of Large Networks

To study the properties of the system (6.7)–(6.8) for large networks, we have to rely on numerical simulations. In particular, such analysis was performed in [33] under the assumption that the cross-immunoreactivity network is scale free. Such an assumption is biologically relevant because the scale-free properties of the HCV cross-immunoreactivity network were discovered in a prior experimental study [7]. The model was also discretized by assuming that a viral variant is completely eliminated by the immune system if its population reaches a sufficiently low level. This assumption makes the outcomes of numerical simulations more realistic because in real life, a population that becomes too small is eliminated and cannot rebound to a higher level at a later time.

The numerical results confirm and reinforce the conclusions that were reached using more rigorous reasoning for small networks. If the cross-immunoreactivity network is an empty graph O_n, all viral variants are eliminated by the immune system. Therefore, the only way for a virus to persist is by constantly generating new variants via mutations. When the cross-immunoreactivity network is not empty, the evolutionary dynamics of the viral population fundamentally change.

The major findings obtained by the model numerical solution for 360 randomly generated scale-free networks are summarized below. The model described by equations (6.7)–(6.8) was solved with the following parameters, which have the same values as in [30]:

$$\alpha = 0.5, \quad \beta = \alpha^K, \quad K \in \{2, 5, 10\}, \quad f = 2.5, \quad p = 2, \quad c = 0.1, \quad b = 0.1.$$

The cross-immunoreactivity networks were generated as random scale-free networks with power-law exponent $\gamma = 1.5$; this value was experimentally obtained in [7]. Viral variants and immune responses are considered to have been eliminated when their population sizes become smaller than the initial conditions of the system (6.7)–(6.8). The numbers reported below are the average values over all simulations.

- In the model settings, the immune system still manages to eliminate the majority of viral variants. However, ca. 11% of variants survive and persist at the system equilibrium.
- The persistent variants exist in a state of local immunodeficiency. Indeed, the specific immune responses against the persistent variants comprise only ca. 0.3% of the overall equilibrium immune response.
- There exists a distinct group of variants with specific properties—these variants are called *altruists* (see Figure 6.1). The altruistic variants themselves are eliminated by the immune system, but their specific immune responses persist throughout the infection because they are stimulated by other variants. Despite the presence of antibodies against altruists, they cannot efficiently neutralize persistent variants. Although the altruists constitute only ca. 1.2% of the total number of viral variants, immune responses against them are dominant at equilibrium and constitute ca. 99.7%. This impedes the production of antibodies that are efficient against persistent variants. Interestingly, the existence of altruists can theoretically be deduced from the presence of local immunodefficiency. This is left to the reader as an exercise at the end of the chapter.

- Altruistic variants are in-hubs; that is, vertices with high in-degree. The average in-degree of altruistic variants was found to be at least 12 times greater than the average in-degree of all other variants.

This section has demonstrated how the analysis of a graph-based differential equation model can reveal novel properties of biological systems. The analysis suggests that even such simple pre-organismic life forms as viruses can develop complex behaviour when they are organized into a complex network.

6.3 Graphs and Outbreak Analyses

In this section, we will move from more theoretical modelling problems described in the previous sections to more application-oriented algorithmic challenges. They arise when researchers use viral genomes to investigate outbreaks and reconstruct the history of virus transmissions.

6.3.1 Some Algorithmic Challenges in Epidemiology

One of the goals of computational molecular epidemiology is to design computational tools that facilitate outbreak investigations and help us to understand the spread of diseases in susceptible populations. The basic algorithmic challenge of computational molecular epidemiology can informally be defined in the following way. Suppose that viral genomes sampled from n infected patients are given.

Problem A *Find transmission clusters, such as groups of patients that are related by direct or indirect chains of virus transmissions.*

Problem B *Infer a transmission network for each transmission cluster.*

Problem A can be tackled significantly more easily, and the approaches to its solution are usually based on the simple assumption that viral genomes sampled from epidemiologically related patients should be significantly closer to one another than the genomes sampled from unrelated patients. Thus, this problem can be solved by first calculating the pairwise distances between viral genomes or intra-host viral populations using the variety of population genetics measures. Then, it is necessary to link the patients with the viral genomes that are at a distance below a particular threshold [8, 38]. The threshold value is usually established using the available training genomic datasets from patients, whose epidemiological relationships are confirmed by epidemiological investigations. The obtained (undirected) *genetic relatedness graph* $\mathcal{G}_r$ may serve as a good approximation for the host network (see Section 6.1), and connected components of $\mathcal{G}_r$ can be identified by transmission clusters.

Problem B is significantly more complicated, so the rest of this chapter is devoted to this problem. It is reasonable to assume that each individual was infected only once. Exceptions to this assumption are possible, but they are usually rare and are, therefore, not considered here. Under this assumption, the transmission network $\mathcal{G}_t$ is a spanning tree (or forest) of the genetic relatedness graph $\mathcal{G}_r$. For simplicity of presentation, we assume that the graph $\mathcal{G}_r$ is connected. We may also assume that the edges $ij \in E(\mathcal{G}_r)$ have weights $w_{i,j}$ representing the corresponding distances. From this, it is quite natural to infer that the transmission network is a minimum spanning tree of $\mathcal{G}_r$. Indeed, one of the first computational tools for outbreak investigation using genomic data was based on this idea [22].

Furthermore, in some situations, we may make reasonable guesses about the possible transmission direction. For example, the fact that patient i was diagnosed earlier than patient j may be indicative of the possibility that, in a case of direct transmission between them, i was the source. Alternatively, higher genetic heterogeneity of the viral population inside patient i in comparison to the population from patient j can also serve as an indicator that i was infected earlier. In this case, $\mathcal{G}_r$ can be considered as a directed graph, and $\mathcal{G}_t$ can be inferred as a minimum spanning arborescence of $\mathcal{G}_r$.

Unfortunately, this simple approach often results in insufficiently accurate solutions, especially when applied to emerging outbreaks of highly heterogeneous viruses such as HIV and HCV. Let us consider, for example, three patients i, j and k such that both i and j were recently infected by k. It was experimentally observed that the viral genomes sampled from i and j can be closer to each other than to the population of k because the latter is usually significantly more heterogeneous and, besides the transmitted viral variant, can contain many variants that are not transmissible due to their inherent biological features [34]. As a result, the minimum spanning tree would contain the edge ij rather than the edges ki and kj.

To resolve this issue, it is necessary to redefine the objective function for the sought-for spanning tree of the relatedness graph $\mathcal{G}_r$. The new objective may be suggested by the epidemiology. It is known that transmission modes of HIV, HCV and a number of other viruses are associated with socio-behavioural risk factors such as drug use or unsafe sexual contact. This means that those viruses are essentially transmitted over the social network of interactions between susceptible individuals. Social networks have been extensively studied for a long time and are known to have particular properties, such as power law degree distribution, presence of hubs (high-degree vertices) and a small diameter [29]. The most common model for such networks is the so-called *scale-free* network, as discussed in Chapter 1. It may be argued that transmission networks mirror the properties of the underlying social networks and, therefore, they should be scale free. This assumption also has an experimental support [4, 38]. The above observation brings to life the following informally defined *scale-free spanning tree problem*: for a given graph $\mathcal{G}_r$, find its most 'scale-free-like' spanning tree.

To measure the 'scale-freeness' of a tree, we may use an idea proposed in [28]. It is based on a graph parameter called *s-metric*:

$$s(T) = \sum_{ij \in E(T)} d_i d_j,$$

where d_i is the (undirected) degree of a vertex i in the tree T. In the statistical ensemble $\mathbb{G}(\underline{d})$ of random graphs with the same expected degree sequence $\underline{d}$, a high s-metric indicates the presence of most of the typical properties of scale-free networks [28]. Moreover, if s^* is the maximal s-metric for graphs from $\mathbb{G}(d)$, then the value

$$\frac{s(\mathcal{G})}{s^*}$$

is proportional to the relative log-likelihood of the graph $\mathcal{G}$ under the generalized random graph model [28]. Thus, we have the following algorithmic problem:

s-SF Spanning Tree Problem
Given: A connected graph G.
Find: A spanning tree T of G such that $s(T)$ is maximal.

Unlike the minimum spanning tree and minimum arborescence problems, this problem is computationally hard:

Theorem 6.1 [32] *The s-SF Spanning Tree Problem is NP-hard.*

Proof: Let us consider the following problem, which is NP-complete [27]:

SPANNING TREE WITHOUT 2-VERTICES PROBLEM
GIVEN: A connected cubic graph G.
QUESTION: Is there a spanning tree of G without vertices of degree 2?

We will reduce this problem to the s-SF SPANNING TREE PROBLEM.

Let G be an instance of the SPANNING TREE WITHOUT 2-VERTICES PROBLEM, and let T be its spanning tree. If n_k denotes the number of vertices of degree k in T, then

$$n_1 + n_2 + n_3 = n.$$

By the handshaking lemma, we have

$$n_1 + 2n_2 + 3n_3 = 2(n - 1).$$

Deriving n_2 and n_3 from the above equalities, we obtain

$$n_2 = n + 2 - 2n_1, \quad n_3 = n_1 - 2. \tag{6.20}$$

Hence, if n is odd then n_2 is odd, and therefore G does not have a spanning tree without vertices of degree 2. Thus, it remains to consider the case when n is even and n_2 is also even. We will show that among all n-vertex trees T ($n \geq 4$ is even) with maximum degree $\Delta(T) \leq 3$, the trees without vertices of degree 2 have the highest s-metric. Indeed, the following claim holds:

Claim 6.1 *If $\Delta(T) \leq 3$ and $n \geq 4$ is even, then $s(T) \leq 6n - 15$. The equality holds if and only if T has no vertices of degree 2.*

Proof: If T has no vertices of degree 2, then $s(T) = 6n - 15$. The proof of this statement is left to the reader as an exercise.

Now suppose that T has $n_2 \geq 2$ vertices of degree 2. Let u and v be two vertices of degree 2. Denote the neighbourhoods of u and v by $N_T(u) = \{u_1, u_2\}$ and $N_T(v) = \{v_1, v_2\}$. We may assume that u_1 and v_1 belong to the path between u and v in T; hence v is not adjacent to u_2 (i.e. $v_2 \neq u_2$). Note that u_1 may coincide with v_1, or u may be adjacent to v (i.e. $u_1 = v$ and $v_1 = u$). Let T' be the tree obtained from T by removing the edge uu_2 and adding the edge vu_2. We have

$$\begin{aligned}
\Delta s &= s(T') - s(T) \\
&= 3\deg(u_2) + \deg(v_1) + \deg(v_2) - 2\deg(u_2) - \deg(u_1) \\
&= \deg(u_2) + \deg(v_1) + \deg(v_2) - \deg(u_1).
\end{aligned}$$

Because v is connected to u through v_1 in T, we have $\deg(v_1) \geq 2$. Hence $\deg(u_2) + \deg(v_1) + \deg(v_2) \geq 4$ and $\Delta s > 0$. By iteratively repeating this procedure for all pairs of vertices of degree 2, we will obtain a tree with higher s-metric and without vertices of degree 2. This proves the claim. ∎

According to Claim 6.1, the graph G has a spanning tree without vertices of degree 2 if and only if it has a spanning tree T with $s(T) = 6n - 15$. The proof of Theorem 6.1 is complete. ∎

6.3.2 Inference of Transmission Trees and Networks

Theorem 6.1 from the previous section implies that the inference of the transmission networks based on the scale-freeness criterion is computationally hard. Furthermore, completely ignoring genetic distances and relying only on the scale-freeness criterion may produce sub-optimal inference results. On the other hand, completely ignoring the structural objective and relying only on genetic distances may result in the poor performance of the minimum-spanning-tree-based and minimum-arborescence-based approaches. Thus, it is desirable to use a synthetic approach that combines both objectives. This was implemented in the tool called QUENTIN [34], which is based on graphs/networks and models the viral evolution and the epidemic spread using the following graphs:

- A *genetic network* $\mathcal{G}_g$ is an undirected graph with vertices corresponding to viral genomes; two genomes are connected by an edge if they differ by a single mutation.
- A connected *host network* $\mathcal{G}_h$ is a tournament whose vertices are infected hosts and links represent possible transmission directions. The link weights in the vector

$$\Omega = (\omega_e)_{e \in E(\mathcal{G}_h)}$$

are equal to genetic distances between corresponding viral populations. An example of a host network is shown in Figure 6.3a.
- A *transmission tree* $\mathcal{T}$ represents the transmission history of an outbreak. It is a rooted binary labelled tree, whose leaves are infected hosts and interior nodes are transmission events. An interior node labelled z and its children labelled z and y signify that host y was infected by host z. For a given transmission tree $\mathcal{T}$, the *transmission network* $\mathcal{G}_\mathcal{T}$ indicates who infected whom. More precisely, the vertices of $\mathcal{G}_\mathcal{T}$ are infected hosts, and

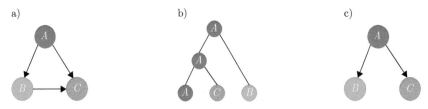

Fig. 6.3 a) Host network $\mathcal{G}_h$; b) transmission tree $\mathcal{T}$; c) transmission network $\mathcal{G}_\mathcal{T}$.

directed edges connect hosts that are linked by transmission. Examples of a transmission tree and a transmission network are depicted in Figures 6.3b and 6.3c.

The algorithm described below infers a transmission tree from the host network $\mathcal{G}_h$ and prior information about the expected structure of the transmission network. The objective is to find the transmission tree $\mathcal{T}$ that maximizes the probability

$$\mathbb{P}[\mathcal{T} \mid \mathcal{G}_h, \Omega]$$

of observing $\mathcal{T}$ given $\mathcal{G}_h$ and Ω. This probability is estimated in a Bayesian fashion as follows:

$$\mathbb{P}[\mathcal{T} \mid \mathcal{G}_h, \Omega] \propto \mathbb{P}[\Omega \mid \mathcal{G}_h, \mathcal{T}] \times \mathbb{P}[\mathcal{T} \mid \mathcal{G}_h], \qquad (6.21)$$

where $\mathbb{P}[\Omega \mid \mathcal{G}_h, \mathcal{T}]$ is the likelihood of the distances Ω given the tree $\mathcal{T}$ and given the possible transmission directions represented by $\mathcal{G}_h$. Also, $\mathbb{P}[\mathcal{T} \mid \mathcal{G}_h]$ is a prior probability of $\mathcal{T}$ given $\mathcal{G}_h$.

Likelihood of Estimated Genetic Distances

The likelihood $\mathbb{P}[\Omega \mid \mathcal{G}_h, \mathcal{T}]$ is estimated by assessing how the distances Ω correlate with the topology of the tree $\mathcal{T}$. We use the least-squares approach [18] under the assumption that the differences between collection times of viral samples are small in comparison with the time elapsed from transmission events.

Let us define the variables x_e as weights of the edges in $\mathcal{T}$ measured in viral generations. Also, let $\bar{x}_{i,j} = \sum_{e \in P_{i,j}} x_e$ be the weight of the path $P_{i,j}$ in $\mathcal{T}$ between the leaves corresponding to the hosts i and j. Note that $\bar{x}_{i,j}$ is well defined for all $ij \in E(\mathcal{G}_h)$ because the leaves of $\mathcal{T}$ represent all the nodes of $\mathcal{G}_h$.

We first solve the following constrained linear least-squares problem:

$$\min_{ij \in E(\mathcal{G}_h)} \sum (\bar{x}_{i,j} - \omega_{i,j})^2 / \omega_{i,j}^2 \tag{6.22}$$

subject to

$$\sum_{e \in P_{i,j}} x_e = \bar{x}_{i,j}, \quad i \neq j, \quad i, j = 1, 2, ..., n \tag{6.23}$$

$$\sum_{e \in P_i} \alpha_e x_e = \sum_{e \in P_{i+1}} \alpha_e x_e, \quad i = 1, 2, ..., n - 1 \tag{6.24}$$

$$x_e \geq 0, \quad e \in E(\mathcal{T}) \tag{6.25}$$

In this problem, the path P_i connects the root of $\mathcal{T}$ and the leaf corresponding to the host i. The terms $\sum_{e \in P_i} \alpha_e x_e$ represent the physical times of sample collections, where the coefficients α_e are used for the adjustment of evolutionary time (in generations) and physical time. In our model, these coefficients reflect the difference in the speed of viral evolution between the donor and recipient intra-host populations following the transmission event represented by the parent vertex of e. The speed of evolution for a recipient is generally higher due to the bottleneck effect and the initial immune response absence [11]. Therefore, assuming that e connects nodes corresponding to the hosts i and j, we set $\alpha_e = 1$ if $i = j$ and $\alpha_e = \alpha \in (0, 1]$ otherwise (α is a constant).

Now, the likelihood $\mathbb{P}[\Omega \mid \mathcal{G}_h, \mathcal{T}]$ is estimated as follows:

$$\mathbb{P}[\Omega \mid \mathcal{G}_h, \mathcal{T}] = \max\{r(\Omega, \bar{\mathcal{X}}), 0\},$$

where $r(\Omega, \bar{\mathcal{X}})$ is the Pearson correlation coefficient between $(\omega_{i,j})_{ij \in E(\mathcal{G}_h)}$ and $(\bar{x}_{i,j})_{ij \in E(\mathcal{G}_h)}$, which are consistently represented as vectors. The approach where a positive Pearson correlation coefficient is interpreted as a probability has been suggested in other studies (e.g. see [17]).

Prior Probability of the Transmission Tree

Let $\mathcal{T}$ be a transmission tree and $\mathcal{G}_\mathcal{T}$ be the corresponding transmission network, which can easily be constructed by adding a link $zy \in E(\mathcal{G}_\mathcal{T})$ for every internal node of $\mathcal{T}$ with label z and two children with labels z and y. The tree $\mathcal{T}$ *agrees* with the host network $\mathcal{G}_h$ if $\mathcal{G}_\mathcal{T}$ is a subgraph of $\mathcal{G}_h$. If $\mathcal{T}$ does not agree with $\mathcal{G}_h$, then

$$\mathbb{P}[\mathcal{T} \mid \mathcal{G}_h] = 0.$$

Otherwise, for an estimation of the prior probability $\mathbb{P}[\mathcal{T} \mid \mathcal{G}_h]$, we use the fact that the typical RNA virus transmission networks are social networks, which are usually scale free [4, 38]. We assume that transmission trees, which agree with $\mathcal{G}_h$, are distributed in such a way that the trees corresponding to the scale-free transmission networks have higher observed probabilities.

To measure the 'scale-freeness' of the transmission network $\mathcal{G}_\mathcal{T}$, we use the s-metric, the graph parameter discussed above:

$$s(\mathcal{G}_\mathcal{T}) = \sum_{ij \in E(\mathcal{G}_\mathcal{T})} d_i d_j,$$

where d_i is the (undirected) degree of a vertex i. Because we have no prior knowledge about degree sequences of real transmission networks, we cannot use the aforementioned statistical ensemble $\mathbb{G}(d)$ of random graphs with the same expected degree sequence $\underline{d}$. Instead, we use a wider statistical ensemble $\mathbb{H}_k$ consisting of graphs with the expected number of hubs (high-degree vertices) equal to k. Inside $\mathbb{H}_k$, the maximal s-metric can be calculated as follows:

$$s^*(k) = \left(\left\lfloor \frac{n}{k} \right\rfloor + k - 2 \right)(n-1) + (k-1)\left(\left\lfloor \frac{n}{k} \right\rfloor - 1 \right)\left\lfloor \frac{n}{k} \right\rfloor. \qquad (6.26)$$

The reader is encouraged to construct the graph H_k for which this bound is achieved.

Let c_1 and c_2 be constants. Based on the above considerations, the prior probability $\mathbb{P}[\mathcal{T} \mid \mathcal{G}_h]$ is estimated as

$$\mathbb{P}[\mathcal{T} \mid \mathcal{G}_h] = \begin{cases} c_1 e^{-c_2 \left| 1 - \frac{s(\mathcal{G}_\mathcal{T})}{s^*(k)} \right|} & \text{if } \mathcal{T} \text{ agrees with } \mathcal{G}_h, \\ 0 & \text{otherwise.} \end{cases} \qquad (6.27)$$

Inference of the Most Probable Transmission Tree

The transmission tree that maximizes the probability in (6.21) can be estimated using the Markov Chain Monte Carlo (MCMC) sampling method. MCMC is a class of algorithms for sampling from a given probability distribution. If the distribution is defined on a set of graphs or other discrete structures, then MCMC can be considered as a randomized heuristic for maximizing a probabilistic objective. Under this perspective, MCMC is related to local search, simulated annealing and other similar metaheuristics based on transformations of local solutions.

In our case, we can utilize the general scheme that is based on the Metropolis–Hastings algorithm. The algorithm stochastically explores the space of transmission trees and proceeds iteratively as follows:

Step 1 Produce a new transmission tree T' from the current tree T using a preset local tree rearrangement operation. There are a number of such operations known in the literature [21], all of which are based on the removal of some internal edges of a tree and consecutive reconnection of the obtained subtrees by adding new edges.

Step 2 Recalculate the labels of the internal nodes of T' in order to make them agree with the host network $\mathcal{G}_h$. This can be done recursively using the following rule: an internal node with children having labels z and y receives the label z if $zy \in E(\mathcal{G}_h)$.

Step 3 Calculate the likelihood $\mathbb{P}[\Omega \mid \mathcal{G}_h, T']$ and the prior probability $\mathbb{P}[T \mid \mathcal{G}_h]$ using the algorithms and formulae described above.

Step 4 Accept or reject the tree T' as the new current tree. This is based on the acceptance ratio ρ that is calculated as follows:

$$\rho = \min \left\{ \frac{\mathbb{P}[\Omega \mid \mathcal{G}_h, T'] \times \mathbb{P}[T' \mid \mathcal{G}_h]}{\mathbb{P}[\Omega \mid \mathcal{G}_h, T] \times \mathbb{P}[T \mid \mathcal{G}_h]}, 1 \right\}. \qquad (6.28)$$

Step 5 Repeat Steps 1–4 until the predefined number of iterations is reached. Among the generated trees, choose the most likely tree.

In order to use (6.27), the expected number of hubs k in the true transmission network should be estimated. We use the following heuristic approach. First, the neighbour-joining tree $\mathcal{T}_0$, which is based on the distances Ω, and the corresponding transmission network $\mathcal{G}_{\mathcal{T}_0}$ are constructed. The degree sequence of $\mathcal{G}_{\mathcal{T}_0}$ is partitioned into two clusters using hierarchical clustering, and the cardinality of the cluster containing the highest degrees is used as an estimation for k.

6.4 Exercises

6.4.1

Consider the four-letter alphabet of nucleotides for RNA:

$$\mathcal{A} = \{A, C, U, G\}.$$

(a) How many seven-letter words are there using the letters from $\mathcal{A}$?
(b) How many four-letter words are there with the condition that none of the letters are repeated?
(c) How many seven-letter words are there with the condition that contiguous letters are different?
(d) How many RNA sequences of length seven are there theoretically with the condition that every sequence contains repeated contiguous nucleotides (e.g. ACAUUGA)?

6.4.2

(a) What is the concept of 'error catastrophe'?
(b) Explain the meaning of the 'survival of the flattest' concept. In particular, give the definition of the vector of eigenvalue centralities.

6.4.3

(a) Give the formal definition of the PageRanks of web pages.
(b) Explain why PageRank is a special case of the quasispecies model.

6.4.4

Address the following questions in the context of the model (6.9)–(6.12):

(a) Explain the phenomenon of 'local immunodeficiency'.
(b) What is meant by the mechanisms of 'antigenic cooperation' and 'immune escape by antigenic cooperation'?

6.4.5

This exercise is devoted to the solutions of the system (6.9)–(6.12).

(a) Let us consider Solution 4 of the system (6.9)–(6.12):

$$x_u^\diamond = 0, \quad x_v^\diamond = \frac{b}{c}\frac{f}{p}, \quad r_u^\diamond = 0, \quad r_v^\diamond = \frac{f}{p}.$$

Find all eigenvalues of the Jacobian $J(x_u^\diamond, x_v^\diamond, r_u^\diamond, r_v^\diamond)$. Hence prove that the solution is unstable.

(b) The system (6.9)–(6.12) has several stationary solutions. Explain why all stationary solutions with $r_u = 0$ are unstable. Hint: use the equation (6.9).

(c) Show that $\Delta_2 > 0$ in the proof of Proposition 6.1.

6.4.6

(a) Give the definition of a cross-immunoreactivity network.

(b) Explain the meaning of the following concepts: 'population transition via indirect variant interactions' and 'enhancement of one variant by another'.

6.4.7

In this exercise, we discuss altruistic and selfish variants in cross-immunoreactivity networks.

(a) Which variants are called altruistic in a cross-immunoreactivity network and what are their properties? Which variants are called selfish?

(b)* Show that the existence of altruists can theoretically be deduced from the presence of local immunodefficiency. Hint: find a relationship between the total viral population X and the total immune response R by summing up the equations in (6.8).

6.4.8

(a) Explain the meaning of Hamming distance and Levenshtein distance in the context of a viral population.

(b) Give the formal definition of a genetic network.

(c) Suppose that the edges in a given genetic network $\mathcal{G}_g$ connect variants at Hamming distance 1. Given a viral variant v and a fixed number k, how can one computationally find all viral variants in $\mathcal{G}_g$ that are at Hamming distance k from v?

6.4.9

(a) Give the definition of a host network and its transmission clusters. How can one computationally find all transmission clusters in a large host network?

(b) Explain what is meant by a genetic relatedness graph, and how it can be used.

(c) Give the definitions of a transmission tree and a transmission network.

(d) Compare and contrast Problems A and B from Section 6.3.

6.4.10

The following questions are devoted to s-metric:

(a) Calculate the s-metric for every tree with six vertices, and hence find the tree that has the highest s-metric among all six-vertex trees.

(b) Construct the graph H_k with k hubs that has the highest s-metric among all n-vertex graphs with k hubs, where k and n are fixed. Hint: the s-metric of H_k must be equal to (6.26).

(c) Prove the 'if' part of Claim 6.1: *If a tree T has no vertices of degree 2, $\Delta(T) \leq 3$ and $n \geq 4$ is even, then $s(T) = 6n - 15$.*

6.4.11*

(a) Formulate the constrained linear least-squares problem (6.22)–(6.25) for the weighted host network $\mathcal{G}_h$ and the transmission tree $\mathcal{T}$ shown in Figure 6.3, where $w_{AB} = 2, w_{AC} = 1, w_{BC} = 6$ are the weights of the links in the host network $\mathcal{G}_h$.

(b) Let $x_{AA} = \tau$ be a parameter. Using the constraints in your formulation from part (a), express all of the variables $x_{AB}, x_{AC}, \bar{x}_{AB}, \bar{x}_{AC}$ and $\bar{x}_{BC}$ in terms of τ.

(c) Without using the objective function of your formulation, calculate the likelihood $\mathbb{P}[\Omega \mid \mathcal{G}_h, \mathcal{T}]$ for the following values of α: 0.1, 0.5 and 1.
(d) Calculate the numerical values of the variables x_{AA}, x_{AB}, x_{AC}, $\bar{x}_{AB}$, $\bar{x}_{AC}$ and $\bar{x}_{BC}$ for $\alpha = 0.1$.

6.5 Solutions

6.4.1

(a) One of four letters can be used in the first position of the word, one of four letters can be used in the second position of the word etc. Therefore, there are $4^7 = 16{,}384$ such words. ∎

(b) One of four letters can be used in the first position of the word, one of three letters can be used in the second position of the word, one of two letters can be used in the third position of the word and one letter can be used in the last position of the word. Therefore, there are $4 \times 3 \times 2 \times 1 = 24$ words. ∎

(c) One of four letters can be used in the first position of the word, one of three letters can be used in the second position of the word, one of three letters can be used in the third position of the word etc. Therefore, there are $4 \times 3^6 = 2{,}916$ words. ∎

(d) The question can be reworded in this way: how many seven-letter words are there with the condition that every word contains repeated contiguous letters? Using parts (a) and (c), we obtain

$$16{,}384 - 2{,}916 = 13{,}468. \quad ∎$$

6.4.2

(a) The concept of *error catastrophe* refers to the phenomenon whereby the effects of having a too high mutation rate (exceeding the so-called error threshold) outweigh the effects of natural selection and cause the population to lose its structural coherence and eventually die out. ∎

(b) According to the *survival of the flattest* concept, the highest population size is achieved by the genomic variants with the highest centrality. Thus, under high mutation rates, natural selection favours variants that are robust against deleterious mutational effects. More precisely, the

distribution of the frequencies of viral variants converges to the eigenvector corresponding to the largest eigenvalue of the genetic network
G. This eigenvector is called the vector of eigenvalue centralities of
G. Therefore, the state of the population of selectively neutral viral
variants converges to the distribution that is completely determined by
the graph-theoretic properties of the sequence space. This observation
is the foundation of the *survival of the flattest* phenomenon.   ■

6.4.3

(a) Consider the web digraph $W = (V(W), E(W))$, where $V(W) = \{1, 2, ..., n\}$ is the set of web pages and $ij \in E(W)$ if and only if there
exists a hyperlink from web page i to web page j. The $n \times n$ edge-weight
matrix $R = (r_{i,j})_{i,j=1}^n$ is defined as follows:

$$r_{i,j} = \begin{cases} \frac{1}{d_i^+} & \text{if } ij \in E(W), \\ 0 & \text{if } ij \notin E(W). \end{cases}$$

These weights are usually interpreted as the probabilities of moving between edges. Let $\alpha \in (0, 1)$ and $s = (s_1, s_2, ..., s_n)^T$, where
$s_i > 0$ and $\sum_{i=1}^n s_i = 1$. The *PageRanks* of the web pages are elements
of the vector $p = (p_1, p_2, ..., p_n)^T$, which is the solution to the following
system of linear equations:

$$p = (1 - \alpha)s + \alpha R p.$$

■

(b) Let $G = (V(G), E(G))$ be a digraph with the vertex set $V(G) = \{1, 2, ..., n\}$. Consider the quasispecies equations (6.1) with $f_i = 1$ for
every $i = 1, 2, ..., n$ and $q_{j,i} = \alpha r_{j,i} + (1 - \alpha)s_i$. Then, the equation
(6.2) takes the form of a linear system of differential equations with
constant coefficients:

$$\frac{dc(t)}{dt} = (Q^T - I)c(t).$$

Its general solution can be written in the form

$$c(t) = \sum_{i=1}^r e^{(\lambda_i - 1)t} \sum_{j=1}^{k_i} P_i^{j-1}(t)\, u_i^j,$$

where $\lambda_1 > \lambda_2 > ... > \lambda_r$ are different eigenvalues of the matrix Q with multiplicities $k_1, k_2, ..., k_r$. Also, $(u_i^1, u_i^2, ..., u_i^{k_i})$ are Jordan chains of vectors corresponding to the eigenvalues λ_i, and P_i^{j-1} are polynomials of degree $(j-1)$, $i = 1, 2, ..., r$. Because Q is a strictly positive stochastic matrix, we have $\lambda_1 = 1$, $k_1 = 1$, u_1^1 is a non-negative eigenvector corresponding to the eigenvalue λ_1 and $\lambda_i - 1 < 0$ for every $i = 2, 3, ..., r$. Therefore, $c(t) \to p = Cu_1^1$ as $t \to \infty$, where $C = P_1^0$. With the proper normalization, we can assume that $\sum_{j=1}^{n} p_j = 1$. Taking into account this property, it follows from the equation $p = Q^T p$ that p satisfies (6.6). Thus, p is the PageRank of G.

∎

6.4.4

(a) *Local immunodeficiency* is a state where there is a reduction in the specific immune responses against particular viral variants, whereas the response against other variants increases or does not change. When this phenomenon is observed, the population of viral variants is typically split into three groups. One of them consists of persistent viruses— their population size is large but there is practically no immune response against them. Thus, persistent variants are not detected by the immune system; this is called *immunodeficiency*. The group in which a variant belongs is determined by its local position in the cross-immunoreactivity network. ∎

(b) Suppose that the phenomenon of local immunodeficiency is happening for the variant v: there is some reduction in the specific immune responses against v, whereas the response against u stays the same. The beneficial effect of cross-immunoreactivity for the variant v can be accompanied by its detrimental effect on the variant u, whose equilibrium population size may decline. Thus, the model suggests differential roles for the variants u and v during the viral evolution—this mechanism is called *antigenic cooperation* between u and v. The variant v acts selfishly and stimulates immune responses against the variant u. This diminishes the capability of the immune responses to produce the v-specific antibodies and allows the variant v to survive. In contrast, the variant u acts altruistically by maintaining its own specific immune response on a level sufficiently high to allow the variant v to survive, even by decreasing its own survival chances. Thus, the viral population as a whole exhibits quasi-social behaviour by the mechanism that can be called *immune escape by antigenic cooperation*. ∎

6.4.5

(a) For Solution 4, the Jacobian of the system is

$$J = J(x_u^\circ, x_v^\circ, r_u^\circ, r_v^\circ) = \begin{pmatrix} f & 0 & 0 & 0 \\ 0 & 0 & -\beta bf/c & -bf/c \\ c & 0 & b(\alpha - 1) & 0 \\ 0 & c & -b\alpha & -b \end{pmatrix}.$$

The characteristic polynomial of J is as follows:

$$\begin{aligned} P(\lambda) &= |J - \lambda E| \\ &= (f - \lambda)\Big(b(\alpha - 1) - \lambda\Big)\Big(-\lambda(-b - \lambda) - c(-bf/c)\Big) \\ &= (f - \lambda)\Big(b(\alpha - 1) - \lambda\Big)\Big(\lambda^2 + b\lambda + bf\Big). \end{aligned}$$

Therefore, the eigenvalues are

$$\begin{aligned} \lambda_1 &= f, \\ \lambda_2 &= b(\alpha - 1), \\ \lambda_3 &= \frac{1}{2}\Big(-b - \sqrt{b^2 - 4bf}\Big), \\ \lambda_4 &= \frac{1}{2}\Big(-b + \sqrt{b^2 - 4bf}\Big). \end{aligned}$$

Because $\lambda_1 = f > 0$, Solution 4 is unstable. ∎

(b) Let $(x_u^*, x_v^*, r_u^*, r_v^*)$ be a stationary solution and $r_u^* = 0$. We have $\dot{x}_u^* = 0$, and the equation (6.9) implies that $x_u^* = 0$. Therefore, the Jacobian of the system is

$$J(x_u^*, x_v^*, r_u^*, r_v^*) = \begin{pmatrix} f & 0 & 0 & 0 \\ 0 & f - pr_v^* & -p\beta x_v^* & -px_v^* \\ c & 0 & \frac{c\alpha x_v^*}{r_v^*} - b & 0 \\ 0 & c & \frac{-c\alpha x_v^*}{r_v^*} & -b \end{pmatrix}.$$

The characteristic polynomial of $J(x_u^*, x_v^*, r_u^*, r_v^*)$ is

$$P(\lambda) = |J - \lambda E| = (f - \lambda)P'(\lambda),$$

where $P'(\lambda)$ is some cubic polynomial. Therefore, $f > 0$ is an eigen-value of the Jacobian $J(x_u^*, x_v^*, r_u^*, r_v^*)$. This means that the stationary solution $(x_u^*, x_v^*, r_u^*, r_v^*)$ is unstable. Thus, all stationary solutions (x_u, x_v, r_u, r_v) with $r_u = 0$ are unstable. ∎

(c) For Δ_2, we have

$$\Delta_2 = b^4 f^2 \frac{(1-\alpha)(1-\beta)}{(\alpha-\beta+1)^2} g_3(\alpha, \beta),$$

where

$$g_3(\alpha, \beta) = (2+\alpha\beta-2\beta)\left(2+\alpha\beta-\alpha-\beta+\frac{b}{f}\frac{(1-\alpha)(1-\beta)}{\alpha-\beta+1}\right)(2+\alpha-\beta)$$

$$-(2+\alpha\beta-2\beta)^2 - (1-\alpha)(1-\beta)(2+\alpha-\beta)^2.$$

Simple rearrangements yield:

$$g_3(\alpha, \beta) \geq (2+\alpha\beta-2\beta)(2+\alpha\beta-\alpha-\beta)(2+\alpha-\beta)$$
$$-(2+\alpha\beta-2\beta)^2 - (1-\alpha)(1-\beta)(2+\alpha-\beta)^2$$
$$= (1+\alpha-\beta)\big((1-\alpha)(1-\beta)-1\big)^2 \geq 0. \qquad (6.29)$$

The inequality (6.29) achieves equality if and only if $\alpha = 0$ and $\beta = 1$, or $\alpha = \beta = 0$. Both cases are impossible due to the original assumptions about the cross-immunoreactivity strength (i.e. $\alpha \geq \beta > 0$), and therefore we have $\Delta_2 > 0$. ∎

6.4.6

(a) A *cross-immunoreactivity network* is a directed graph with vertices corresponding to viral variants. Two variants a and b are connected by the link ab if a elicits antibodies that can bind to b. This network represents interactions between viral variants and the host's immune system. ∎

(b) Various interactions (e.g. antigenic cooperation) occur between adjacent vertices in cross-immunoreactivity networks. Such interactions are usually complex, and the viral variants can influence other variants even if they are not adjacent. The model discussed in Section 6.2.2 suggests that a new vertex joining the network can affect even distant variants and either trigger their extinction or boost their frequencies. Such effects are called *population transition via indirect variant interactions*.

Suppose that, in a given cross-immunoreactivity network, the variant u goes extinct, whereas the variant v survives. Assume that the network is modified by adding a new variant w, for example as shown in Figure 6.2. Now, it is quite possible that the variant u will survive, but the variant v will be eliminated. This change is mediated by the addition of the variant w to the network, such that w is able to stimulate immune pressure on the variant v, thus relieving the pressure on the variant u that was stimulated by v. Therefore, the survival of u is due to the effect of the variant w—this phenomenon is called an *enhancement of one variant by another*—in this case an enhancement of u by w. This phenomenon can be viewed as another form of cooperation among viral variants, in our scenario between w as a 'benefactor' and u as a 'beneficiary'.   ∎

6.4.7

(a) There exists a distinct group of variants with specific properties—these variants are called *altruists* (see Figure 6.1). The altruistic variants themselves are eliminated by the immune system. However, their specific immune responses persist throughout the infection because they are stimulated by other variants, which are called *selfish*. Despite the presence of antibodies against altruists, they cannot efficiently neutralize persistent variants but still impede the generation of specific responses against those variants.

Although the altruists constitute only ca. 1.2% of the total number of viral variants, immune responses against them are dominant at equilibrium and constitute ca. 99.7%. This impedes the production of antibodies that are efficient against persistent variants. Altruistic variants are in-hubs, that is, vertices with high in-degree. The average in-degree of altruistic variants was found to be at least 12 times greater than the average in-degree of all other variants.   ∎

(b) Let $X = \sum_{i=1}^{n} x_i$ be the total viral population, and let $R = \sum_{i=1}^{n} r_i$ be the total immune response. By summing up the equations in (6.8), we obtain

$$\sum_{i=1}^{n} \dot{r}_i = \sum_{i=1}^{n} \left(c \sum_{j=1}^{n} \frac{u_{j,i}^{+} r_i}{\sum_{k=1}^{n} u_{j,k}^{+} r_k} x_j - b r_i \right)$$

$$= c \sum_{j=1}^{n} \frac{\sum_{i=1}^{n} u_{j,i}^{+} r_i}{\sum_{k=1}^{n} u_{j,k}^{+} r_k} x_j - \sum_{i=1}^{n} b r_i$$

$$= c \sum_{j=1}^{n} x_j - b \sum_{i=1}^{n} r_i$$

$$= cX - bR.$$

Therefore,

$$\dot{R} = cX - bR.$$

At equilibrium, $\dot{R} = 0$, so the equilibrium solutions

$$x = (x_1, x_2, ..., x_n) \quad \text{and} \quad r = (r_1, r_2, ..., r_n)$$

are related as follows:

$$bR = cX.$$

Thus, the relationship between the total viral population X and the total immune response R is linear. Therefore, the existence of variants with smaller r_i and larger x_i can theoretically imply the existence of variants with larger r_j and smaller x_j. ∎

6.4.8

(a) The Hamming distance between two viral variants of equal length is the smallest number of mutations required to obtain one variant from the other. For example, Hamming distance 1 means that they are different by a single mutation. The Levenshtein distance between two viral variants is the smallest number of single-character edits required to obtain one variant from the other. The possible operations are substitutions (i.e. mutations), insertions and deletions. ∎

(b) A *genetic network* $\mathcal{G}_g$ is an undirected graph with vertices corresponding to viral genomes, and two genomes are connected by an edge if they are at a distance from each other that is below a certain threshold (e.g. at Hamming distance 1).

(c) In terms of graph theory, it is necessary to find all nodes (variants) in $\mathcal{G}_g$ that are at a distance k from the node v, where the distance between nodes x and y is the length (number of edges) of the shortest (x, y)-path. The required nodes can be found by the Breadth-First Search method or using Dijkstra's algorithm, which are discussed in Chapter 1. Notice that for finding distances in a graph with Dijkstra's algorithm, the edge weights should be equal to 1.   ■

6.4.9

(a) A *host network* represents epidemiological relationships between infected individuals. Its vertices are infected hosts, and two vertices are adjacent whenever they are epidemiologically linked; that is, when there is evidence that one of the hosts directly infected the other or they are linked by a sufficiently short chain of viral transmissions between observed and, possibly, unobserved individuals. Connected components of a host network are called *transmission clusters* or *outbreaks*. They can be found by the Breadth-First Search or Depth-First Search methods, as discussed in Chapter 1.   ■

(b) We first calculate the pairwise distances between viral genomes or intra-host viral populations using the variety of population genetics measures. Then, it is necessary to link the patients with the viral genomes that are at a distance below a particular threshold. The threshold value is usually established using the available training genomic datasets from patients, whose epidemiological relationships are confirmed by epidemiological investigations. The obtained (undirected) *genetic relatedness graph* $\mathcal{G}_r$ may serve as a good approximation for the host network, and connected components of $\mathcal{G}_r$ can be identified by transmission clusters.   ■

(c) A *transmission tree* $\mathcal{T}$ represents the transmission history of an outbreak. It is a rooted binary labelled tree, whose leaves are infected hosts and interior nodes are transmission events. An interior node labelled z and its children labelled z and y signify that host y was infected by host z. For a given transmission tree $\mathcal{T}$, the *transmission network* $\mathcal{G}_{\mathcal{T}}$ indicates who infected whom. More precisely, the vertices of $\mathcal{G}_{\mathcal{T}}$ are infected hosts, and directed edges connect hosts that are linked by transmission.   ■

(d) Problem A can be tackled significantly more easily, and the approaches to its solution are usually based on the simple assumption that viral genomes sampled from epidemiologically related patients should be significantly closer to one another than the genomes sampled from unrelated patients. Thus, this problem can be solved by constructing the corresponding genetic relatedness graph $\mathcal{G}_r$, whose connected components can be identified by transmission clusters.

Problem B is significantly more complicated. Indeed, its solution usually requires finding an optimal spanning tree of $\mathcal{G}_r$ that maximizes or minimizes some predefined objective function. If the objective is just the total weight of the optimal tree, then the problem can be solved using Prim's or Kruskal's algorithm. However, the objectives are usually more complicated; this makes the corresponding problems NP-hard. ∎

6.4.10

(a) There are six non-isomorphic trees with six vertices. We have:
  (1) $s(P_6) = 16$;
  (2) $s(T) = 18$ for the tree with degree sequence $(3,2,2,1,1,1)$ where the two vertices of degree 2 are adjacent;
  (3) $s(T) = 19$ for the tree with degree sequence $(3,2,2,1,1,1)$ where the two vertices of degree 2 are not adjacent;
  (4) $s(T) = 21$ for the tree with degree sequence $(3,3,1,1,1,1)$;
  (5) $s(T) = 22$ for the tree with degree sequence $(4,2,1,1,1,1)$;
  (6) $s(K_{1,5}) = 25$.
  Thus, the star $K_{1,5}$ has the highest s-metric among all six-vertex trees. ∎

(b) This bound is achieved on the graph H_k obtained by taking the disjoint union of k stars

$$K_{1,\lfloor n/k \rfloor - 1}$$

and connecting the centre of one of them to the centres of all the others. ∎

(c) If T has no vertices of degree 2, then (6.20) implies that $n_1 = \frac{n+2}{2}$. Furthermore, taking into account that $n \geq 4$, we obtain

$$s(T) = 3m_1 + 9m_3,$$

where m_1 is the number of pendant edges and m_3 is the number of edges with both ends having degree 3. Clearly,

$$m_1 = n_1 = 0.5n + 1$$

and, taking into account that $T \neq K_2$,

$$m_3 = n - 1 - n_1 = 0.5n - 2.$$

Thus,

$$s(T) = 6n - 15. \qquad \blacksquare$$

6.4.11

(a)
$$\min \ \nu = \sum_{ij \in E(\mathcal{G}_h)} \left(\frac{\bar{x}_{i,j}}{\omega_{i,j}} - 1 \right)^2$$
$$= (\bar{x}_{AB}/2 - 1)^2 + (\bar{x}_{AC} - 1)^2 + (\bar{x}_{BC}/6 - 1)^2$$

subject to

$$2x_{AA} + x_{AB} = \bar{x}_{AB}$$
$$x_{AA} + x_{AC} = \bar{x}_{AC}$$
$$x_{AC} + x_{AA} + x_{AB} = \bar{x}_{BC}$$
$$2x_{AA} = \alpha x_{AB}$$
$$\alpha x_{AB} = x_{AA} + \alpha x_{AC}$$
$$x_{AA} \geq 0$$
$$x_{AB} \geq 0$$
$$x_{AC} \geq 0 \qquad \blacksquare$$

(b) From the fourth and fifth constraints in the above formulation, we obtain

$$x_{AB} = 2x_{AA}/\alpha = 2\tau/\alpha$$

and

$$x_{AC} = x_{AB} - x_{AA}/\alpha = 2\tau/\alpha - \tau/\alpha = \tau/\alpha.$$

Now, using the first three constraints, we have

$$\bar{x}_{AB} = 2\tau\left(\alpha^{-1} + 1\right),$$
$$\bar{x}_{AC} = \tau\left(\alpha^{-1} + 1\right),$$
$$\bar{x}_{BC} = \tau\left(3\alpha^{-1} + 1\right).$$

∎

(c) The likelihood $\mathbb{P}[\Omega \mid \mathcal{G}_h, \mathcal{T}]$ is equal to the Pearson correlation coefficient $r(\Omega, \bar{\mathcal{X}})$ if it is positive. By definition, we have

$$r(\Omega, \bar{\mathcal{X}}) = \frac{\mathrm{cov}(\Omega, \bar{\mathcal{X}})}{\sigma_\Omega\,\sigma_{\bar{\mathcal{X}}}}.$$

Therefore,

$$r(\Omega, \bar{\mathcal{X}}) = \frac{\mathbb{E}[\Omega\,\bar{\mathcal{X}}] - \mathbb{E}[\Omega]\,\mathbb{E}[\bar{\mathcal{X}}]}{\sqrt{\mathbb{E}[\Omega^2] - (\mathbb{E}[\Omega])^2}\,\sqrt{\mathbb{E}[\bar{\mathcal{X}}^2] - (\mathbb{E}[\bar{\mathcal{X}}])^2}}.$$

We obtain

$$\mathbb{E}[\Omega\,\bar{\mathcal{X}}] = \frac{\tau}{3}\left(23\alpha^{-1} + 11\right), \quad \mathbb{E}[\Omega] = 3, \quad \mathbb{E}[\bar{\mathcal{X}}] = \frac{\tau}{3}\left(6\alpha^{-1} + 4\right).$$

Also,

$$\mathbb{E}[\Omega^2] = \frac{41}{3} \quad \text{and} \quad \mathbb{E}[\bar{\mathcal{X}}^2] = \frac{\tau^2}{3}\left(5\left(\alpha^{-1} + 1\right)^2 + \left(3\alpha^{-1} + 1\right)^2\right).$$

Thus,

$$r(\Omega, \bar{\mathcal{X}})$$
$$= \frac{\frac{\tau}{3}\left(23\alpha^{-1} + 11\right) - \tau\left(6\alpha^{-1} + 4\right)}{\sqrt{\frac{41}{3} - 9}\,\sqrt{\frac{\tau^2}{3}\left(5\left(\alpha^{-1}+1\right)^2 + \left(3\alpha^{-1}+1\right)^2\right) - \frac{\tau^2}{9}\left(6\alpha^{-1}+4\right)^2}},$$

which can be rewritten as

$$r(\Omega, \bar{\mathcal{X}}) = \sqrt{\frac{3}{14}}\,\frac{5\alpha^{-1} - 1}{\sqrt{15\left(\alpha^{-1}+1\right)^2 + 3\left(3\alpha^{-1}+1\right)^2 - \left(6\alpha^{-1}+4\right)^2}}.$$

Therefore,

$$\mathbb{P}[\Omega \,|\, \mathcal{G}_h, \mathcal{T}] = 0.924 \quad \text{if} \quad \alpha = 0.1,$$
$$\mathbb{P}[\Omega \,|\, \mathcal{G}_h, \mathcal{T}] = 0.817 \quad \text{if} \quad \alpha = 0.5,$$
$$\mathbb{P}[\Omega \,|\, \mathcal{G}_h, \mathcal{T}] = 0.655 \quad \text{if} \quad \alpha = 1. \qquad \blacksquare$$

(d) For $\alpha = 0.1$, we have

$$x_{AA} = \tau, \quad x_{AB} = 20\tau, \quad x_{AC} = 10\tau$$

and

$$\bar{x}_{AB} = 22\tau, \quad \bar{x}_{AC} = 11\tau, \quad \bar{x}_{BC} = 31\tau.$$

Hence, the objective function takes the form

$$\nu = 2(11\tau - 1)^2 + \left(\frac{31}{6}\tau - 1\right)^2.$$

This objective function is minimized when τ satisfies the following equation:

$$\nu' = 44(11\tau - 1) + \frac{31}{3}\left(\frac{31}{6}\tau - 1\right) = 0.$$

We obtain $\tau = 0.1011$ (to 4 dp), and therefore

$$x_{AA} = 0.101, \quad x_{AB} = 2.022, \quad x_{AC} = 1.011$$

and

$$\bar{x}_{AB} = 2.224, \quad \bar{x}_{AC} = 1.112, \quad \bar{x}_{BC} = 3.134. \qquad \blacksquare$$

Acknowledgements

Section 6.3.2 is based on the article in *Bioinformatics*, **34** (1), P. Skums et al., QUENTIN: reconstruction of disease transmissions from viral quasispecies genomic data, 163–170, © 2017, with permission from Oxford University Press.

References

[1] M. Ajtai, J. Komlós and E. Szemerédi, Largest random component of a k-cube, *Combinatorica*, **2** (1)(1982), 1–7.

[2] B. Bollobás, *Random Graphs*, Cambridge studies in advanced mathematics, Volume 73, Cambridge: Cambridge University Press, 2001.

[3] S. Brin and L. Page, The anatomy of a large-scale hypertextual web search engine, *Proceedings of the Seventh International Conference on World Wide Web 7*, Amsterdam, The Netherlands, 1998, 107–117.

[4] A. J. Leigh Brown, S. J. Lycett, L. Weinert, G. J. Hughes, E. Fearnhill and D. T. Dunn, Transmission network parameters estimated from HIV sequences for a nationwide epidemic, *Journal of Infectious Diseases*, **204** (9)(2011), 1463–1469.

[5] L. Bunimovich and L. Shu, Local immunodeficiency: Minimal networks and stability, *Mathematical Biosciences*, **310** (2019), 31–49.

[6] J. D. Burtin, The probability of connectedness of a random subgraph of an n-dimensional cube, *Problemy Peredachi Informatsii*, **13** (1977), 90–95.

[7] D. S. Campo, Z. Dimitrova, J. Yokosawa, D. Hoang, N. O. Perez, S. Ramachandran and Y. Khudyakov, Hepatitis C virus antigenic convergence, *Scientific Reports*, (2)(2012), Article 267.

[8] D. S. Campo, G.-L. Xia, Z. Dimitrova, Y. Lin, J. C. Forbi, L. Ganova-Raeva, L. Punkova, S. Ramachandran, H. Thai, P. Skums, S. Sims, I. Rytsareva, G. Vaughan, H.-J. Roh, M. A. Purdy, A. Sue and Y. Khudyakov, Accurate genetic detection of hepatitis C virus transmissions in outbreak settings, *Journal of Infectious Diseases*, **213** (6)(2016), 957–965.

[9] F. Chung, PageRank as a discrete Green's function, *Geometry and Analysis I ALM*, **17** (2010), 285–302.

[10] F. Chung and W. Zhao, PageRank and random walks on graphs, *Proceedings of Fete of Combinatorics and Computer Science Conference*, Berlin: Springer, 2010, 43–62.

[11] N. De Maio, C.-H. Wu and D. J. Wilson, SCOTTI: Efficient reconstruction of transmission within outbreaks with the structured coalescent, *PLoS Computational Biology*, **12** (9)(2016), 1–23.

[12] E. Domingo, Biological significance of viral quasispecies, *Viral Hepatitis Reviews*, **2** (1996), 247–261.

[13] E. Domingo, J. Sheldon and C. Perales, Viral quasispecies evolution, *Microbiology and Molecular Biology Reviews*, **76** (2)(2012), 159–216.

[14] M. Eigen, Selforganization of matter and the evolution of biological macromolecules, *Naturwissenschaften*, **58** (10)(1971), 465–523.

[15] M. Eigen, J. McCaskill and P. Schuster, The molecular quasi-species, *Advances in Chemical Physics*, **75** (1989), 149–263.

[16] P. Erdös and J. Spencer, Evolution of the n-cube, *Computers and Mathematics with Applications*, **5** (1)(1979), 33–39.

[17] R. Falk and A. D. Well, Many faces of the correlation coefficient, *Journal of Statistics Education*, **5** (3)(1997), 1–18.

[18] W. M. Fitch and E. Margoliash, Construction of phylogenetic trees, *Science*, **155** (3760)(1967), 279–284.

[19] F. R. Gantmacher, *Matrix Theory*, New York, NY: Chelsea, 1959.

[20] S. Gavrilets, *Fitness Landscapes and the Origin of Species*, MPB-41, Vol. 41, Princeton, NJ: Princeton University Press, 2004.

[21] D. H. Huson, R. Rupp and C. Scornavacca, *Phylogenetic Networks: Concepts, Algorithms and Applications*, Cambridge: Cambridge University Press, 2010.

[22] T. Jombart, R. M. Eggo, P. J. Dodd and F. Balloux, Reconstructing disease outbreaks from genetic data: a graph approach, *Heredity*, **106** (2)(2011), 383–390.

[23] W. O. Kermack and A. G. McKendrick, A contribution to the mathematical theory of epidemics, *Proceedings of the Royal Society of London. Series A, Containing Papers of a Mathematical and Physical Character*, **115** (772)(1927), 700–721.

[24] M. Kimura, Evolutionary rate at the molecular level, *Nature*, **217** (5129)(1968), 624–626.

[25] A. V. Kostochka, Maximum matching and connected components of random spanning subgraphs of the n-dimensional unit cube, *Diskretny Analis*, **48** (1989), 23–29.

[26] A. V. Kostochka, A. A. Sapozhenko and K. Weber, Radius and diameter of random subgraphs of the hypercube, *Random Structures and Algorithms*, **4** (2)(1993), 215–229.

[27] P. Lemke, The maximum leaf spanning tree problem for cubic graphs is NP-complete, *IMA Preprint Series*, University of Minnesota, Minneapolis, 428, 1988.

[28] L. Li, D. Alderson, J. C. Doyle and W. Willinger, Towards a theory of scale-free graphs: Definition, properties, and implications, *Internet Mathematics*, **2** (4)(2005), 431–523.

[29] M. E. J. Newman, *Networks: An Introduction*, Oxford: Oxford University Press, 2010.

[30] M. A. Nowak, *Evolutionary Dynamics: Exploring the Equations of Life*, New Haven, CT: Harvard University Press, 2006.

[31] M. A. Nowak and R. M. May, *Virus Dynamics: Mathematical Principles of Immunology and Virology*, Oxford: Oxford University Press, 2000.

[32] Y. Orlovich, V. Keibel, K. Kukharenko and P. Skums, Scale-free spanning trees: complexity, bounds and algorithms, *arXiv preprint arXiv:2005.13703*, 2020.

[33] P. Skums, L. Bunimovich and Y. Khudyakov, Antigenic cooperation among intrahost HCV variants organized into a complex network of cross-immunoreactivity, *Proceedings of the National Academy of Sciences*, **112** (21)(2015), 6653–6658.

[34] P. Skums, A. Zelikovsky, R. Singh, W. Gussler, Z. Dimitrova, S. Knyazev, I. Mandric, S. Ramachandran, D. Campo, D. Jha, L. Bunimovich, E. Costenbader, C. Sexton, S. O'Connor, G.-L. Xia and Y. Khudyakov, QUENTIN: reconstruction of disease transmissions from viral quasispecies genomic data, *Bioinformatics*, **34** (1)(2017), 163–170.

[35] A. Soshnikov and B. Sudakov, On the largest eigenvalue of a random subgraph of the hypercube, *Communications in Mathematical Physics*, **239** (1–2)(2003), 53–63.

[36] H. Tolou, J. Nicoli and C. Chastel, Viral evolution and emerging viral infections: what future for the viruses? A theoretical evaluation based on informational spaces and quasispecies, *Virus Genes*, **24** (3)(2002), 267–274.

[37] E. Van Nimwegen, J. P. Crutchfield and M. Huynen, Neutral evolution of mutational robustness, *Proceedings of the National Academy of Sciences*, **96** (17)(1999), 9716–9720.

[38] J. O. Wertheim, A. J. Leigh Brown, N. L. Hepler, S. R. Mehta, D. D. Richman, D. M. Smith and S. L. Kosakovsky Pond, The global transmission network of HIV-1, *Journal of Infectious Diseases*, **209** (2)(2014), 304–313.

[39] C. O. Wilke, Quasispecies theory in the context of population genetics, *BMC Evolutionary Biology*, **5** (1)(2005), 44.

Index